# The Life of Structures

# The Life of Structures
## Physical Testing

Edited by
**G.S.T. Armer**
Building Research Establishment, UK

**J.L. Clarke**
British Cement Association, UK

**F.K. Garas**
Taylor Woodrow Construction Ltd, UK

**Butterworths**
London Boston Singapore Sydney Toronto Wellington

PART OF REED INTERNATIONAL P.L.C.

First published 1989

© **The Informal Study Group for Model Analysis as a Design Tool, Institution of Structural Engineers, 1989**

---

**British Library Cataloguing in Publication Data**

The Life of Structures.
  1. Structures. Durability. Effects of design
  I. Armer, G. S. T.  II. Clarke, J. L. (John Lowther),
  *1941–*  III. Garas, F. K.
  624.1'71

ISBN 0-408-04245-1

---

**Library of Congress Cataloging-in-Publication Data**

The Life of structures : physical testing / edited by
  G.S.T. Armer, J.L. Clarke, F.K. Garas.
    p. cm.
    Includes bibliographical references.
    ISBN 0-408-04245-1
    1. Structural engineering. 2. Buildings—
Testing. 3. Building materials—Service life.
I. Armer, G. S. T. II. Clarke, J. L. III. Garas, F. K.
TA636.L53  1989
624.1—dc20

---

Printed in Great Britain at the University Press, Cambridge

# Preface

The decision to build or purchase a particular building is usually controlled by economic factors related to both financial strategies for the future and to a current situation. In the past, buildings were often expected to last "forever". This sometimes led to a very high ratio of building/plant capital investment and, in consequence, to high demolition costs when the function disappeared long before the structure itself failed. The natural progression in design towards more efficient use of materials and labour, fast-track construction etc, exposes the designer's need to consider the life of a new building. Concurrently, the rehabilitation of old stock requires an assessment of remaining life to justify particular courses of action. In many circumstances therefore, an estimate of the useful life of the building could clearly have an effect on the levels of investment in its fabric. Currently however, there are no viable methods for its evaluation. Against this background, the development of a method of useful life prediction would be a significant asset for the designer.

The physical testing of full and large scale structures and elements and modelling life-time environments combine to form the only approach which will support a realistic method of structural life prediction. Modern building research has generated a large body of knowledge on the physical properties of construction materials, the physical environment of structures and the performance of existing structures in service. There has however, been scant attention paid to the synthesis of this data to provide a soundly based process by which an engineer may predict the useful life of a structure he is designing or assesing.

This book is based upon the proceedings of an international seminar entitled "The Life of Structures". The objective of the seminar was to provide a forum in which experts on:

- actions on structures and the performance of existing populations of structures,
- the properties and performance of building materials,
- and the internal and external environments of buildings,

could address the common goal of predicting the life of structures by interpreting their own special knowledge as part of a design process.

Fifty one papers are reproduced in this volume together with written discussion and commentaries. The subjects covered which are related to the life of structures are:

- Basic principles of design and assessment
- Structural environments
- Materials and test methods
- Nuclear and special structures
- Bridges
- The performance of populations of structures

Together these represent a state-of-the-art examination of design for durability in the built environment.

G S T Armer
J L Clarke
F K Garas

# Contents

# Introduction

**P L Campbell,** President, Institution of Structural Engineers

This conference has been organised by the Institution of Structural Engineers Study Group entitled "Model Analysis as a Design Tool". There are twelve such study groups, but this one chaired by Dr Garas is probably the most productive group; one that has done a great deal to advance the understanding of the art and science of structural engineering; achievements of which this Institution is justly proud.

It is stating the obvious when one observes that the structure in the context of buildings is but one, albeit a very important, factor in the life expectancy of that building. When one embarks on the creation of a new facility one must take into account the climate, environmental and geotechnical conditions that exist in each considered case; the nature of the architectural aesthetic and the materials that are required to achieve that. Then there is the detailing of the use of those materials; the anticipated useful life that the promoter expects, and the funds he is prepared to commit to the project. There is the question of maintenance, both short and long term, the serviceability requirements dictated by the use that will be made of the structure, and the need to build in flexibility to allow for changing uses with the passage of time.

When contemplating the life of structures one is inevitably going to look at the performance of modern buildings and not so modern buildings. It is worthwhile using the Second World War as the reference point to look at structures pre and post war. When this is done it is sad to find that there is overwhelming evidence that pre-1939 buildings and structures have performed generally much better over a longer period of time than post 1939 buildings and structures. It would appear that greater understanding, advancing technology, new materials, and the need to build both quickly and cheaply has been the undoing of the quality of buildings and therefore their life expectancy.

I believe we should know about where we have been in order to decide where we are going and learn from the designers and builders that have gone before, and to appreciate that they had no alternative but to use a very limited variety of materials that they understood very well, like brick, stone, metal, timber, lead and glass. I would like to organise a competition for the design of a 21st Century building using only six primary materials and I wonder what would be produced.

It is a strange reality that, for example, we undoubtedly have much greater understanding of every aspect of the material concrete, but appear to produce a material inferior in many respects to that produced in earlier times. Is this so, and if so, why?

One does not forget that buildings and structures post-war are inevitably far more complex than pre-war buildings and structures, but I suspect that in general the basic structures have changed little, but nowadays the external envelope of the building is frequently part of the structure and therefore subject to far more stringent

1

conditions in use, and a greater potential for deterioration.

A post-war feature in the development of structures has seen the arrival of offshore structures, and nuclear radiation containment structures, where the ability to confidently design and build for a specific useful life becomes critical in conditions that can only be described as hazardous.

It would therefore appear that since the Second World War we have been busy designing and building structures with a life span rarely specified at the design stage, made of a complex of materials that we do not understand fully and often have no proof of their durability; to resist forces and loads that are unreliable, in conditions that are almost impossible to achieve and then wonder why we find the prediction of the life expectancy of our buildings and structures difficult.

I have, and do, publicly register my deep concern about the profligate way in which we use our material and financial resources, being only concerned about the bottom line of the next balance sheet and showing little evidence of any real concern for the quality of the built environment that others will inherit.

On two occasions in the past I asked clients with collections of largely mediaeval buildings how long they wished their new buildings to last. On the first occasion I eventually was given a considered reply - 500 years with a minimum of external maintenance, and on the second occasion an answer that amounted to the same - 250 years. (Ask a silly question and you sometimes get an answer that concentrates the mind).

The resulting overtly modern buildings were designed and constructed using a minimum number of proven materials with high quality structural materials used at two thirds the normal working stress, a detailed specification, careful detailing and quality construction. The clients' requirements implicitly imposed all the key factors affecting the life of the buildings. High quality primary structure, limited number of materials, all of which are well understood and have proven durability, very low short term and long term maintenance which favourably affected the life cycle costs even when allowance was made for increased first cost.

By contrast it is worthwhile visiting docklands. One cannot be other than amazed by the level of activity there, but I am one of those depressed by the tackiness of the vast majority of buildings, there being, seemingly, a competition for designers that use the greatest number of materials with the most complicated detailing in conditions of maximum exposure, who will then be surprised when it all goes wrong.

Longevity in buildings and structures is a direct function of design appraisal, a detailed understanding of the sub-soil conditions, the use of proven materials, and a profound understanding of building and construction.

I must add that the subject of sub-soil investigation, interpretation and the effect of substructure design on the life of structures is conspicuous by its absence in this conference.

However, I am confident that the papers presented during these three days will add greatly to our knowledge of many of these matters, and will enable us to serve society better by more responsible use of our resources.

# 1 Methodology for the prediction of the life of existing structures

**James A O'Kon,** PE, M-ASCE

## I.  SUMMARY

Throughout the built environment of our planet the numbers of buildings constructed prior to the advent of modern building systems greatly exceed the number of buildings which were constructed with modern internal environments and are now serving the purpose for which they were constructed.

The high costs of new construction have made the preservation and restoration of these older buildings quite attractive.  Older structures that are to be given a renewed life are likely to contain flaws that must be corrected or proven not to adversely affect the predicted life of the structure; or if an older building is to be used for different functional purposes then the structure may need to be strengthened.

In addition to the upgrading of commercial structures, most of the world's renowned structures that comprise part of our historical heritage are showing signs of wear and age; these buildings must be given an extended life span.

Older buildings which suffer from the stresses and strains of age can be given new life by the new profession of "building doctors" who can prescribe corrective action by taking the building's pulse.  The demand for building diagnosticians is now greater than ever; and the professionals who are the practitioners of this new field must use all the tools that are available including:  efficient management techniques, new technology, deductive skills and common sense.

The purpose of this paper is to provide an overview of viable methodologies that can be used to diagnose the conditions of existing structures; identify deficiencies in performance, recommend remedial repair programs and provide a quantitative prediction of the life of the structure.

The techniques presented in this paper are based on years of experience, knowledge of structural behavior, state-of-the-art building technology and archeological principals.

## II.  PLANNING THE INVESTIGATION

The implementation of a successful Life Prediction effort begins with the development of a logical operating plan. The diagnostic engineer should assume the role of team leader and develop a plan of operations that will trigger a systematic process that will predict the life of the structure.

The planning effort includes client contact, selection of the investigative team, site observation and testing, document collection, modeling, synthesis of data and development of hypotheses, as well as the budgetary and scheduling requirements for each phase.

A) Client Interface: When the investigative plan has been outlined, the client should be apprised of the scope of work that is required for a proper investigation. The scope should be expressed in the form of a written proposal, along with a contract for engineering services. The proposal should include a budget which should reflect the cost of the operation including professional services, consultants, testing, travel, communications and printing.

The schedule submitted in the proposal should extend from the initial conferences with the client to the completion of the final report and include all the salient elements of the investigation. The client's representative should be identified at this stage.

B) The Investigation Team: When commissioned to carry out the investigation of a large or complex building, the diagnostic engineer must retain other professional disciplines. The prime requirement of the interdisciplinary team member is a high level of technical qualifications in an area of expertise which is germane to the project. To implement a comprehensive investigation the diagnostic investigation may require experts from the following fields:

| | |
|---|---|
| Geotechnical Engineer/Testing | Structural Engineering |
| Materials Technology/Testing | Mechanical Engineering |
| Chemical Technology/Testing | Electrical Engineering |
| Metallurgy Technology/Testing | Civil Engineering |
| Dynamics Technology | Surveying |
| Accoustical/Vibration Eng. | Photography |
| Construction Management | Photogrammetry |
| Construction Equip. Operators | Video Camera Operator |
| Maintenance/Operations/Expert | Financial Analysis Experts |

III. DOCUMENT/LITERATURE SEARCH

Documents which are related to the design and construction of the facility will be an invaluable aid in the investigation process. When investigating older buildings the construction documents will probably be unavailable, however, a literature search to identify building techniques used in the era of the original construction will be helpful.

A. Document Search: The body of documents generated during the development of a project will provide an overview of the building's construction history.  Documents which are generated during the construction of a project which can be used in the investigation could include the following:

1) Construction Drawings
2) Contract Specifications
3) Contractural Agreements
4) Contract Changes
5) Shop Drawings
6) As-Built Drawings
7) Test Reports
8) Field Reports
9) Inspection Reports
10) Design Calculation
11) Building Dept. Permits
12) Maint./Modif. Records

B. Literature Search: An extensive body of published works exists dealing with historical accounts of construction techniques as well as techniques that can be used in the investigation of constructed facilities.  Sources for identifying research literature, in addition to the bibliographies contained in the papers presented at this seminar, include:  publications of professional societies and trade associations, proceedings of conferences and symposiums, engineering libraries and computer data bases.

C. Historical Documents: The collection and review of data from historical sources can assist in developing a body of information that can provide graphic or narrative accounts of the site prior to construction, the building during construction or the configuration of the building prior to renovations.  The best sources of visual information are historical volumes, records of historical societies, newspaper files, insurance companies records, building owners and the local building department.

IV.  SITE OBSERVATIONS

The goals of building site visits will be to conduct overall visual examinations, prepare graphic and narrative surveys and carry out testing programs.

A. INITIAL SITE VISIT: The initial site visit should be carried out to evaluate the overall condition and geometry of the building.  The initial site visit should be used to facilitate the following aspects of the investigation:

1) Outline the investigative plan
2) Assess equipment required for further site visits
3) Outline field and laboratory testing programs
4) Develop a list of interdisciplinary experts required for the investigation.
5) Determine the extent of physical survey required, if design drawings are not available.

7

B. OVERALL VISUAL EXAMINATIONS: The development of graphic and narrative records of the investigation should be implemented in order to provide a complete survey of the condition of the facility. Field data can be collected in the form of measured drawings, sketches, verbal descriptions, video recordings, and photographs.

1. Drawings/Sketches: Well conceived and competent drawings and sketches are vital to the investigation. Measured drawings of the overall configuration and detailed sketches of specific areas will augment other types of data collected at the site. A reference system of coordinates should be established for spatial relationships of the entire building.

2. Photographs: The use of photography is valuable as a tool for collecting data in the investigation and for use as a visual recording mechanism in the on-going effort. The use of wide-view and detailed photos will record data that is germane to the effort.

3. Videotaping: The use of videotaping with audio backgrounds can provide an invaluable visual and narrative overview of the site.

4. Verbal Description: An effective technique for verbal description is to use a pocket recorder and succinctly record thoughts or impressions as they occur. Later the thoughts can be reviewed in a controlled environment which will provide an opportunity for examination and consideration.

V. TESTING PROGRAMS

The diagnostic engineer and the investigative team will be responsible for the implementation of testing programs. The program should include field and laboratory testing. The goals of the testing program include acquisition of data relative to material quality, workmanship, unstable materials and exposure to deleterious substances. The testing program could include physical and chemical tests on materials of construction, testing of the various elements of equipment, structural load testing, and model testing.

Some tests may be made in the field, however, the majority of tests will be carried out in the testing laboratory utilizing samples acquired in the field. Testing laboratories familiar with material sampling and testing which adhere to acceptable standards should be engaged as part of the team. Following is a summary of each type of testing and it's applications that may be used in the investigation.

A. FIELD TESTING: Field testing of materials is used to provide specific characteristics of the materials of construction, while the subject materials are in-situ. The following tests could be carried out in the field:

1) TESTS OF METAL COMPONENTS

a) Ultrasonic Testing: For flaws in metal components or welds.
b) Magnetic or Eddycurrent Testing: Location and dimensions of cracks in metals or welds.
c) Radiographic Testing: Locate and detect flows in welds or parent metal.
d) Hardness Testing: Indicates strength of materials.

2) TESTS OF REINFORCED CONCRETE COMPONENTS

a) Swiss Hammer: Measures the strength of concrete.
b) Radar Probing: To evaluate density and voids
c) X-Ray Testing: To ascertain size and amount of reinforcing.

3) TESTS OF TIMBER COMPONENTS

a) Boring Tests: Drilling and rate of penetration may indicate low density or voids resulting from decay, fungi, insect or bacterial action.

4) TESTS ON MASONRY COMPONENTS

Tests may be carried out on masonry units, steel reinforcing, sealants, coatings.

B) LABORATORY TESTING: Laboratory testing uses controlled conditions to carry out a wide range of tests. Following is a listing of frequently employed tests for various materials of construction:

1. METAL TESTING
a) Strength
b) Charpy V-Notch
c) metallurgical exams
d) chemical tests
e) microscopic tests
f) electron microscope
g) x-rays

2. CONCRETE TESTING
a) compressive strength
b) long term creep tests
c) Long term skrinkage/ expansion tests
d) aggregate matrix micro cracking
c) moisture cont. tests
f) petrographic studies
g) split tensile strength tests
h) modulus of elasticity tests
i) air content tests

9

3.  TIMBER TESTS
a) organic chemistry          c) strength of material
b) moisture content

4.  MASONRY TESTS
a) prism tests: compression, flexure, shear
b) masonry tests: compression, water absorption, freeze-thaw
   resistance, thermal expansion
c) mortar tests: cement and air content

C) STRUCTURAL LOAD TESTS:  Structure load tests may be
performed on an individual part or a complete structural
system in order to verify the load carrying capabilities of
the structure.  Load tests may be performed on an existing
structure when analytical methods alone cannot accurately
reflect the actual structural response of a building. This
is often done when verifying the actual load carrying capabilities of an existing building.

The methodology for conducting structural load tests should
be developed by the testing laboratory using acceptable
standards. A structural analysis should preceed all load
tests.

D) MODEL TESTING: In the event that a structural load test
or analytical analysis cannot verify the load carrying
capabilities of a structure then model tests can be used to
simulate the behavior of the structure.

VI.  SYNTHESIS OF THE INVESTIGATION

The diagnostic engineer will work closely with the team to
synthesize the results of the investigation and promote the
intellectual process that will establish the predictable
life of the building.

The synthesizing process is carried out using a systematic
appraisal of investigative data combined with technology,
deductive reasoning and practicality.

A. INVESTIGATIVE SCENARIOS:  The scope of the synthesis
process can vary depending on the condition of the building
and it's intended usage and alterations.  The following
investigative scenarios will require different scopes of
data synthesis:

1) Review of sound building system that will utilize the
   same loading after renovation

2) Review of building with degraded structural system, that
   maintain the same loading after renovation

3) Review of building with sound structural systems: with
   intent to change usage and increase floor loadings

4) Review of building with sound structural systems, increased floor loading with intent to increase the of floors

5) Review building with degraded systems: increase floor loadings, and/or increase number of floors

6) Review of building or monument with degraded structural systems for purpose of historical preservation

B. SYNTHESIS PROGRAM: The complexity and scope of the program will vary with the investigative scenario. The effort can range from a nominal survey and structural review of a sound building whose renovation will not change it's functional use; to a complex effort required of a building with a degraded structural system that is anticipated to experience changes in loadings and additional floor levels.

To assess the load carrying capability of the structure and to identify modifications required to fulfill the renovation program. The investigative data can be used to develop a simple mathematical model as in the case of the review of a sound structure, or used to develop and plot a three dimensional computer model in the case of a complex investigation.

Unique methods of repairing and/or strengthening of structures have been developed using space-age materials combined with traditional building materials. Some of these techniques and materials include epoxies and high strength steel that increase the strengths of steel reinforced concrete and timber structures; repairs using the post-tensioning of concrete, structural steel, and timber structures; injection of high strength epoxy grout to strengthen structures, high-tech cladding used to achieve look-alikes of traditional materials, and transfer of stresses from primary load paths to secondary systems using computer models.

The synthesis of the investigation should evaluate the condition and capabilities of the base structure and identify remedial measures for repair and strengthening as required.

VII. LIFE PREDICTION HYPOTHESIS

The establishment of a hypothesis for the life span of a building should be based on the work carried out during the synthesis of the investigaton. Consideration should be given to the condition of the building after repairs, the extent of the repair program, the internal and external environment at the building which could degrade the structure and the projected program of maintenance.

The prediction of the useful life of a structure should be expressed in terms of a time period related to the retirement of the mortgage or bond issued. A definitive period not to exceed thirty years should be used in the case of a historic structure. A complete financial analysis including life cycle costing for each of the building's components should be developed as part of the program.

The hypothesis should condition the structure's life prediction contingent upon a period of contiunous maintenance of the structure and equipment, with recommendations for an engineering review of the building systems at least every five years. It is recommended that a maintenance manual reflecting these recommendations be prepared and presented to the owner for his use.

## VIII. THE REPORT

The culmination of a life prediction analysis should be the preparation of a comprehensive written report which details the fact-finding mission that resulted in the useful life prediction of a building.

The report should contain the salient aspects of the investigation and the conclusions including the life expectancy of the building and recommendations for a remedial program. The maintenance manual for the building should be included in this document.

The comprehensive report is intended to provide the complete history of the investigation, with conclusions and caveats under one cover, so that the document can be used by the client for funding purposes and as a guide through the renovation process. This document will serve to protect the diagnostic engineer in the event that misuse or neglect of the renovated structure leads to litigation.

# 2 Serviceability and its role in predicting structural life

**A Scanlon,** The Pennsylvania State University, Pennsylvania, USA
**D R Green,** Glasgow University, Glasgow, Scotland

Efforts to predict structural life must consider the serviceability of
the structure.  This paper examines some of the issues related to
serviceability and its impact on projected design life.  Particular
emphasis is placed on concrete structures for which serviceability
usually relates to control of deflections, vibrations, and cracking.
Consideration is given to uncertainty involved in predicting structural
response for new and existing structures and the question of
establishing serviceability limits.  In situ monitoring to detect
damage is also discussed.

## INTRODUCTION

Serviceability refers to the ability of a structure to adequately
perform its intended function in service.  Loss of serviceability may
occur for example if floor slab deflections become excessive causing
damage to non-structural elements, or causing alarm to occupants who
may associate perceptible sagging with imminent collapse.  Other
examples include liquid retaining structures that leak as a result of
concrete cracking, and floor beams and bridges that are subject to
unacceptable levels of vibration.  The service life of a structure is
clearly linked to its serviceability so that to predict service life
one must be able to identify serviceability characteristics and the
manner in which they may change with time.

In the general sense it is necessary to deal with two sides of the
issue.  On the one hand, one must be able to predict structural
response under prescribed conditions of loading and environment.  It is
also necessary to define what is and is not acceptable in terms of
structural response.  Both issues require treatment either implicitly
or explicitly on a probabilistic basis since there will be considerable
uncertainty associated with both structural response prediction and
specification of appropriate acceptability limits.

## TREATMENT OF UNCERTAINTY

Limit states design provides a framework for considering uncertainty in
loading and response through the use of load and resistance factors
that account for expected variability in design parameters.  Extensive

research has been conducted to determine suitable factors for codification to provide for reasonably uniform safety levels for certain classes of structures. Relatively little effort has been expended to date on applying probabilistic concepts to serviceability limit states.

There appear to be two main difficulties. Design parameters affecting serviceability show significantly higher levels of variability than is the case for strength limit states. For example long term deflections of concrete structures depend on creep and shrinkage characteristics, modulus of elasticity, and concrete cracking, all of which exhibit high levels of variability compared for example to the yield strength of reinforcing steel which is the main factor affecting flexural strength of under-reinforced beams. A Monte Carlo study conducted by Ramsey et al [1] indicated coefficients of variation for the prediction of short term deflections of beams of the order of 30% due largely to variability in concrete tensile strength. Given the high levels of variability of creep and shrinkage characteristics [2] even higher uncertainty can be expected in predicting long-term deflections.

At the design stage, when the actual characteristics of the structure are unknown there is a high degree of uncertainty associated with the expected behavior of the structure in service. The fact that for the most part structures in service perform in a satisfactory fashion suggests that current design rules are reasonably conservative. However, given the empirical nature of most serviceability design rules one might expect a wide range in the "serviceability margin" for a given class of structures.


## FLOOR SLAB DEFLECTIONS

With the increasing use of sophisticated analysis and design computer programs, structures are being designed with less "reserve" of strength and stiffness than was previously the case. If insufficient attention is paid to design for serviceability the end product may be a structure with adequate strength but poor serviceability characteristics. A prime example is found in the design of two-way floor slabs. A great deal of attention is usually paid to proportioning the flexural reinforcement whereas the slab thickness is given cursory attention and is usually selected on the basis of empirical minimum thickness rules. Problems often arise when the slab thickness is reduced to minimize dead load or overall structure height. There is usually no difficulty in providing the flexural reinforcement required for strength but the slab may experience excessive deflections due to the reduction in flexural rigidity.

On the other hand, analytical tools are being developed that will increase our ability to predict long term deflections of concrete structures with greater accuracy. Combining the results of these computational techniques with a probabilistic assessment of the expected variability of response parameters will put the assessment of deflection serviceability on a stronger footing. Along with the development of design and analysis tools, well documented data on field measurements are required to provide a data base that can be used to quantify the variability of deflections of structures in service (3,4,5).

## DEFLECTION LIMITS

So far we have addressed the problems of predicting structural response. Design for serviceability must also consider the "demand" on the structure. In other words, how does one define acceptability in terms of serviceability. Design codes (eg ACI 318) generally specify allowable computed deflections in terms of a fraction of the span. These design rules are empirical in nature and have developed over the years based on experience. Since they are somewhat limited in scope they can be considered as a means of keeping designers out of trouble in routine design. Unusual structures or design conditions may require more stringent limitations since it would be impractical to cover every conceivable situation in the design code. A review of the historical development of deflection requirements is given by Warwaruk (6) while an extensive bibliography related to static and dynamic deflections is included in a report by Galambos et al (7).

An attempt to incorporate serviceability considerations in a probabilistic format was made by Reid and Turkstra (8) for the case of floor slab deflections. They noted that since there is no objective absolute measure of structural serviceability, the measurement of serviceability involves value judgments that can only be expressed with respect to a subjective scale of relative merit. By converting such utility scales onto a monetary scale, they indicate that objective economic assessment of situations involving subjective evaluations can be made. The methodology involves combining probability density functions of structural response parameters (eg deflections) with utility functions associated with the response parameter. Considerable research is still required to generate appropriate probability density functions and utility functions to make this approach viable.

## CRACKING

Crack control is generally considered a serviceability measure in that cracking may occur at service load levels. Damage due to cracking includes effect on appearance, loss of liquid retaining ability, loss of stiffness, and enhancement of corrosion of reinforcement due to chloride and moisture penetration. Corrosion damage may be limited to local spalling although eventually a significant loss of strength may be caused by a reduction in steel area (particularly prestressing steel) or loss of anchorage due to excessive spalling. Crack mechanisms and procedures for crack control and methods for evaluation and repair of cracks are described in reports by ACI Committee 224 (9,10).

Cracking can clearly limit the service life of a structure and significant costs may be involved to counteract the effects of cracking.

15

EXISTING STRUCTURES

For existing structures, uncertainty in predicting future response is considerably reduced because the past performance will give an indication of the future performance of the structure. In addition it is possible to take appropriate measurements to determine in-situ structural characteristics. For most serviceability conditions, and problems that develop will become apparent early in the life of the structure, and prediction of the future response will be improved by observations of the structure in service.

Bayesian statistical procedures provide a formal method for improving the prediction of long term deflections if one or more measurements are available. The procedure has been applied by Bazant and Chern (11) to predict creep and shrinkage deformations given a set of actual measurements prior to the time that the prediction is being made. Some work is required to develop the procedure in a form that can be applied by structural engineers in practice for predicting future deformations given a set of deflection measurements.


VIBRATION RESPONSE

Vibration is a serviceability criterion that may be a significant consideration in certain structures. Serviceability issues related to vibration have been examined recently by Tallin and Ellingwood (12). The response of a structure to dynamic loading depends largely on its mass and stiffness characteristics which together determine the natural frequencies of the system. In situ test procedures are available to determine natural frequencies. In addition, these procedures can be incorporated into a monitoring program to identify the onset of damage as detected by a change in natural frequency. Such damage might be significant in terms of both serviceability and strength.


RESPONSE ANALYSIS

All elastic structures have a large number of natural frequencies of vibration which are dependent upon the geometry of the structure and the distribution within it of stiffness and mass. Random forces acting on such a structure will cause it to respond by vibrating predominantly at its natural frequencies. Although the distribution of the amplitudes of vibration at the different frequencies depend on the amplitude of the exciting force and its point of application, the natural frequencies themselves depend only on the distributed structural properties of stiffness and mass. In addition the relative amplitudes of vibration of all points on the structure at each of the natural frequencies are independent of the exciting force. This set of relative amplitudes of natural frequencies is termed the mode shape determined solely by the distributed structural properties. Any structure which remains unchanged will always have the same mode shape. Conversely an isolated change in stiffness or mass will affect some frequencies and modes of vibration more than others, the extent to

which each is affected depending on how involved that stiffness or mass
component is in the different modes of vibration.

During the lifetime of a structure it is possible for changes to occur
both to the magnitude and distribution of mass and stiffness of the
structure. These changes and their effect may not be visually apparent
but may affect structural integrity. Monitoring mode shape provides
evidence of changes in structural stiffness and mass such as might be
caused by local failure and damage.

Whereas visual methods of monitoring structural integrity infer
soundness from lack of evidence of damage, response monitoring provides
positive evidence of structural soundness by confirming that stiffness
and mass are unchanged.

Since 1975 Response Analysis and Monitoring has been used commercially
to monitor the structural integrity of offshore platforms excited by
wave action (13).

In 1981 Structural Monitoring Ltd. were commissioned by TRRL to monitor
four bridges excited by normal traffic flow to examine the viability of
extending the response analysis technique to bridge structures (14). A
complementary SERC sponsored theoretical study was undertaken by Green
and Burns (15) at the University of Glasgow at the same time. The
conclusions from these studies may be summarized as follows:

1. Bridge vibration modes can be simply and accurately measured using
   normal traffic flow to excite the structure. Modern vibration
   analysis equipment makes the identification of natural frequencies
   and geometries a relatively simple procedure, and the accuracy with
   which frequencies and vibration amplitude ratios can be determined
   is quite high so that small changes in frequency can be detected.
2. The most accurate way to use the monitoring technique is on a basis
   of historical comparison. If a baseline measurement of natural
   frequencies is made while the structure is new and probably sound,
   subsequent measurement will show any deviation from that condition.
   Care must be taken to ensure that initially sufficient tests are
   made so that the effects of a variety of traffic loadings, ambient
   temperatures and other variables can be isolated from the results.
3. Correlation between changes in frequency and changes in structural
   form can be achieved through finite element modelling. An initial
   numerical model may be developed and tuned into the initial mode
   shapes of the bridge. Modelling accuracy must be sufficient to
   calculate the effect of damage at various points and of different
   types on the set of frequencies and geometries. Such a series of
   calculations, simulating the effects of the most likely failure
   modes, can provide a failure dictionary so that, if a significant
   shift in natural frequency and mode geometry is measured,
   indicating possible damage, it may be possible to identify the type
   of damage by comparing the changes with those obtained by finite
   element simulation.
4. The costs of Response Monitoring do not at present compare
   favorably with visual monitoring techniques. Vibration monitoring
   equipment is expensive and the analytical techniques require a
   fairly powerful dedicated computer. However, modern bridges are

17

becoming more sophisticated and new exacting disciplines on bridge inspection methods, reporting procedures and structural evaluation may make response analysis more competitive.

CONCLUSION

Modern computational techniques and probability based procedures show promise for placing serviceability assessment on a more rational basis than is presently the case. Much work remains, however, particularly in the area of selecting appropriate serviceability limits. Since such limits are often related to human perception and are therefore highly subjective it may be necessary to combine structural engineering and psychology to arrive at appropriate means of dealing with the subject. Structural monitoring tools are being developed that will provide valuable information under service conditions that will be useful in assessing both strength and serviceability conditions.

REFERENCES

1. Ramsay, R. J., Mirga, S. A., and MacGregor, J. G. "Monte Carlo Studies of Short-Time Deflections of Reinforced Concrete Beams." ACI Journal, V. 76, No. 8, August 1979, pp. 897-918.

2. Bazant, Z. P., and Panula, L. "Creep and Shrinkage Characterization for Analyzing Prestressed Concrete Structures." PCI Journal, V. 25, No. 3, May - June 1980, pp. 86-122.

3. Jokinen, E. P. and Scanlon, A. "Field Measured Two-Way Slab Deflections." Canadian Journal of Civil Engineering, V. 14, No. 4, December 1987, pp. 807-819.

4. Jokinen, E. P., Scanlon, A., and Burghardt, G. J. "Comparison of Field-Measured Deflections for Two Reinforced Concrete Buildings." Proc. of Symposium on Serviceability of Buildings. National Research Council Canada, May 1988, pp. 112-123.

5. Montgomery, C. J., Brockbank, R. G., Stephens, M. J., and Kryviak, G. "Design, Construction and Deflection of Two-Way Slabs." Proc. of Symposium on Serviceability of Buildings, National Research Council Canada, May 1988, pp. 101-111.

6. Warwaruk, J. "Deflection Requirements - History and Background Related to Vibrations." ACI Special Publication SP-60, 1979, pp. 13-41.

7. Galambos, T. V., Gould, P. L., Ravindra, M. R., Suryoutomo, H., and Crist, R. A. "Structural Deflections - A Literature and State-of-the-Art Survey." Building Science Series 47, National Bureau of Standards, Washington, D. C., October 1973, 104 pp.

8. Reid, S. G. and Turkstra, C. J. "Serviceability Limit States:Probabilistic Description." Report ST 80-1, Department of Civil Engineering and Applied Mechanics, McGill University, June 1980.

9. ACI Committee 224. "Control of Cracking in Concrete Structures (ACI 224R-80, Revised 1984) American Concrete Institute, 1984, 43 pp.

10. ACI Committee 224. Causes, Evaluation, and Repair of Cracks. "ACI Journal, Vol. 8, No. 3, May-June 1984, pp. 211-230.

11. Bazant, Z. P. and Chern, J. C. "Bayesian Statistical Prediction of Concrete Creep and Shrinkage." ACI Journal, V. 81, No. 4, 1984, pp. 319-330.

12. Ellingwod, B. and Tallin, A. "Structural Serviceability: Floor Vibrations." Journal of Structural Engineering, ASCE, V. 110, No. 2, February 1984, pp. 401-418.

13. Begg, R. D. and Mackenzie, A. C. "Overall Monitoring of Offshore Structures by Vibration Analysis." Oceanology International, 78.

14. Structural Monitoring Ltd. "Investigation of Vibration Methods for Monitoring the Integrity of Bridges." Report Nos. 205 and 283, 1982.

15. Burns, J. P. A. and Green, D. R. "Dynamic Reanalysis." Civil Comp 85, Proc. 2nd International Conference and Civil and Structural Engineering Computing, London, 1985.

# 3 Reliability of service-proven structural systems

**W Brent Hall and Maolin Tsai,** The University of Illinois at Urbana-Champaign, Urbana, Illinois, USA

The reliability of a structure in service is analysed as a function of successful past performance, including the effects of proof testing on correlated failure modes, and resistance of past service loads. For many failure modes not involving wearout, successful past service of a structure can be likened to a proof load test with uncertainty, and results in smaller probabilities of low strength and increased reliability estimates for older structures. After-service strength is found to be relatively insensitive to assumptions on the distribution of initial strength, and service-proven structural systems are found to have decreased likelihood of gross errors. Compared to single components, these beneficial effects of survival age are found to be more pronounced in non-redundant systems and less pronounced in redundant systems. Examples include chain and parallel ductile structural sytems.

## INTRODUCTION

Not all effects of time and service on a structure are negative. After all, the existing structure is a tested structure. For many failure modes not involving wearout, the successful past performance of a structure can be likened to a proof test with uncertainty, resulting in increased reliability estimates and decreased likelihood of gross errors for older structures. This paper is concerned with these and other positive effects of proven structural performance. It extends the work in [1] to include the influence of system type, redundant or non-redundant, on service-proven structural reliability.

The paper begins with a brief description of proof load testing for single and multiple modes of failure. This is followed by a reliability analysis of structures that have survived past service loadings, and an illustration of the effects of survival age on reliability and gross error. The analysis is then applied to basic weakest-link and ductile-parallel systems and a comparison of service-proven reliability is made for the two types of systems.

## PROOF LOAD TESTING

The basic idea of proof load testing is a simple one: The structure is tested at a known proof load, and if successful, the undamaged structure is assumed to have a strength greater than the proof load for the failure mode tested. One analysis approach for this type of test is based on structural reliability theory, as outlined briefly in the following.

### Before-Test Strength and Reliability

The failure mode to be tested is represented by a performance function

$$Z = g(X_1, X_2, ... X_n) \geq 0 \tag{1}$$

in which the random variables $X_i$ represent uncertainties in the structural geometry, material strength, and other variables. Often it is possible to write this condition in the form

$$Z = R - Q \tag{2}$$

wherein R represents strength and Q load. Violation of the inequality represents failure. That is, the probability of failure before the test is

$$P_F = P(Z < 0) \tag{3}$$

or, in the case of load versus strength, the probability that Q is greater than R. Reliability is defined to be

$$Rel = 1 - P_F = \int_{-\infty}^{\infty} F_Q(r) f_R(r) \, dr \tag{4}$$

in which $F_Q(q)$ is the cumulative probability distribution of load Q and $f_R(r)$ the probability density function of strength R. Further details on these and other methods of structural reliability analysis can be found in many good references including [2].

## After-Test Strength and Reliability

After a proof test at a known load q*, the strength of a successful structure has a probability distribution that is truncated at the proof load. The test screens out low values of strength, giving

$$f''_R(r) = \frac{f'_R(r)}{1 - F'_R(q^*)} \; ; \quad r \geq q^*$$
$$f''_R(r) = 0; \qquad\qquad r < q \tag{5}$$

in which the notation " indicates an after-test (posterior) condition and ' a before-test (prior) condition. The effect of a proof test is illustrated in Figure 1, which shows before- and after-test strength distributions for a normal prior strength with a mean of 200 units and a standard deviation of 40 , tested at a proof load of q*=160. The screening out of low strengths produces increased reliability estimates for proof-tested structures.

## Multi-Mode Proof Testing

A proof test will have an effect on all failure modes that are correlated to the test mode. If the correlation is strong it may be possible to verify safety in several failure modes with a single test (or with a small number of tests). The analysis of a single test mode with correlated service modes is as follows. Let $S_S$ represent successful performance in service for one or more failure modes and $S_T$ represent survival of the test. The conditional reliability for after-test service is then

$$P(S_S | S_T) = \frac{P(S_S \cap S_T)}{P(S_T)} = \frac{P(S^+)}{P(S_T)} \tag{6}$$

$P(S^+)$ is the reliability of a fictitious system with the service failure modes of the untested system in "series" with the test failure mode. Thus, the conditional reliability of a proof-tested system can be found from two system reliability calculations, one for an augmented system including the test and the other for the test alone.

The above result has practical significance, since the work that has been done on system reliability analysis can be applied to proof testing problems, and commercial software packages are beginning to ease the burden of calculations. (See [3] for a good review of system reliability methods.) One can imagine that someday it may be possible to design a *battery* of tests to prove a system with many failure modes, and then to obtain the revised reliability estimate effortlessly from a system reliability package. However, this is not yet possible and we offer here only the following simple illustration.

*Illustration:* A simply-supported beam has a rectangular cross section of dimensions B and H, a material strength of Y, and a span length of L. It has two service modes. The first resists a central load $Q_1$ in bending and the second resists a load $Q_2$ in tension. The strengths associated with the two modes are

$$R_1 = \frac{2YBH^2}{3L}, \quad R_2 = BHY \tag{7}$$

It is assumed that tension and bending modes do not occur simultaneously. Suppose that a test is performed in the bending mode at a proof load of $q_1^*$ and is successful. The after-test two-mode reliability is

21

$$\text{Rel}'' = P(S_1 \cap S_2 \mid S_T) = P[\ (Q_1<R_1) \cap (Q_2<R_2) \mid (R_1>q_1^*)\ ] \tag{8}$$

or, using Equation 6,

$$\text{Rel}'' = \frac{P(S^+)}{P(S_T)} = \frac{P(S_1 \cap S_2 \cap S_T)}{P(S_T)} \tag{9}$$

The numerator is evaluated as a three-mode series system reliability problem and the denominator is readily obtained from single-mode reliability caculations. Because the modes share the common random variables Y, B and H, they will be correlated (which is accounted for in the system analysis), and the bending proof test can have a strong influence on both service modes.

## SERVICE-PROVEN STRUCTURES

A service load acts as a proof test on the structure. A past service load with uncertainty can be modelled as a random proof load, Q*. The "after-service" strength distribution is then

$$f'_R(r) = \frac{F_{Q^*}(r) f_R(r)}{\int_{-\infty}^{\infty} F_{Q^*}(r) f_R(r)\, dr} \tag{10}$$

For a successful structure with unknown but maintained strength (wearout controlled or not yet significant) Q* is the maximum load the structure has survived to age T, giving

$$f'_R(r) = \frac{F_Q^T(r) f_R(r)}{\int_{-\infty}^{\infty} F_Q^T(r) f_R(r)} \tag{11}$$

where $F_Q^T(r)$ is the cumulative probability function for the T-period maximum load. A simple case is statistically independent load maxima for each of T periods. Then T can be regarded as an exponent on the single-period maximum load distribution, $F_Q(q)$. The effect of survival age (T) on undamaged structures is to decrease the probablility of low strengths much as in a proof load test, except that there is not a sharp truncation of the lower tail of the strength distribution. This is shown in Figure 2 for the same initial strength distribution as in Figure 1, and normally-distributed independent single-period load maxima with a mean of 100 and a standard deviation of 30. As can be seen, the estimate of strength improves with survival age T.

### Insensitvity to Assumptions on the Distribution of Initial Strength

The choice of initial strength distribution is not likely to be an important factor (beyond the mean and variance, say) for re-assessment of a structure in service. One of the reasons for this is illustrated in Figure 3, in which four different prior distributions $f_R(r)$ are compared using Equation 11: normal, lognormal, extreme-value, and bimodal representing gross errors in strength. For all of these cases, after T = 20 periods of service loading as above (Figure 2), the after-service strength distributions $f'_R(r)$ are remarkably similar, particularly in their critical lower tails. In fact, it is evident that the lower tail of strength for an existing structure is determined much more by the type of loading it has survived than by its initial strength distribution.

### Time-Dependent Reliabilty

The above approach can be used to find the reliablility for a period t,

$$\text{Rel}(t) = \int_{-\infty}^{\infty} F_Q^t(r) f_R(r)\, dr \tag{12}$$

and the conditional reliability for an additional service period t starting from a survival age of T:

$$Rel(t|T) = \frac{Rel(T + t)}{Rel(T)} = \frac{\displaystyle\int_{-\infty}^{\infty} F_Q^{T+t}(r) f_R(r)\, dr}{\displaystyle\int_{-\infty}^{\infty} F_Q^{T}(r) f_R(r)\, dr} \tag{13}$$

This conditional reliability improves with age; that is, during early and useful life -- before wearout -- the used structure is better than new. This result is particularly true for structures with gross errors, as illustrated below.

## Gross Error in Strength

Many structural failures occur not because of expected variation in strength, but because of a mistake or human error. Here we consider a simple model of gross error in strength. The before-service strength distribution is assumed to be a composite of two distributions, one for error-free structures with unreduced strength R and one for defective structures with reduced strength gR. The occurrence of an error has probability $P(g)$. The result is a distribution of strength values similar to the bimodal distribution in Figure 3(d).

Equation 13 can then be used to find conditional reliability for a structure with survival age T. This is done in Figure 4(a), for a normally distributed strength with a mean of 200 and a standard deviation of 40, as before, but with a possibility of a gross error with a magnitude of g=0.6 and probability P(g)=0.10. The service loading conditions are the same as in earlier examples. As seen in the figure, the resulting conditional reliability estimate is higher for older structures. This indicates a failure rate that is decreasing with age, which is typical of a "burn-in" or "early life" process, in which defective components are screened out in service.

The probability of a gross error also depends on survival age:

$$P(g \mid T) = \frac{Rel(T \mid g)\, P(g)}{Rel(T)} \tag{14}$$

This is illustrated in Figure 4(b) for the above example with g=0.6 and g=0.8. The figure shows a rapid decrease in the likelihood of gross error for a structure in service for time T, and the larger the error (the lower the value of g) the more rapid is the reduction. Excepting wearout, older structures are less likely to contain gross errors in strength than newer structures.

## BASIC STRUCTURAL SYSTEMS

The foregoing analysis applies mainly to a single-mode load versus strength situation. Real structures can be far more complex, involving many compononts and failure modes. Despite their complexity, however, structural systems fall into two basic categories: non-redundant and redundant systems. Here we consider a representative example of each type and examine the effects of system structure and redundancy on service-proven reliability.

## Non-Redundancy: "Chain" Systems

A basic non-redundant system is the "weakest link" series-type system, which includes all statically determinate structures. The simplest case is the "chain" system in which the chain strength C is the minimum of "n" link strengths $L_i$. For statistically independent and identically distributed link strengths, the probability density function of chain strength C is,

$$f_C(r) = n[1 - F_L(r)]^{n-1} f_L(r) \tag{15}$$

in which $F_L(r)$ is the cumulative distribution of link strength and $f_L(r)$ is the corresponding probability density function. Using this in place of $f_R(r)$ in Equations 12 and 13 extends the foregoing analysis to chain systems.

Figure 5 shows the conditional reliability of a chain system for several cases of the number of links n and survival age T. Link strengths are normally distributed with a mean of 200 and a standard deviation of 40, and

23

the single-period load maxima are independent and normal with a mean of 100 and standard deviation of 30. For a fixed number of links, the conditional reliability Rel(t/T) is higher for older structures, as for the single-component case. And not surprisingly, chain reliability is lower for larger values of n, since the greater the number of links the greater the chance of a weak link. But as can be seen in the figure, the improvement in after-service reliability Rel(t/T) increases with the number of elements in the system. In other words, the proof load effects of survival age are intensified for non-redundant systems.

### Redundancy: "Ductile-Parallel" Systems

A basic redundant system is the "ductile-parallel" system, in which system strength D is the sum of element strengths. Thus, for independent and identically-distributed element strengths, the ductile system has a mean and variance in terms of element values (i) as follows:

$$\mu_D = n\,\mu_i, \qquad \sigma_D^2 = n\,\sigma_i^2 \tag{16}$$

Figure 6 shows the effect of the number of ductile elements, n, on conditional reliability for the same conditions as the chain example, except that the element mean strength and standard deviation are taken as 200/n and 40/n respectively. The figure illustrates an opposing influence to that of the chain sytem. As the number of elements increases, the differences in conditional reliablity between new and old structures reduce. The beneficial effects of survival age are attenuated in redundant systems.

### Gross Error

The model of gross error in strength used for single components (Figure 4) can be applied to the basic systems discussed above. Each element has a chance of gross error, and reduced strength, with probability P(g). The probability of all elements being error-free reduces rapidly with the number of elements n, which is most significant for the chain system, since even one defective link can weaken the system considerably. Numerical results for the two example systems are given in Figures 7 and 8. As expected, for both chain and ductile parallel systems the probability of a gross error in the system is initially higher for larger numbers of elements, and reduces with survival age. The systems differ, however, in how rapidly this reduction takes place. Compared to single components, the likelihood of a gross error reduces more rapidly for non-redundant systems and less rapidly for redundant sytems.

## CONCLUSION

The successful past performance of a structure is evidence of its strength and safety. Such evidence may take the form of a formal proof load test of the system in one or more failure modes, or it may arise from the resistance of loads in service. Methods of system reliabilty analysis can be used to solve these and other conditional reliability problems with little or no modification. Service-proven structural systems not subject to wearout are found to have increased conditional reliability estimates and decreased likelihood of gross errors for older structures. These beneficial effects of survival age are intensified in non-redundant systems and attenuated in redundant systems.

## ACKNOWLEDGEMENTS

This research is supported by the National Science Foundation under a grant on reliability models of load testing. The assistance of Mr. David Wears is also acknowledged with appreciation.

## REFERENCES*

[1] Hall, W. B., 'Reliability of service-proven structures,' *Journal of Structural Engineering*, ASCE, Vol. 114, No. 3, March (1988).

[2] Madsen, H. O., Krenk, S., and Lind, N. C., *Methods of Structural Safety*, Prentice-Hall, (1986).

[3] Ditlevson, O., and Bjerager, P., 'Methods of structural systems reliability,' *Structural Safety*, 3 (1986).

* A more extensive list of references can be found in [1].

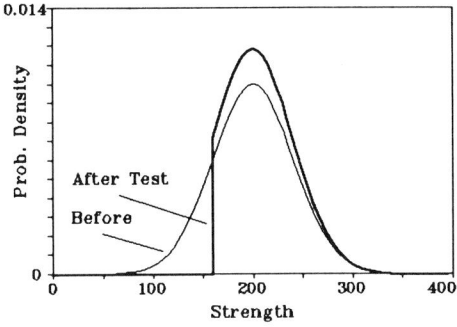

Figure 1 — A Proof Load Test

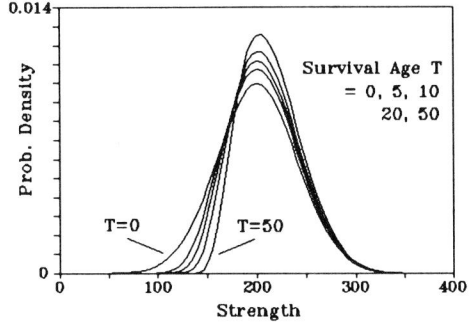

Figure 2 — Effect of Survival Age
on After—Service Strength

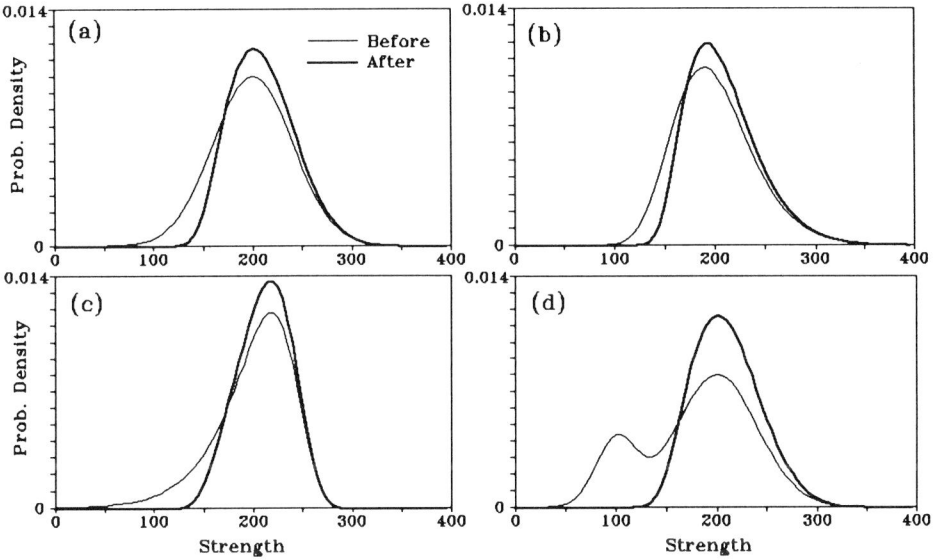

Figure 3 — After—Service Strength Distributions for Four
Initial Strength Distributions (T = 20).

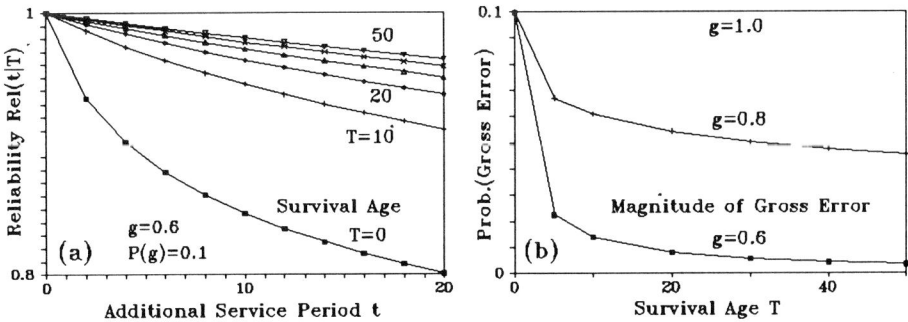

Figure 4 — Effect of Survival Age on (a) Conditional Reliability
and (b) Probability of Gross Error in Service

25

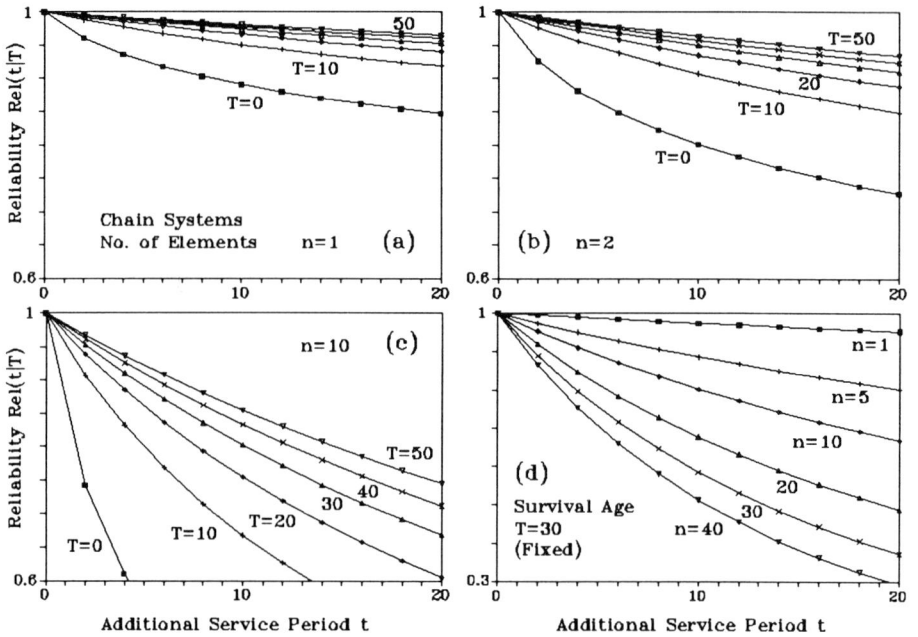

Figure 5 — Effect of Survival Age on Reliability of Chain Systems

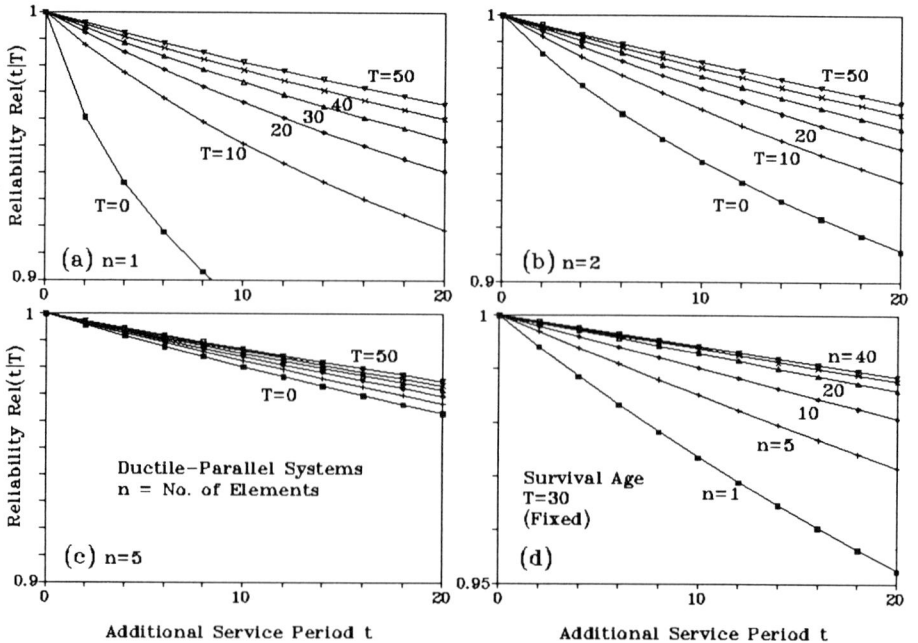

Figure 6 — Effect of Survival Age on Reliability of Ductile Systems

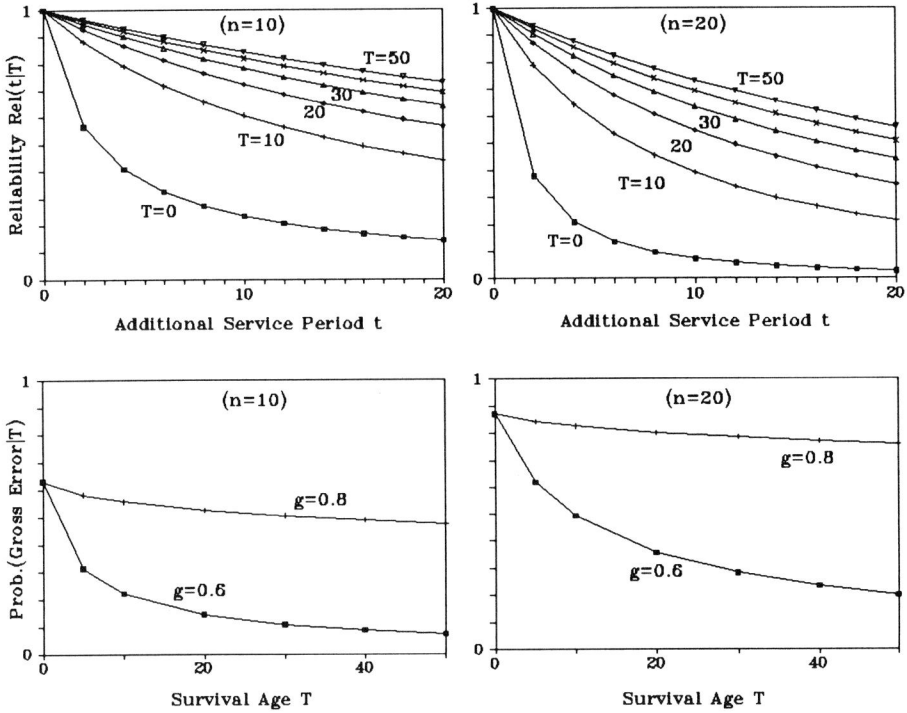

Figure 7 — Service—Proven Chain Systems: Reliability (Top)
and Probability of Gross Error (Bottom) in Service.
(See Figure 4 for the Single—Component Case)

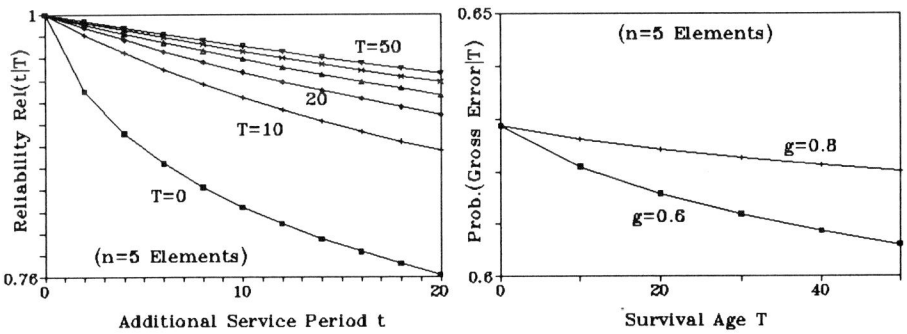

Figure 8 — Service—Proven Ductile—Parallel Systems:
Reliability and Probability of Gross Error for n=5
(See Figure 4 for the Single—Component Case)

# 4 Life cycle financial analysis for bridge rehabilitations

**R S Reel and C Muruganandan,** Ministry of Transportation, Ontario, Canada

The paper presents the methodology of carrying
out financial analysis to compare the life
cycle costs of different rehabilitation
options. A computer program, utilizing Lotus
1-2-3, version 2.01, is developed for the
analysis. Actual life cycles of different
techniques employed on bridge rehabilitation
projects are given. The application of the
technique is illustrated by an actual example
of a bridge rehabilitation project.

## PRINCIPLES OF FINANCIAL ANALYSIS

The technique used for the financial analysis is the present
value analysis.

### Present Value Analysis

The present value analysis technique involves the calculation
of the cost of alternative schemes in present day monetary
terms, i.e. the amount that are required in today's value to
obtain goods and services at any future date.  This is based
on a simple investment principle.

A capital or principal (P) invested for n years at an
interest rate (r),  compounds to a sum (C) such that:

$$C = P(1 + r)^n$$

Stated in another way the net present value (P) of a sum (C)
spent in year n at a discount rate (r) is:

$$P = \frac{C}{(1 + r)^n}$$

This is the basis of net present value analysis as it allows
for the comparison of alternative schemes on an equitable
basis. In general, the alternative with the least present
value is the preferred alternative.

Consider two options for a bridge rehabilitation project,
such that:

Option (1): Time frame to be considered = $N_1$ years
Cost Estimate                    = $C_{1n}$  (n = 1, .....$N_1$)
$C_{1n}$ = the cost of Option (1), over the time frame considered.

Option (2): Time frame same  as for Option (1) = $N_1$ years
Cost Estimate                      = $C_{2n}$  (n = 1, .....$N_1$)
$C_{2n}$ = the cost of Option (2), over the time frame considered.

The present values for option 1 and 2 are

$$PV_1 = \sum_{n=1}^{N_1} \frac{C_{1n}}{(1+r)^n} \qquad\qquad PV_2 = \sum_{n=1}^{N_1} \frac{C_{2n}}{(1+r)^n}$$

## Parameters Required for Financial Analysis

The following parameters are required to perform the financial analysis.

Parameters related to the proposed rehabilitation or replacement of each alternative.
-  Capital costs
-  Life cycle
-  Residual life
-  Future maintenance costs

Parameters related to the existing condition of the structure.
-  Estimated residual life without remedial work.

Parameters related to discount rate.
-  Rate recommended by the agency or the government.
-  Variations in discount rate.


Capital Cost

The following should be estimated for each alternative in constant monetary terms.
-Engineering design cost
-Construction cost
-Miscellaneous costs such as, demolition, right-of-way, approaches, utilities, stream-diversion, detours, etc.


Life Cycle

The life cycles of the alternatives should be estimated for the financial analysis.  Usually, it is the time between two successive replacements or rehabilitations.

The assumed life cycle values given in Table 1 are based on experiences in Ontario. These may be varied based on local experiences.

TABLE 1  Life Cycles

| Treatment and Treatment Elements | Life Cycle |
|---|---|
| **Paving** | |
| 1. Asphalt | 15 |
| 2. Mastic waterproofing and asphalt | 15 |
| 3. Hot applied rubberized waterproofing | 30 |
| . Mill and replace top course of asphalt | 15 |
| . Replace waterproofing and asphalt | 30 |
| | |
| **Bare Concrete Deck slabs** | |
| 1.   0 mm to  75 mm deck slab - plain reinf. | 5 |
| 2.  75 mm to 125 mm deck slab - plain reinf. | 10 |
| 3. 125 mm to 180 mm deck slab | |
| . Plain reinforcement | 10 |
| . Top layer, epoxy coated | 20 |
| 4. 180 mm to 225 mm deck slab | |
| . Plain reinforcement | 15 |
| . Top layer, epoxy coated | 25 |
| 5. Solid thick deck slabs | |
| . Plain reinforcement | 10 |
| . Top layer, epoxy coated | 20 |
| 6. Voided thick slabs | |
| . Plain reinforcement | 10 |
| . Top layer, epoxy coated | 20 |
| | |
| **Coating Systems for Structural Steel** | |
| 1. Alkyd System | 10 |
| 2. High Build Alkyd System | 10 |
| 3. Inorganic-Zinc / Vinyl System | 20 |
| 4. Epoxy-Zinc / Vinyl System | 20 |
| 5. Aluminum filled Epoximastic System | 10 |
| 6. Hot Dip galvanizing | 20 |
| 7. Zinc Metallizing | 20 |

Residual Life and Value.

The various alternatives considered may have useful lives at
the end of the time-frame. This is termed as the residual
life.  There are no specific methods of assessing this. A
thorough knowledge of the performance of past rehabilitations
and experienced engineering judgement are probably the best
way of assessing the useful residual life. From the residual
life the residual value of the structure is calculated. There
are several methods available for calculating the residual
value. The method used here is the 2nd cycle replacement
method.

Let the 2nd replacement of Option (1) be $N_1$ years.
and 2nd replacement of Option (2) be $N_2$ years.

The calculations for residual values are shown in Table 2:

TABLE 2   Residual Value

| Option | Year of Replacement (2nd cycle) | Replacement cost | Residual Years | Value at Year $N_1$ | Differential Value (option 1 as base) | Residual Value at Year 0 |
|--------|----------|----------|----------|----------|----------|----------|
| 1 | $N_1$ | C | 0 | $C_1$ | $C_{D1}$ | $C_{R1}$ |
| 2 | $N_2$ | C | $N_2-N_1$ | $C_2$ | $C_{D2}$ | $C_{R2}$ |

Where r = Discount rate          C = Replacement cost.

$$C_1 = C$$

$$C_2 = C_1/(1+r)^{N2-N1}$$

$$C_{D1} = C_1-C = 0$$

$$C_{D2} = C_2-C$$

$$C_{R1} = 0$$

$$C_{R2} = C_{D2}/(1+r)^{N1}$$

Maintenance Costs

Costs associated with maintenance are the routine maintenance costs. These would include minor repairs, maintenance, touch up painting, etc., carried out on a regular basis to minimize the incidence of costly and dangerous situations that may arise from unexpected deterioration of the bridge.

The impact of maintenance costs is not very significant on the majority of projects, but may be useful to consider when the present value of two alternatives are close to each other. However, if these costs are not available the analysis could be performed without them.

Probabilities of Occurrences

In an environment of uncertainties determining cost estimate are not accurate. The degree of uncertainty is reduced by assigning probabilities to various estimated costs. If $C_1, .C_2, \ldots \ldots C_n$ are n estimated costs with probabilities of occurrence, $p_1, p_2, \ldots \ldots \ldots p_n$, then the expected costs $c^1, c^2 \ldots \ldots c^N$ are given by:

$$c^1 = p_1C_1 + p_2C_2 + \ldots \ldots \ldots + p_nC_n$$

## Discount Rate

Benefits and costs of government expenditure may be realized over different time periods. To allow projects to be compared on an equitable basis the costs and benefits should be multiplied by a discount factor.

$$\text{Discount factor} = \frac{1}{(1+r)^n}$$

Where r is the discount rate and n is the number of the years. The appropriate discount rate for government projects will depend on several factors[1], such as magnitude of investment return, tax rates, capital market conditions, preferences for current and future consumption, methods used to finance projects etc. A discount rate of 6% is recommended in this paper with sensitivity analysis using 3% and 9% for projects owned by government agencies. These rates may be different for other agencies.

## Effects of Inflation

During inflation there is a difference between current and real prices, and similarly between nominal and real discount rates. The effects of inflation or deflation may be treated as follows[2]:

### Constant Inflation Rate

If the rate of inflation is constant at f then the relation between nominal cost and real cost is given by:

$$C_n = C_o (1 + f)^n \quad \ldots \ldots (a)$$

where $C_n$ is the cost at year n, and $C_o$ is the real cost at year 0.

The relation between the real and the nominal discount rate is given by:

$$(1+R) = (1+r)(1+f) \quad \ldots (b)$$

where R = nominal discount rate and r = real discount rate

Substituting these in the equation for present values gives:

$$PV = \sum_{n=1}^{N} \frac{C_n}{(1+R)^n} = \sum_{n=1}^{N} \frac{C_o(1+f)^n}{[(1+r)(1+f)]^n} = \sum_{n=1}^{N} \frac{C_o}{(1+r)^n}$$

From the above it is evident that f has no effect on the analysis.

Relative change in Inflation Rate

When there is a relative change in inflation rate the general prices move at an inflation f and the construction costs move at $f^1$.

The equations (a) and (b) become:

$$C_n = C_o(1 + f^1)^n \ldots\ldots(a^1)$$

$$(1+R) = (1+r)(1+f) \cdots\cdots (b^1)$$

Substituting these in the equation for present value gives:

$$PV = \sum_{n=1}^{N} \frac{C_n}{(1+R)^n} = \sum_{n=1}^{N} \frac{C_0(1+f^1)^n}{[(1+f)(1+r)]^n} = \sum_{n=1}^{N} \frac{C_0}{\left[(1+r)\frac{(1+f)}{(1+f^1)}\right]^n}$$

for small values of $r, f$ and $f^1$, this is simplified as:

$$PV = \sum_{n=1}^{N} \frac{C_o}{[1+r-(f^1-f)]^N}$$

Where the adjusted discount rate $= r - (f^1 - f)$

The difference between the discount rate r, and the adjusted discount rate is small and therefore inflation has no effect on the analysis.

## Present Value Analysis Using PRVAL Program

PRVAL is a template overlay developed to perform financial analysis for bridge rehabilitation projects using Lotus 1-2-3, Version 2.01, on a worksheet format. There are four different options available to carry out the financial analysis at different levels of sophistication. Sensitivity analysis can be carried out for each level by varying the discount rates by ±3%.

The four options available are:

PRVAL01.WK1 - Level 1 Analysis - Capital Costs.
PRVAL02.WK1 - Level 2 Analysis - Capital Costs and Residual Values.
PRVAL03.WK1 - Level 3 Analysis - Capital Costs, Residual Values and Maintenance Costs
PRVAL04.WK1 - Level 4 Analysis - Probability Analysis

# Examples of Financial Analysis

The example is the selection of the most economical option for a deck rehabilitation project. The project engineer has come up with the following data and options as part of his engineering assessment.

a) The time frame to be considered is 50 years.

b) The life cycles for the various treatments are as follows:

1) Deck replacement                                    50 years
2) Rehabilitation - Patch, Waterproof and Pave         15 years
3) Milling and replacing top coat of asphalt           15 years
4) Replace waterproofing and asphalt                   30 years
5) Residual life without rehabilitation                 5 years

c) The three options considered are:

Option (1): Immediate rehabilitation: patch, waterproof and pave. This would extend the useful life of the deck by 15 years, after which the deck will be replaced. In year 30 mill and replace the top asphalt and in year 45 replace waterproofing and asphalt. The deck will require a second replacement in year 65.

Option (2): Do nothing now and replace the deck, waterproof and pave in 5 years, at the end of the useful life of the bridge. In year 20 mill and replace the top asphalt surface. In year 35 replace waterproofing and asphalt. In year 50 mill and replace the top asphalt. The bridge will need to be replaced in year 55.

Option (3): Replace the deck now, waterproof and pave. In year 15 mill and replace the top asphalt surface. In year 30 replace waterproofing and asphalt. In year 45 mill and replace top asphalt surface. In year 50, replace deck.

The results of the financial analysis performed at the fourth level, for the above example are shown below.

Financial Analysis. - Level (4).

TABLE 3   Cost data (Can. $)

| Year | Option 1 | Option 2 | Option 3 |
|------|----------|----------|----------|
| 0    | 600000   |          | 1000000  |
| 5    |          | 1000000  |          |
| 15   | 1000000  |          | 100000   |
| 20   |          | 100000   |          |
| 30   | 100000   |          | 200000   |
| 35   |          | 200000   |          |
| 45   | 100000   |          | 200000   |

Name of Bridge     : ABC
Site number        : 12-04
Number of options  : 3
Discount rate      : 0.06
Number of entries  : 7
Life cycle         : 50

TABLE 4   Output

| | Option 1 | | Option 2 | | Option 3 | |
|-------|---------|---------|---------|--------|---------|---------|
| Years | Cost | P.V | Cost | P.V | Cost | P.V |
| 0  | 600000  | 600000 | 0       | 0      | 1000000 | 1000000 |
| 5  | 0       | 0      | 1000000 | 747258 | 0       | 0       |
| 15 | 1000000 | 417265 | 0       | 0      | 100000  | 41727   |
| 20 | 0       | 0      | 100000  | 31180  | 0       | 0       |
| 30 | 100000  | 17411  | 0       | 0      | 200000  | 34822   |
| 35 | 0       | 0      | 200000  | 26021  | 0       | 0       |
| 45 | 100000  | 7265   | 0       | 0      | 200000  | 14530   |

|  | Option 1 | Option 2 | Option 3 |
|--|----------|----------|----------|
| Total P.V. | $1,041,941 | $ 804460 | $1,091,079 |
| Res.Value | (31,636) | (13721) | 0 |
| Net P.V | $1,010,305 | $ 790,739 | $1,091,079 |
| Net P.V (Adjusted for uncertainty) | $1,030,511 | $ 802,600 | $1,112,900 |

References

1. Robert Haveman and Julius Margolis. Public expenditure
   and Policy Analysis. Third edition. Houghton Mifflin
   Company Ltd. Boston, 1983, Chapter 12.
2. Robert Sugden and Alan Williams. The principles of
   practical Cost-benefit analysis. Oxford University
   Press, Walten Street, Oxford, U.K. 1978.

# 5 Life prediction and optimum safety

**T G Kármán,** Building Science Institute, Hungary

This paper presents practical design life categories to be used for designing load-bearing structures and some special considerations concerning
optimum safety of temporary structures and monumental edifices respectively. It takes into
account changing in structural resistances,varian-
ce in maximum load distributions and difference
in discount coefficients all of them as results
of changing in design life.

The service of a load-bearing structure - it means the life
of this structure - will be finished when it comes

- either to a technical condition
- or to a functional situation

which proves to be inconvenient - that is to say unprofitable,
unreliable or unable - to ensure further exploitation of
structure. Consequently in connection with a load-bearing
structure the concepts of technical service life and of functional one have to be distinguished. The technical service
life depends first of all on structure itself, on its material
feature, safety level and maintenance conditions, while functional service life will be determined by process of changing
in use of building.

Designing new load-bearing structure the assumed and expected
functional service life has to constitute design requirement
which must be satisfied in course of design by selecting appropriate structure, namely convenient structural material,
righs safety level and good directions for maintenance of
structure. This can be demonstrated by following formel design
demand:

$$FSL \leq TSL$$

... 1

To apply this simple Formula 1 is just as sophisticated as
in case of any other similar design demand because of following essential reasons:

- both sides of inequality and so realization of it
  too mean events being random variables,

- the probability level on which demand should be
  satisfied - the so called optimum safety - depends
  on hazy economical considerations,

- realization of inequality on this optimal level
  generally needs considerations for extreme frac-
  tiles of    not-well-revealed    probability distributions.

To estimate the left side of demand, that is to say to predict
functional service life of a building or any other structure
to be designed, moreover to predict its probability distribution is not the business of structural engineer. The life
requirement should have to be supplied by expected utilizer
of structure. However in order to avoid over-autocracy of
customers it seems to be necessary to create official directions for expected service life - what is called design life -
of most frequent types of buildings. Studying a lot of national codes of practice it can be found that most generally used
official categories of life-time are as follow:

- temporary structures such as fair stands, site
  buildings and the like with a design life of 3-5
  years,

- permanent buildings as the most common category
  which includes usual residental and office buildings,
  etc. having a required life-time of 40-50 years,

- monumental edifices, churches, theatres and so on
  with an expected life-period of 150-250 years.

As an over-simplification let us suppose that according to
Figure 1 designing new load-bearing structure means that -
knowing all about load distribution  F(L) - engineer has to
select appropriate structure resistance distribution  F(R)
of which assures that convolution of both distributions
  $\Sigma P(L > R)$ will be equal to or lower than optimal probability
of failure. In this case life requirement controls design of
load-bearing structures that is to say estimate of the right
side of inequality in the following three main respects:

- it is a common knowledge that resistance characteristics and maintenance demand of the various
  structural materials generally depend on its
  service life,

- it can be proved that load characteristics, first
  of all the very important maximum load distribution
  change with changing of life-time,

- it isn't well-known but it must be assumed that
  optimum safety will also be influenced by design
  life.

In respect to the first group we suppose - at least we hope -
that structural materials including their durability characteristics are well-known and engineers are qualified to select
the appropriate one for structures of whichever life-time
category without any difficulties. In addition it is a generally used design approach for benefit of safety that final
resistance or rather its distribution can be taken into
account as being structural characteristic during the whole
life-time. Further technical details of the problem of material-
degradation isn't matter of our investigations at issue.

In a most general meaning each load of structure varies in
time and these changes create stochastic processes. A lot of
empirical analyses concerning snow-, wind-, traffic-, floor-
load etc. show that the I type of Gumbel extremum distributions can be assumed as being assipmtotic distribution of maximum loads occuring in a given time-period.
Formula 2 shows this distribution:

$$F = \exp\left[-\exp\left(-\frac{\tilde{\pi}}{\sqrt{6}}\frac{L-m}{s} - c\right)\right] \qquad \dots 2$$

where m means mean value of
        L load being random variable,
        s means standard deviation and
        c = 0.577216 the Euler-constant.

For a time-period being t-time longer than given one a similar distribution can be obtained which coincides in all respects with original one only that it will be slided with a
load difference showed in Formula 3:

$$\Delta L = \frac{\sqrt{6}}{\pi} s \ln t \qquad \dots 3$$

Let us adopt this Formula taking into account design load
of permanent buildings as being basic load. Then the design
load difference to be used to temporary structures and to
monumental edifices respectively will be as follow:

$$L_T = \frac{\sqrt{6}}{\pi} s \ln (4/45) = -1.89 s \text{ and}$$

$$L_M = \frac{\sqrt{6}}{\pi} s \ln (200/45) = +1.16 s$$

In such a way load and resistance of structure belonging to
any life-time category can be obtained. But how to determine
the right safety level which ought to ensure economical optimum of load-bearing structure belonging to a given life-
time category?

Let us admit that economical optimum can be realized by
structure the total expenses of which will be the lowest one,
that is to say that all costs of creation of structure
including design, production and construction costs as well
as losses caused by accidental failures during the whole

life-time will be minimum as it is shown in Formula 4:

$$TOTAL = COST + P \times D \times LOSS \rightarrow MIN \qquad \dots 4$$

where P means probability of failure and
D means discount coefficient being used for reduceing
losses by time of construction.

In Formula 4 it is easy to recognise that optimal probability of structural failure depends on design life. Namely in this regard design life has an influence on discount coefficient which is shown in Formula 5:

$$D = \frac{q^N - 1}{N(q-1)} \qquad \dots 5$$

where $\qquad q = \dfrac{1}{1+p} \qquad \dots 6$

Here $\qquad$ N means design life and
p means discount rate.

Taking into account a discount rate p = 0.05 and in turn N = 4, 45 and 200 years as average design lives for each life-time category the ratios of discount coefficients are in turn D = 2.24, 1.0 and 0.25. Optimal probability of structural failure has to be inversely proportional to discount coefficient as it can be recognised in Formula 4.

On theoretical considerations and on practical estimations as well it can be assumed that probability of failure depends on structural resistance as it is shown in Formula 7:

$$(P1/P2)^b = (R2/R1)^{100} \qquad \dots 7$$

where $\qquad$ R1, R2 mean resistance of structures 1 and 2 both
used in the same load condition,

$\qquad$ P1, P2 mean probability of failure of structures
1 and 2,

$\qquad$ b $\qquad$ exponent depends on feature of load and of
resistance distributions and has its value
between 2 and 5 with an average value of 3.

Adopting Formula 7 the average supplementery safety factors to be used to temporary structures and to monumental edifices respectively are as follow:

$$\gamma_T = 2.24^{3/100} = 1.025 \text{ and}$$

$$\gamma_M = 0.25^{3/100} = 0.960.$$

Summarized our results we can show design demand for any life-time category in the following manner:

- for temporary structures:

$$L + \Delta L_T \leq R_T / \gamma_T$$

- for permanent buildings:

$$L \leq R_P$$

- for monumental edifices:

$$L + \Delta L_M \leq R_M / \gamma_M$$

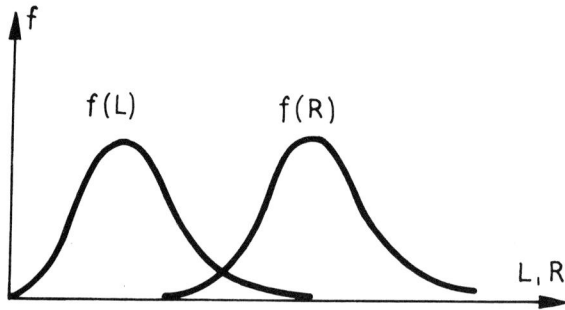

Figure 1

Load and resistance density functions

# 6 The life of structures

**W R Millard and M J Robinson,** Thermalite Ltd, Station Road, Coleshill, Birmingham, UK

This paper illustrates the ways in which Autoclaved Aerated
Concrete manufacturers have attempted to predict and control
the lives of buildings constructed with their material.
The life of a building is seen as dependent on a number of
factors and examples are given of how a building materials
producer has succeeded in influencing these factors through
physical testing.

The life of a building depends on the following factors:

1. THE DESIGN OF THE BUILDING

2. THE STANDARDS OF CONSTRUCTION

3. THE QUALITY OF THE INDIVIDUAL COMPONENTS

4. ECONOMICS AND TRADITION

The building materials producer can influence all of these factors to various
extents, and the following describes how autoclaved aerated concrete (AAC)
producers throughout the world have gained confidence in their products through
experience and testing, allowing them to predict and control the Life of
Structures built from the material.

## 1. THE DESIGN OF THE BUILDING

By having full awareness of current regulations and the capabilities and limits of
performance of their products, manufacturers can influence building design
through arguments along economic and efficiency lines. The life of a building is
governed by its design, whether the end of the life of the building is caused by the
building's being no longer suitable for its original purpose or the failure of the
building to survive environmental conditions.

The following are 8 examples of manufacturers discovering and verifying their
products' properties and helping their users to take advantage of special product
properties to the benefit of guaranteed life of their products and structures.

### 1.1.   Resistance to Freeze Thaw Damage

AAC can withstand repeated cycles of freezing at -20°C in the saturated state
and thawing. This property is attributed to the pore structure of the material
which consists of large pores which do not fill with water, and smaller pores.
It is believed that in the very fine pores water does not solidify at -20°C.
However, in areas where ice is formed, expansion is permitted into the larger
pores of the material without damage to the cellular structure.

The photographs Fig. 1. show specimens of AAC which have undergone 60 cycles of freezing and thawing and have suffered no noticeable deterioration.

Figure 1

The compressive strength of AAC after freeze thaw cycling is unaltered. The presence of ice in frozen AAC, contributes significantly to the compressive strength and values as high as 15 N/mm$^2$ have been measured in frozen material.

This property of AAC is particularly beneficial below d.p.c. where many other materials including clay bricks and dense concrete blocks, suffer frost damage when wet and frozen during the Winter. Special sized blocks are manufactured for use below d.p.c.

## 1.2. Resistance to Chemical Attack

The resistance of AAC to chemically hostile soils has been amply demonstrated in the Northwick Park Project (1) where samples of various building materials were submerged in sulphate bearing soils for a period of 15 years.

Fig.2. The block on the left was submerged in sulphate bearing soil of class 4. The block on the right was stored in 15g/litre SO$_3$ as Mg SO$_4$ (at BRE) for 15 years. No deterioration was suffered and in some cases a strength increase was measured.

Figure 2

Agrement certification for the use of Autoclaved Aerated Concrete in sulphate bearing soils was achieved in 1968 (2).
This is further evidence of the suitability of AAC for use below d.p.c. at foundation level. The manufacturer's own research findings, together with those of the BRE, have been used here to provide confidence in the ability of structural blockwork to survive the ravages of thermal and chemical attack and hence contribute to confidence in the life of a building.

## 1.3.    Variation of Strength with Age

The variation of compressive strength with age has been monitored both in Europe and Britain.   This was achieved by storing blocks, exposed to normal climatic conditions in the U.K., near Birmingham and in Germany near Munich.   Blocks were then tested for compressive strength at regular intervals.   Typical strength age graphs are shown in Fig.3.

The graphs show no reduction in strength and in some cases an increase can be seen.

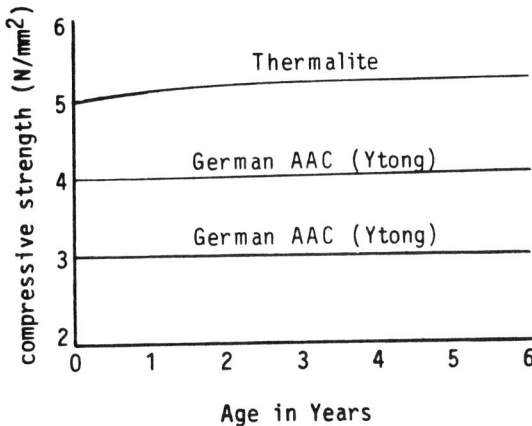

Figure 3

Recently a 60 year old house constructed in Sweden from AAC was demolished, and when blocks from the walls were tested, the compressive strength was found to be unchanged from its original value.

## 1.4.    Fire Resistance

The low thermal conductivity of AAC contributes to its ability to withstand high temperatures in fire conditions.

Numerous loadbearing and non-loadbearing panel tests at the Fire Research Station, have proved the ability of the AAC at all densities to withstand temperatures of up to 1000°C for up to 6 hours (3).
Work carried out at BRE in 1986, (4) proved the ability of AAC to contribute to the fire resistance of steel structures.   The diagrams in Fig.4. show how AAC blocks protect the web of the steel beam and how similar fire resistance times can be obtained with smaller cross section steel beams insulated by AAC blockwork.

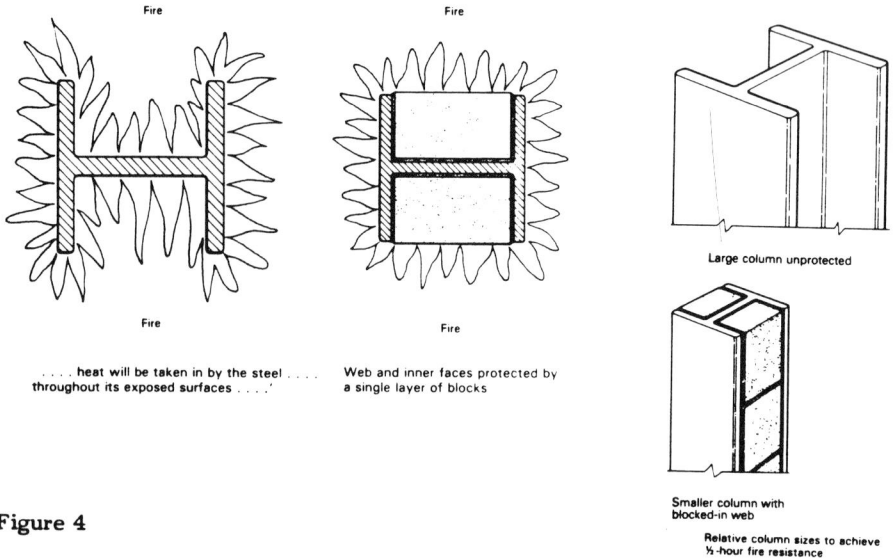

Fire     Fire

Fire     Fire

'. . . . heat will be taken in by the steel . . . .
throughout its exposed surfaces . . . .'

Web and inner faces protected by
a single layer of blocks

Large column unprotected

Smaller column with
blocked-in web

Relative column sizes to achieve
½-hour fire resistance

**Figure 4**

## 1.5.     Sound Insulation

The sound performance of AAC is significantly superior to that which would be
predicted by Established Mass Laws. This has been proved in numerous field and
laboratory tests (5). The graph in Fig.5. shows sound reduction related to density
predicted by the Mass Law (6) compared with the actual performance of
Autoclaved Aerated Concrete at various densities.

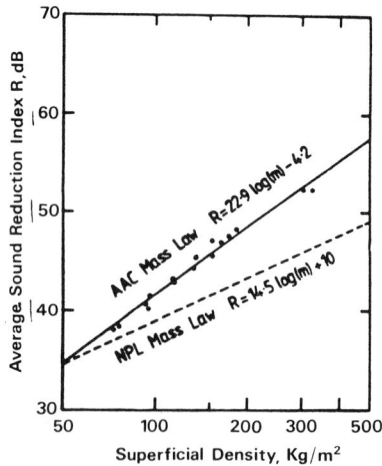

**Figure 5**

Superficial Density, Kg/m²

## 1.6.     Education of Specifiers and Users

The larger AAC manufacturers produce technical literature and organise
symposia (7), in which renowned, independent experts in particular fields,
present papers to Architects, Specifiers and Builders. On a smaller scale,
presentations by experts in structural design and testing give audio visual
presentations and demonstrations to meetings of Building Control
Officers, Architects, Builders and Specifiers.

1.7.    Involvement with Building Regulations, British and European Standards

AAC manufacturers are active through their trade association, the Autoclaved Aerated Concrete Products Association on British and European standard committees. The aim of this work is to ensure that any changes in regulations are forseen and can be catered for, whilst ensuring that the special benefits of AAC are fully exploited. Committees of current interest include those concerning fire regulations, acoustics, thermal insulation, structural regulations and testing.

The foregoing properties can all be seen to contribute to the life of a building, whether by sustained performance in normal conditions or by the ability to withstand exceptional conditions.

1.8.    Technical Services

AAC manufacturers are well equipped to carry out testing in accordance with British Standards under the NAMAS Accreditation Service. Specialist testing is also carried out in association with such bodies as BBA and NHBC.

A computer aided structural advisory service is also available to recommend the most suitable product to comply with current regulations in accordance with customers' requirements.

## 2. THE STANDARDS OF CONSTRUCTION

This is perhaps the most difficult area for the materials producer to influence, because of the ways in which building construction is administered and the use of independent traditional trades.  The correct use of the correct mortar, for example, will significantly contribute to the life of blockwork, and there has been cooperation between AAC block manufacturers and mortar producers (7).  The correct use of suitable fixings (9) in construction, will have a bearing on the maintenance necessary throughout the life of a building which can in turn influence the life of a building.

Technical Advice and site visits as well as liaison with such bodies as NHBC, BRE, CIOB, RIBA aimed at influencing correct use of AAC is frequently organised by manufacturers.

## 3. THE QUALITY OF THE INDIVIDUAL COMPONENT

This is the area were the producer can have the greatest influence. AAC is at an advantage here, since its manufacture has always involved rather more sophisticated technology than is necessary for the production of other building materials. This is borne out by the following history of the development of the process.

The first method of producing aerated concrete was patented by E Hoffmann in 1889 (10). The aeration was produced by carbon dioxide generated in the reaction between hydrochloric acid and limestone.
In 1917, a Dutch patent was registered using yeast as an aerating agent. Later patents involved reaction between zinc dust and the alkalis in the cement mixture, hydrogen peroxide and air foaming.

The first documented attempt at autoclaving aerated concrete was in 1927 (11) in Sweden. The discovery was almost accidental in that a lecturer (Dr Axel Eriksson) decided to autoclave a porous mass of burnt shale limestone, water and aluminium powder. The result was that the porous mass survived autoclaving with a much increased strength. This discovery soon led to the development of an autoclaved aerated concrete industry using local raw materials, producing what were known as "warm stones".

In the late 1940's, research was carried in Britain in the use of pulverised fuel ash from coal burning Power Stations together with sand, cement and water to produce autoclaved aerated concrete. The production of Thermalite commenced at Reading in 1951.

Only one size and density of block was produced and although the properties of the block were intended to combine the strength of stone with the workability of wood, many of the other special properties of the material were not discovered until later. The insulating properties of autoclaved aerated concrete were subsequently to become its most important properties. At this time Building Regulations required that the 'U' values of walls were at least 1.7 W/m²k and this could be achieved using two leaves of brick in a cavity wall. Constructing the inner leaf from 100 mm autoclaved aerated concrete reduced the 'U' value to 1 W/m²k. When a 'U' value 1.0 W/m²k became a mandatory requirement, autoclaved aerated concrete was already an established product and when the regulations changed to requiring a 'U' value of 0.6 W/m²k in domestic dwellings, manufacturers were quick to provide numerous solutions for Architects and Builders. The nature of the manufacturing process permitted easy control of strength and density whilst wire cutting facilitated easy production of a wide range of block sizes from 50 mm thick 275 mm thick.

Increased production and an increase in the variety of uses to which AAC was put (e.g. Trenchblocks used in foundations and Party Wall blocks used in Semi-detached houses) together with the imposition of more stringent regulations have brought about the need for increased sophistication and frequency of testing. To this end, Thermalite became a BCS Accredited Laboratory for thermal conductivity testing to BS874 (12) using the plain and guarded hot-plate methods in 1981 and a NAMAS (13) accredited block testing laboratory for testing compressive strength, dimensional accuracy, density and drying shrinkage in 1988. This ensures that the calibration and quality systems within the company are subject to the closest scrutiny by an independent national body and that claimed laboratory test results are beyond dispute.

The basic standard to which blocks are manufactured is BS6073: 1981. But in order to satisfy Building Regulation requirements a number of additional standards such as BS874, Thermal Conductivity, BS2750 Sound Insulation, BS476 Fire Testing have to be complied with. Compliance with the above standards has to be demonstrated by regular and frequent quality testing.

The widely accepted definition of the word "quality" is "fitness for purpose". Fitness for purpose is conveniently defined for structural blocks in BS6073: 1981. A number of AAC block factories are now able to offer "Quality Assurance" as defined in BS5750. This means that the quality systems of the factories have been subjected to third party assessment and have been found to comply with the requirements of the British Standards Institution. The essential features of the quality system are:-

3.1.    that a Senior member of staff is responsible for the operation of the system and is independent of sales of production pressures.

3.2.    a well documented system of procedures and instructions is in use.

3.3.    records of all inspections are maintained.

3.4    adequate training of staff is carried out.

3.5.    rejected products are segregated to prevent their accidental use.

3.6.    adequate packaging and delivery arrangements are provided.

Further guarantees of the fitness for purpose of products are provided by British Board of Agrement Certification (14), which exists for a wide range of AAC blocks and covers performance requirements including structural, thermal, acoustic, sulphate attack, fire resistance and freeze thaw properties.


## 4. ECONOMICS AND TRADITION

The life of a building is not necessarily a function of how long it stands up before it falls down, it may depend upon positive or negative decisions about maintenance. A decision may be made to demolish a building which is unsuitable for other than structural reasons. High rise flats provide a particularly striking examples of this. It has also been known for multi-storey buildings to be demolished for purely economic reasons before construction was complete. This example was seen in the early 1980's in Wanchai, Hong Kong when a 25 storey block of flats, which had been built up as 21 storeys was demolished before its construction was complete. The decision was made on purely economic grounds because in a three month period the rental charges for office space doubled, making flats a less attractive investment for the developer than offices. The solution was to demolish the flats and build offices.

In the area of social factors affecting the life of buildings, sound insulation tests have been carried out on a large number of buildings containing Autoclaved Aerated Concrete materials in separating and flanking walls. In buildings where tests have been repeated after a number of years, no change in the results has been detected. This should ensure that AAC blocks will continue to provide adequate sound insulation and reduce the likelihood of buildings being demolished as unsuitable.


## CONCLUSION

Over its 50 year History, AAC has been subjected to close scrutiny, particularly with regard to its longevity. In order to find a place in an industry steeped with tradition, this relatively novel product has been subjected to prejudice and preference by non users and users respectively. The use of AAC in the world has grown from around 1 million $m^3$ in 1955 to approximately 15 million $m^3$ in 1988. Thermalite declared its complete confidence in its products by issuing a 100 Year Guarantee in 1985.

The aim of AAC manufacturers is to provide sufficient information for specifiers and designers to be confident that the material will survive normal and adverse conditions for what is likely to be a longer period than the Life of the Building.

# REFERENCES

1)  B.R.E. Report - "Sulphate resistance of buried concrete; second interim report on long-term investigation at Northwick Park." W.H. Harrison and D.C. Teychenne (1981).

2)  British Board of Agrement Certificate 68/17.

3)  Loss Prevention Council (formerly Fire Research and Testing Organisation) Test Reports TE5216 , TE5217, TE5218, TE5219 *.

4)  B.R.E. Digest 317 - "Fire resistant steel structures" (December 1986).

5)  Acoustical Research Investigation Organisation Limited. Report AJJ2661/2 *.

6)  Bazley, E.N. - "The airborne sound insulation of partitions":HMSO (1966).

7)  Proceedings of Thermalite Masonry 88 Symposium *.

8)  Mortar Producers Associations Limited.

9)  "Fixings into A.A.C."; W.R. Millard, Proceedings of Thermalite Masonry 88 *.

10) Dampfgehartete Baustoffe: Bauverlag GmbH p5.

11) Eriksson, J.A.        DRP 404 677 vom 17.3.1923
                          DRP 447 194 vom 12.6.1924
                          DRP 454 744 vom 19.8.1925

12) B.C.S. Laboratory No. 0112 Accredited November 1981.

13) NAMAS Laboratory No. 0640 Accredited June 1988.

14) B.B.A. Certificates Nos. 86/1709, 87/1934, 86/1689, 87/1933, 84/1338, 87/1882, 87/1932, 87/1949.

* Information available on request from:-

> Thermalite Limited
> Technical Advisory Department
> Station Road
> Coleshill
> Birmingham
> B46 1HP

A L Gilbertson, W S Atkins, UK

This Conference is concerned with Design Life and I wish to draw your attention to the different meanings 'Life' can have for different people. Taking three definitions given in the draft British Standard, document 87/15323, we find (in my words):-

Required life - the client's assessment of the period the building should last without excessive maintenance or repair.

Design life factor - the designer's assessment which will include a of safety on the required life.

Expected life - the life predicted by experts as in books.

These definitions bring out two points. Firstly, designers are unwise to take the client's expression of life without applying a safety factor. Secondly, the basis upon which the designer can test that the structure's expected life exceeds the design life is in many instances very weak. I hope this conference will help us to strengthen our knowledge base about 'expected life'.

## Reply by W Brent Hall and M Tsai

We agree with the thoughtful comments of A L Gilbertson. It is a most ambitious notion to try to estimate the life of any real structure, with or without testing. Nevertheless, engineers must often wrestle with decisions involving structural lifetime, directly or indirectly. Seminars such as this one are therefore very necessary if we are eventually to make good progress on a difficult problem.

An interesting fact is that most structural failures, perhaps more than 70%, are caused by human errors in design and construction. In fact, the life of a structure is not likely to be determined by the anticipated deviations in strength, fatigue or corrosion performance allowed for in design. Rather, it is more likely to be cut short by some mistake. Prevention, control and screening of structural errors is an important topic that rivals any other relating to the life of structures, but has not received the attention it deserves. Thankfully, it has received some consideration at this seminar, although not always under the same name, and undoubtedly it will receive more in the future.

A good overview of some of the research on human error can be found in a recent paper by C B Brown and X Yin ("Errors in Structural Engineering," Journal of Structural Engineering, Vol. 114, No. 11, November 1988).

Dr G Somerville, C&CA Services, British Cement Association, UK

These 5 papers come under the general heading of Basic Principles for Design and Assessment. I am not going to attempt to sum up all the contributions made, but I will begin with the general observation that these are two areas where much more work is needed if we are to develop a coherent approach in the future.

Certain key words occurred to me as I listened to the presentations and discussion. These were as follows :

Life :    This was qualified in various ways by different speakers using words such as "required", "predicted", "service", "useful", "design", residual". It occurs to me that these are all used loosely at present and each means different things to different people. Identifying what we really mean is not just a question of being pedantic over definitions, but more a question of getting them into a context that will make it much clearer how they relate to each other. This, in turn, will help develop a framework which will be the basis of future strategies in two key areas. There areas are :

          1) Managing and maintaining existing structures
          2) Designing better new structures.

          I see these as two separate, and distinct, issues, requiring different approaches, but both based on the data that is being so assiduously collected at the present time and presented at seminars such as this.

The second key word that occurred to me was :

Function   As structural engineers, we relate mainly and naturally to the structure, but that is generally only part of the final product or artifact that we are trying to produce. It is the totality of the building or bridge or off-shore structure that is crucial and its proper function, at least overall cost. There is no point in designing, detailing and constructing a perfect bridge for a given set of imposed loads if we then foul up on the drainage provided or insert quite inadequate joints or bearings. The influence of fabric, furniture and fittings on a structure is as important to the well-being of that structure as is the attention given to the structure itself.

The third key word - and not a very good one - was :

Issues     What I mean here are the various factors involved in the totality of what we are trying to do. We must of course have predictive models on how a structure actually behaves. These are key tools in developing design and detailing procedures which are compatible with construction standards. However, we are tending to concentrate on this at the moment because it is comparatively easy to do. There are other elements that make up the total package : these include the setting of performance criteria, the need to identify and quantify the loadings that we have to take into account. In durability terms, these are of enormous importance. We are all too familiar with the extensive use of de-icing salts on motorways over the last 20 or 30 years and the detrimental effect that this can have on the serviceability of bridges and pavements. However, there are other examples in other areas ; e.g. if, in the interests of energy conservation, we use cavity fill in masonry construction, then this can alter the temperature regime and temperature profiles in the outer leaf of the masonry itself - possibly leading to frost damage.

Additional items in the list of **Issues** include methods for the proper development and evaluation of materials and components and the general need to more consciously provide maintenance and overall management of the investment that any structure represents, during its entire life.

In relating to all these key words and all these issues, I believe that those people involved in R&D have a key role to play. However, that role will only be significant if research has recognised the need to be involved at the sharp end with what happens in design offices, on construction sites and relating to actual structures in service.

There are encouraging signs that this is being recognised. When I first became involved in engineering, there were very clear distinctions between the roles of researchers, consultants, contractors and material producers. Now there are individual organisations in each of these categories who are widening their fields of expertise in response to the needs that I have defined above.

Finally, the ISE Informal Study Group is to be congratulated in organising an event of this type. It is timely and it presents an opportunity to pool our knowledge in different areas. However, there is still much work to do and I would encourage the Study Group to continue with this particular theme with a follow up seminar in a few years' time.

# 7 Fatigue life estimation of offshore tubular joints using test data and service experience

**M Lalani,** The Steel Construction Institute, Ascot, UK

This paper reports on the development of design guidance for tubular joints which form part of offshore jacket structures. The design guidance has been generated through interpretations of experimental data on tubular joints. Current guidance, extent of research, implications of new data, probabilistic methods and future research areas are defined.

## INTRODUCTION

Steel offshore jacket structures consist primarily of tubular joints which are formed by the intersection of brace and chord members, with the outside diameter of the brace less than or equal to the outside diameter of the chord. The complex intersection gives rise to severe stress concentrations; wave loading on a tubular structure cause fluctuations in these stress levels at the joints, leading to fatigue crack growth and eventual failure. Fatigue failure is defined as the number of stress cycles (and hence the time) taken to reach a pre-defined failure criterion. This paper is primarily concerned with the S-N approach and addresses the fatigue limit state of tubular joints from the resistance standpoint. The S-N approach relies on empirically derived relationships between applied stress ranges and fatigue life. Examination of design codes(1,2) and other published literature(3) give the following parameters which affect fatigue strength:

(a)   The geometry of the tubular joint and weld.
(b)   The type, amplitude, mean level and distribution of applied loads.
(c)   The fabrication process.
(d)   Post-fabrication processes.
(e)   The environment.

In addition to the above parameters, the fatigue strength of a tubular joint may be influenced by the R ratio (ratio of trough applied stress to peak applied stress), seawater temperature and frequency of testing compared with wave frequency. Where appropriate, these parameters are discussed herein. Aspects related to the calculation and use of SCFs in fatigue design are not addressed due to space restrictions; Reference 3 gives a detailed assessment of local stresses.

52

## EXISTING DESIGN GUIDANCE

A number of documents present design guidance for the fatigue life assessment of tubular joints(1,2). Figure 1 presents the family of S-N curves recommended by both API RP2A(1) and the UK Guidance Notes(2).

API RP2A presents two curves, the X and X' curve. The X curve is recommended for joints in which the welds merge smoothly with the adjoining base metal. For welds without such profile control, the X' curve is recommended. The so-called 'thickness and size effect' is taken by API to be appropriately reflected in these two curves, i.e. it is assumed that the curves are sufficiently devalued so that the effect of wall thickness is largely negated if profiled welds are specified. An endurance limit at $2.10^8$ cycles is specified, and it is assumed that all stress cycles below $35N/mm^2$ or $23N/mm^2$ for the X and X' type weld profiles, respectively, are non-damaging. The UK Guidance Notes on the other hand recommend specific size correction factors, as shown in Figure 1. This figure illustrates S-N curves for various wall thicknesses.

## EXTEND OF RESEARCH EFFORT

In 1973, the UK Department of Energy set up a test programme to investigate the problems of fatigue and fracture behaviour of the complex welded structures required for the exploitation of North Sea oil and gas fields. The programme was called the United Kingdom Offshore Research Steel Project I (UKOSRP I). This phase was completed in the late 1970's, and a second phase, which drew on industrial sponsorship, was initiated in 1982 to examine particular problem areas identified from UKOSRP I. This UKOSRP II project is now complete, and many of the results were presented at the Delft Conference(4). Collectively, the UKOSRP I and II projects were executed at a cost exceeding £10 million. In parallel with the UKOSRP projects, a series of research projects funded by the European Coal and Steel community (ECSC) have been undertaken.

The Norwegian national research programme began in 1974 with the first phase completed in 1978/79. A new programme was initiated in 1979, and was completed in 1986/87. In 1981, the Canadian Department of Energy, Mines and Resources initiated a programme of research and development with the objective of ensuring that appropriate regulations, standards and design data are generated to govern the selection and performance of welded steel structures. In 1984, the author of this paper brought together all the data available at that time in a major design manual for tubular joints(3), published by UEG in 1985. These data were screened and tabulated in the UEG design guide, and encompassed the UKOSRP I results as well as the results available from the ECSC, Norwegian and Canadian Programmes, together with data generated from several ad hoc test programmes.

## SCREENED DATABASES

The pre-1984 data has been reported in full by the author in Reference 3; the database comprise of data on 118 T/Y joints, 32 DT joints, 4 H joints and 9 K/KT joints. The post-1984 data have been reported in full by the author in Reference 5, and this database comprise of data on 19 T joints,

7 DT joints, 8 H joints and 3 K joints. In addition, a large amount of data on weldments have been generated; References 3, 4 and 5 indicate a number exceeding 500. The intent of the weldment data has been to generate information in areas where large scale experimentation becomes prohibitively expensive.

Space restrictions here has precluded the full definition of test parameters (eg. geometry, hot-spot stress ranges, cycles to failure etc) or screening procedures adopted. These details can be found in References 3 and 5. The available data are used below to assess the accuracy of design code recommendations. Reference to new data implies the post-1984 data(5), and reference to UEG data implies the pre-1984 data(3).

COMPARISON OF CODE RECOMMENDATIONS WITH TEST DATA

Base Data

Figures 2 and 3 present the new as-welded data (Table 2) in the standard S-N format (hot-spot stress range against life to through-thickness cracking). Superimposed on the graphs are the data from the UEG design guide. Data are separately identified according to the specimen wall thicknesses (16 mm and 32 mm in this case) to enable the size corrections recommended by the UK Guidance Notes to be assessed. The relevant design curves from both API RP2A and the UK Guidance Notes are also shown in these figures. It can be seen that all the data fall on the safe side of the design curves. It is of interest to note the safety margin provided by the API X' curve. For 32 mm thick joints, at a stress range of say 200 N/mm$^2$, the allowable cycles using the API X' curve is approximately 50,000 under constant amplitude conditions; the corresponding value using the UK Guidance Notes 32 mm curve gives 200,000 cycles, a four-fold increase. Similar comparisons using the 16 mm data give an even greater difference. This is due to the differences in slopes adopted in the two documents and the size correction factors introduced in the UK Guidance Notes (see Figure 1). The evidence presented in Figures 2 and 3 indicates that the data are approximately parallel to the -3 slope adopted in the UK Guidance Notes.

Influence of Size

Figure 4 shows the new tubular joint as-welded data plotted in the S-N format. The data cover joints with wall thicknesses for 16 mm to 76mm. Examination of the figure reveals the following:-

-    There is a distinct indication that fatigue strength reduces with increasing size. This is in line with previous observations(3).

-    All the data lie above the relevant API X' curve, including the 76mm data.

Insufficient tubular joint data are available to quantify the size effect throughout the range of thicknesses which occur in practice. However, reference can be made to the wealth of weldment data which are available, many of which are new and were generated recently(5). In order to present all the available data (both tubular joint and weldments) on a common basis, relative strengths have been derived and are shown in Figure 5. It can be seen that the tubular joint data and the weldment data fall within the same scatterband. Further, the figure shows that the size effect is adequately represented by the UK Guidance Notes recommendation of an inverse slope of 0.25 for relative fatigue strengths normalised to 32 mm wall thickness.

The size effect issue has been the subject of considerable research and debate over the past decade. Significant advances have been made in the understanding of the mechanisms which govern this apparent reduction in fatigue strength with increasing size. Figure 5 demonstrates categorically that a size effect exists. Further, it has been shown that weld profiling can improve the fatigue performance of a given joint. This is not surprising since one reason for size effect is the increase in local stress concentration at the weld toe ('notch' stresses) with increasing plate thickness. This 'notch' stress level is a function of the main plate thickness, attachment plate thickness, weld leg length, weld toe radius, weld toe angle and the dihedral angle. The 'notch' stress effect relates to the crack initiation phase which can account for upto 40% of total fatigue life. Once crack propagation commences the crack growth rate will no longer be influenced by notch stresses but by the effects of joint size and thickness on through thickness stress distribution and crack tip opening displacements (and hence crack tip stress intensity factors).

Influence of Environment

Figure 6 shows the new data plotted in standard S-N format. Although the data are limited, tests with cathodic protection clearly indicate that, unexpectedly, the fatigue life to through thickness cracking is reduced when compared to joints tested in air. Included in Figure 6 is the relevant UK Guidance Notes curve recommended for joints with cathodic protection. Whilst both the air and protected data points lie above the design curve, a reduced degree of safety for the joints with cathodic protection can be noted. Also included in Figure 6 is the UK Guidance Notes curve for un-protected joints. Use of this design curve for protected joints would reinstate the desired level of safety for these joints. This behaviour is in marked contrast to that expected from an analysis of welded plate specimens, and brings into question the applicability of plate data to identify environmental effects. It should be noted that the previously held belief in industry that adequate protection restores the in-air behaviour was essentially based on an assessment of welded plate data.

Complex joints, effects of weld improvement techniques and variable amplitude loading

The limited space here precludes a detailed discussion of these aspects. Reference 5, present a detailed overview of the current status in these areas.


ACCOUNTING FOR SERVICE EXPERIENCE

Once commissioned and operating, offshore installations in UK waters must satisfy the inspection requirements of Statutory Instrument No. 289 which states, as one requirement, that a major underway survey preceding the typical 5-year recertification submission should be undertaken. Inspection methods and intervals are usually decided by operators on a year-by-year basis, in consultation with the certifying authority. Selection of joints for inspection is often based on subjective appraisal of predicted fatigue life, stress level and service history, perhaps with a notional weighting for consequence of failure. Given the uncertainties associated with fatigue life predictions, the use of probabilistic methods represents a suitable technology through which an optimised inspection plan can be developed for a given structure. The objective in developing an optimised inspection plan is to ensure a reduction of either risk or cost, without compromising the level of confidence in the fitness-for-purpose of the installation. The ability to inspect provides a unique opportunity with respect to intervention. It has been demonstrated that, given inspection opportunity, the lifetime risk of catastrophic total collapse due to attaining a fatigue limit state can be reduced to substantially less than 0.1%(6).

Within the framework of developing optimised inspection plans, probabilistic methods can readily accommodate the need to account for service experience. Service history and inspection history can be taken into account in the analysis by means of Bayesian corrections whereby the service experience is used to gradually reduce the initial uncertainties. These corrections take the form of a classic saw tooth effect representing the diminishing risk as inspection exposure increases.

The development of probabalistic-based inspection plans is a vast subject area, and discussion is necessarily restricted here due to space constraints. It is perhaps noteworthy to recognise that probabilistic models representing fatigue decay processes with time are beginning to find increasing favour within the offshore industry, and relevant technologies are being developed to enable the implementation of these methods(6, 7).


RECOMMENDATIONS FOR FUTURE RESEARCH EFFORT

Whilst significant advances have been made in the area of fatigue, a number of issues warrant further study, and these are outlined below:-

i)    Hindcast forecasting studies, correlating predicted hot spot stress ranges using parametric equations with measured fatigue lives for simple and overlapping joints.

ii)   Generation of further tubular joint data on overlapping joints, stiffened joints and joints in seawater with cathodic protection.

iii)  Development of non-destructive techniques to determine residual stress levels in tubular joints. Development of inspection techniques and criteria for stiffened joints and further development of probabilistic-based inspection plans.


CONCLUDING REMARKS

A significant amount of research and design development work has been undertaken for the fatigue life assessment of tubular joints. This paper has presented and discussed the status of current design practices with reference to the vast amount of data now available in this field. Whilst space restrictions has precluded the inclusion of all available data and appraisals, appropriate documents have been cited for further reference. Notwithstanding the substantial amount of data available, it is clear that a number of issues warrant further study. Therefore, one specific section in this paper has been dedicated to the important topic of defining areas where future research effort should be directed. It is hoped that this definition will minimise any unnecessary duplication of future research work.

REFERENCES

(1)   American Petroleum Institute.   'Recommended practice for planning, designing and constructing fixed offshore platforms'.   API RP2A, Seventeenth Edition, (April 1987).

(2)   UK Department of Energy.   'Offshore Installations:   guidance on design and construction'.   HMSO, London, (April 1984).

(3)   Underwater Engineering Group.   'Design of tabular joints for offshore structures'.   UEG publication, UR33, (1985).

(4)   Proceedings of the 3rd International ECSC Offshore Conference on Steel in Marine Structures (SIMS 87), Delft, Netherlands, (1987).

(5)   Tolloczko J. A. and Lalani M.   'The implications of new data on the fatigue life assessment of tubular joints'. Paper OTC 5662 of Offshore Technology Conference, Texas, (May 1988).

(6)   The Steel Construction Institute.   'Probabilistic approach for the inspection of steel jacket structures. A technical note.'   Report SCI/104/87, (December 1987).

(7)   UEG.   'Underwater inspection project'.   Report to be issued in 1988/ 1989.

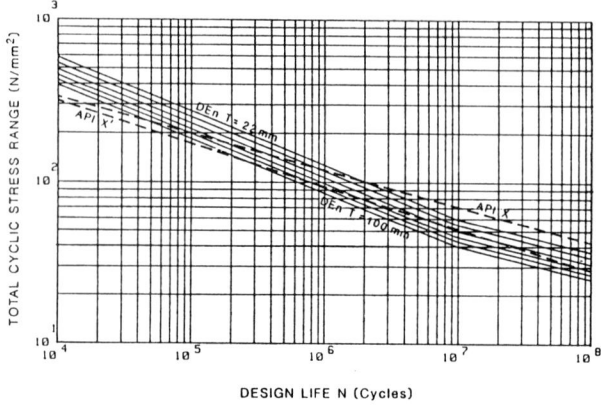

Figure 1. S-N CURVES RECOMMENDED BY
API RP2A AND THE UK GUIDANCE NOTES

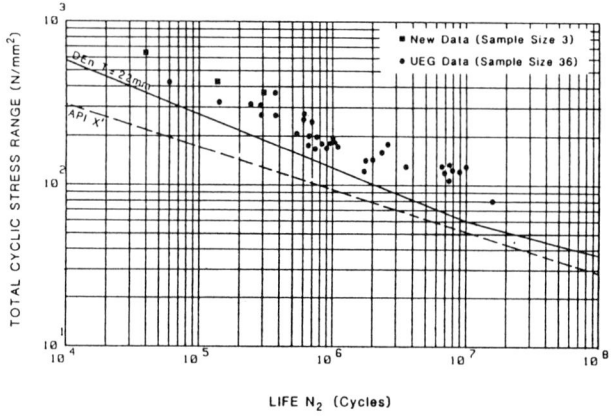

Figure 2. 16mm TUBULAR JOINT FATIGUE DATA

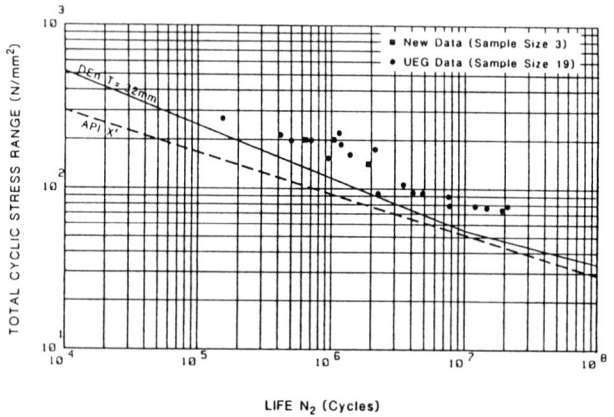

Figure 3. 32mm TUBULAR JOINT FATIGUE DATA

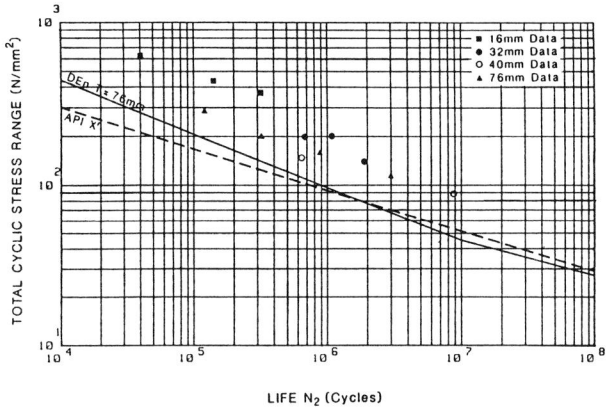

Figure 4. INFLUENCE OF THICKNESS USING THE
NEW AS-WELDED TUBULAR JOINT DATA

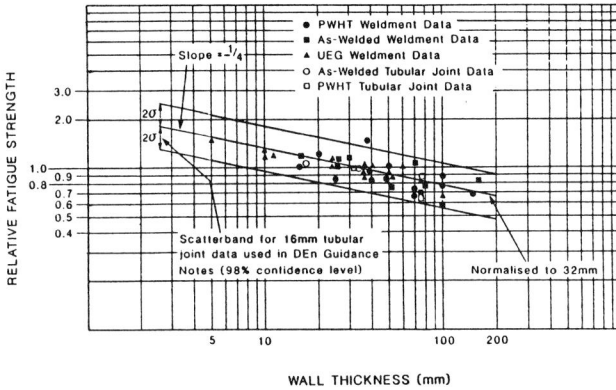

Figure 5. INFLUENCE OF THICKNESS - RELATIVE FATIGUE
STRENGTH AT 2 MIILLION CYCLES FOR ALL DATA

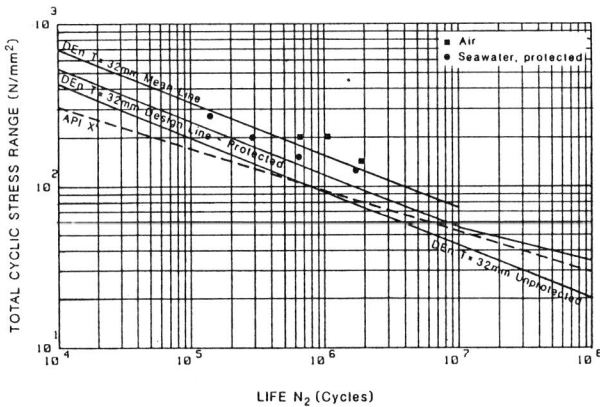

Figure 6. INFLUENCE OF ENVIRONMENT USING
THE NEW 32mm TUBULAR JOINT DATA

59

# 8 Some effects of micro-environment on materials

**J R Southern,** Building Research Establishment, Scottish Laboratory, East Kilbride, Scotland

Water has the greatest effect on the long term deterioration
of most structural materials.  Examples of moisture content
of building materials are given.   Real environmental
conditions can be different from those usually considered.
On one site moisture contents were higher than assumed for
design purposes after 12 years.  Condensation formed in
summer under natural exposure conditions.  Lower moisture
contents were found in the timber frame of cavity filled
houses compared with a small control sample.

INTRODUCTION

Structural materials account for a small part of the cost of housing,
but for a house to maintain its value the main structure must be in an
acceptable condition.  The life expectancy of housing is dependent on
viewpoint, and the designer, new owner, subsequent owners and
financiers all have different expectations of useful life, the current
owner probably expecting the longest life for the building.

Some structures are designed to have a short life but others deteriorate
to a stage where they require extensive repair or renewal before the end
of their projected life, either because they are unsatisfactory or
because a decision has been made to extend their life beyond the design
period.  One cause of reduced life expectancy is deterioration of
important materials in the structure.  These materials may be included
in such minor components as fixings and bearings as well as major
structural elements.

Materials deteriorate by various mechanisms but if we were able to fully
control the moisture content of structural materials we would have
greater control over the long term deterioration process in most
structures.  Deformation and cracking by shrinkage or expansion may be
directly caused by moisture changes, whereas the onset or rate of
deterioration by rot, frost damage, corrosion, carbonation, sulphate
attack and reversion of high alumina cement, are all dependent on
included or contact water.

The chemical and biological, and some of the physical aspects of
deterioration due to moisture are generally understood but there is a
lack of information on in-service moisture conditions.  Standardised
moisture contents are sometimes quoted [1] and laboratory testing of

materials can help in understanding the mechanisms of failure, but the examination of actual examples must set the standard for test conditions and assumptions for calculations. A great deal of information is likely to be collected during the course of investigations but little is published in a form useful to the specifier. Failures, long term changes, test procedures and standardised assumptions need to be constantly re-assessed in the light of present knowledge and site information. We have only to consider the shortened life expectancy of many examples of concrete structures to remove confidence in standard assumptions about site performance. Until there is more information about the micro-environmental conditions to which materials are subjected there are likely to be examples of unexpectedly short life spans of structures.

Climate plays a large part in generating the moisture condition of structures, but there is a difficulty in deciding which parts of the climate have the major effect. Annual, monthly or daily data on rainfall, wind and temperature are used for some purposes. Meteorological station results can be modified[2,3] to estimate local conditions. Further refinement, such as considering shelter belts, can be made to determine micro-climate close to buildings[4]. Interior environmental conditions vary seasonally and under the influence of the occupier. Standardised environmental conditions or the presumed worst case may be used when predicting the influence of interior climate on the structure.
Interior and exterior climates are used as inputs to model the likely effect within the depth of material, but until recently these models have been based on steady state climates. There is a dearth of knowledge on actual climates within the depth of structures and how these climates affect the moisture distribution within the elements.

The following examples illustrate measured water contents which can affect the useful life of building elements. They illustrate real situations where the problem is obvious once it has been recognised, but the problems were not recognised at the time of design or construction.

EXAMPLES

Moisture content of external masonry between dpc and eaves is usually assumed to be 5% by volume[1]. Measurements on site have shown that this moisture content can be exceeded by as much as 8 times or can be much lower for superficially similar constructions. Higher than expected moisture contents can lead to acceleration of damage from frost, wall tie corrosion and loss of mortar strength by sulphate action. Appendix 1 gives a table summarising moisture contents in masonry determined by BRE Scottish Laboratory and brief details of the circumstances under which they were measured as a start in assembling a data base of real moisture contents.

On one site in the North of Scotland cavity fill was injected into rendered walls of two leaves of lightweight aggregate concrete block. The rendering had cracked prior to injection of the cavity fill and within three years of filling the cavity, the outer leaf of blockwork in some of the houses had deteriorated to a stage where rendering was falling off because the block was so damaged that it could be easily rubbed away by hand. Moisture contents in the blockwork on this site ranged between 5% and 74% with a mean of 24%. Some of the walls had to be pulled down and rebuilt. Walls with cracked renderings should not be cavity filled but even so it is likely that the small change in temperature and water content of the external leaf brought about by the cavity fill reduced the life of the blockwork by several years.

The moisture contents of solid autoclaved aerated concrete walls in 3-storey housing have been determined over a period of 12 years. The rendering on the surface of these walls has remained intact but the average moisture content has not fallen below 9% and it appears to have stabilised at approx 9%, see Figure 1. The distribution of the moisture within the thickness of the wall and the high moisture content found in sheltered areas suggests that drying by solar radiation and wetting by interstitial condensation may influence seasonal changes[5]. Up to now the excess moisture has not caused any noticeable early deterioration of the walls.

## High moisture contents in roof timbers

There are standard recommendations for ventilation in roof spaces and a recommended level of ventilation is now given in various publications, including UK Building Regulations and Standards. An estate of houses was built in Scotland to an English design before these recommendations had become widely publicised. A minor change in the design from facing brick to rendering meant that the ventilation gap at the eaves was sealed by the rendering. This caused a major change in the environment at the interface between the sarking and the sarking boarding. The measured moisture content of the sarking board was 130% by weight. Timbers adjacent to the sarking boarding were off-scale on electrical moisture meters, moulds were actively growing on the sarking board and spreading onto the roof timbers, and the nail plates had started to corrode. The relatative humidity continuously recorded in the roof void, was often 95% during night time and the expected life span of the roof probably dropped to 10-15 years. The problem was discovered when a valuation survey was being undertaken and other houses in the estate were subsequently found to be in a similar condition. Ventilation reduced the moisture content to below 25%, within two months.

## High moisture contents in timber with dry-lined solid walls

Following complaints of moisture appearing under skirtings and being found behind dry linings of solid walls during summer, a series of large scale tests was undertaken on dry-lined solid walls exposed to natural conditions[6]. Similar experiments had been carried out concurrently in Denmark[6] and in Canada during the 1960s[7]. Standard calculations, indicate that with dry-lined solid walls there is a risk of interstitital condenstion within the thickness of the wall under winter conditions unless a vapour control layer is placed on the inside of the insulation. Steady state calculations also indicate that there is no risk of interstitial condensation with this construction during summer conditions. In reality, there can be a problem during summer caused when the sun shining on the outside of the wall raises the temperature of the internal surface of the wall above that of the vapour control

layer, see Figure 2. If the masonry is damp then water vapour may
tranfer across the insulation and condense on the back of the vapour
control layer where it can trickle down and possibly wet up battens,
sole plates and joist ends[9]. In this type of construction the timber
may be much wetter in summer than it is in winter even though the
masonry is at a much lower moisture content, see Figure 3. Under the
experimental conditions it was found that this type of condensation
could be present in some of the panels continuously from the end of May
to the beginning of October, this being a period when the building was
unheated. Even when the internal environment was thermostatted at 17°C
there were some days when condensation was deposited on the back of the
vapour control layer. Within one month of the onset of the condensation
black mould formed on the horizontal members at the bottom of some of
the panels. One problem with this condensation is that its effects are
hidden by the internal lining to the wall and the first signs of
problems may be rotting skirting boards or joist ends.

## Low moisture contents in cavity filled timber framed houses

Older timber framed houses have lower thermal insulation standards than
is currently expected. The walls of these houses can be difficult to
insulate. In North East Scotland, a substantial number have been
insulated by cavity filling between the sheathing and the brickwork.
Fifteen of these cavity filled houses were inspected, using the methods
described in Reference 10, and no high moisture contents attributable to
the cavity fill were found. In a small number of controls the moisture
contents were higher in uninsulated compared with the insulated houses.
Since these inspections were carried out, eleven houses have been
monitored whilst the external leaf was being taken down to remove the
cavity fill. Again no high moisture contents were found.

## Rusting in steel framed, steel clad houses

Alterations to longstanding structures can sometimes affect life
expectancy. Steel framed, steel clad houses inspected after 45 years
showed no serious rust problems. Upgrading which included the addition
of internal insulation to the walls led to condensation on the back of
the steel sheet and rust stains appeared from the joints between the
sheets of steel. This occurred even though a vapour control layer had
been placed on the inside of the insulation. The reason for the problem
was that the steel was a more complete vapour control layer than the
internal polythene sheet and the steel was colder after the insulation
had been installed than when the wall construction consisted of a steel
sheet with an internal lining of 20 mm fibreboard. In this instance the
life expectancy reduced from in excess of 30 years to probably between 10
and 15 years.

CONCLUSIONS


Life span can be altered and especially reduced by relatively small
changes in design or construction.

Not enough is known about moisture contents in real situation and these
can be widely different from those taken as standard and those predicted
by present methods of calculation. Refinement of calculation procedures

can help to remove some of the uncertainties but more information on moisture conditions and environmental conditions both surrounding and within elements are required so that expected life spans are not reduced because of inappropriate design assumptions.

REFERENCES

1. Building Research Establishment Digest 108, BRE Garston

2. British Standards Institute. Draft for Development 'Methods for Assessing Exposure to Wind Driven Rain' BS DD93:1984 BSI London

3. Keeble E. J. 'Macroclimatic data and its intepretation for problems of building Deterioration'. Building Deteriorology Seminar, Soc Chem Ind & BBA, London, April 1986

4. Newman A. J. et al 'Microclimate and the environmental performance of buildings'. Seminar, BRE Garston, April 1988

5. Southern J. R. 'Moisture in solid blockwork walls at Glenrothes'. RILEM Symposium Autoclaved Aerated Concrete, Lausanne, 1983

6. Christenson G. 'Summer condensation in post-insulated exterior walls'. Paper for CIB/W40 Meeting, Holzkirchen, September 1985

7. Wilson A. G. 'Condensation in insulated masonry walls in summer'. RILEM/CIB Symposium, Helsinki, 1965, Vol 1 Sec 2 Paper 7.

8. Southern J. R. 'Summer condensation within dry-lined solid Walls'. BSERT, Vol 7, No 3, 1986

9. Southern J. R. 'Summer condensation on vapour checks'. BRE IP 12/88, Garston, November 1988.

10. Southern J. R., High G. 'Surveying the moisture contents of cavity filled timber framed dwellings'. BRE IP/85, Garston, January 1985.

ACKNOWLEDGEMENTS

This work forms part of BRE Scottish Laboratory's Research Programme and acknowledgment is made of contributions from all members of the Materials Section.

# Appendix 1

Table 1(a)  Summary of moisture contents of external leaves of masonry walling (dpc - eaves) - Survey buildings

| Location and year houses built | Wall Material sampled | Sample date | Number of samples | Moisture Content % by volume | | |
|---|---|---|---|---|---|---|
| | | | | Mean | Range | |
| Glenrothes 1976 | Rendered 225 mm a.a.c. solid wall | 09/79 | 40 | 10.6 | 4.2 | 16.8 |
| | | 02/80 | 51 | 13.0 | 8.6 | 20.6 |
| | | 10/80 | 48 | 9.8 | 4.9 | 16.8 |
| | | 03/81 | 47 | 10.4 | 5.5 | 15.0 |
| | | 11/81 | 39 | 8.6 | 5.1 | 16.3 |
| | | 01/82 | 37 | 9.0 | 3.9 | 14.4 |
| | | 02/88 | 30 | 9.4 | 3.1 | 16.8 |
| Aberdeen 1969-1974 | Rendered 200 mm lightweight aggregate block solid wall | 09/79 | 48 | 3.7 | 1.4 | 7.5 |
| | | 02/80 | 47 | 4.4 | 1.8 | 14.0 |
| Stirling mid 1960s | Rendered 102 mm Common brick | 05/81 | 20 | 1.1 | 0.2 | 2.6 |
| | | 09/81 | 22 | 0.7 | 0.2 | 2.6 |
| | | 02/82 | 22 | 1.5 | 0.5 | 4.8 |
| Livingston 1967 | Painted 200 mm a.a.c. solid panels | 06/70 | 32 | 9.3 | 2.6 | 18.6 |
| | | 10/70 | 8 | 5.0 | 1.6 | 13.8 |
| | | 01/71 | 16 | 5.1 | 1.3 | 11.2 |
| Edinburgh 1966 | Painted 203 mm a.a.c. panels | 06/82 2 | 23 | 3.5 | 2.3 | 7.4 |
| | | 10/82 | 5 | 2.3 | 1.5 | 3.6 |
| Glasgow ca 1900 | 600 mm solid stone wall | 07/80 | 7 | 2.6 | 1.4 | 5.8 |
| Glasgow ca 1900 | 600 mm thick sandstone with rubble infill | 09/80 | 56 | 7.2 | 0.2 | 56.8 |
| Lichfield ca 1960 | Concrete panel with eps insulation core | 04/86 | 5 | 3.8 | 1.3 | 5.7 |
| Machrihanish 1964 | Concrete panel with eps insulation core | 12/86 | 18 | 3.9 | 1.5 | 15.2 |
| Whitburn 1961 | Rendered 102 mm Common brick Outer leaf | 06/81 | 18 | 1.6 | 0.6 | 3.3 |
| " | "      " Cavity filled | 06/81 | 19 | 1.8 | 0.4 | 4.5 |
| E.Kilbride 1973 | Rendered 102 mm Common brick Cavity filled wall | 04/82 | 4 | 1.7 | 0.6 | 4.1 |

Table 1(b)   Summary of moisture contents of external leaves of masonry walling
(dpc - eaves) - Render failure investigations

| Location and hyear houses built | Wall material sampled | Sample date | Number of of Samples | Moisture Content % by volume | | |
|---|---|---|---|---|---|---|
| | | | | Mean | Range | |
| Leuchars 1962 | Rendered 100 mm Lightweight aggregate concrete block Cavity filled wall | 03/80 | 64 | 23.8 | 5.4 | 74.0 |
| Fort William 1965 | Rendered 100 mm Dense concrete block | 06/83 | 25 | 3.5 | 1.6 | 6.0 |
| Gilmourton 1975 | Rendered 102 mm Common brick | 10/80 | 25 | 12.4 | 1.0 | 34.8 |
| Glasgow 1960s | Rendered 102 mm Common brick | 11/80 | 15 | 8.6 | 0.5 | 28.3 |
| Cumbernauld 1973 | Rendered 102 mm Common brick | 06/81 | 11 | 2.0 | 0.1 | 6.1 |
| Kilmarnock ND | Rendered 102 mm Common brick | 06/81 | 6 | 7.0 | 0.8 | 27.5 |
| St Andrews 1971 | Rendered 102 mm Outer leaf | 08/81 | 44 | 2.2 | 0.7 | 6.3 |
| Peebles 1971 | Rendered 102 mm Common brick Cavity filled wall | 09/81 | 14 | 9.1 | 0.4 | 24.7 |
| Glasgow 1960s | Rendered 102 mm Common brick | 04/82 | 27 | 3.3 | 0.5 | 13.2 |
| Lochgilphead ND | Rendered 102 mm Common brick | 11/82 | 20 | 20.8 | 3.5 | 34.5 |
| Carstairs 1940s | Rendered 102 mm Common brick | 03/83 | 7 | 6.9 | 2.3 | 14.7 |
| East Kilbride 1975 | Rendered 102 mm Common brick | 08/84 | 6 | 22.6 | 0.2 | 33.9 |

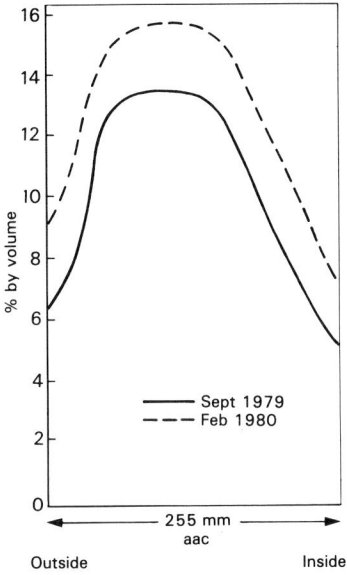

**Figure 1** Variation of moisture content across solid aac wall at Glenrothes

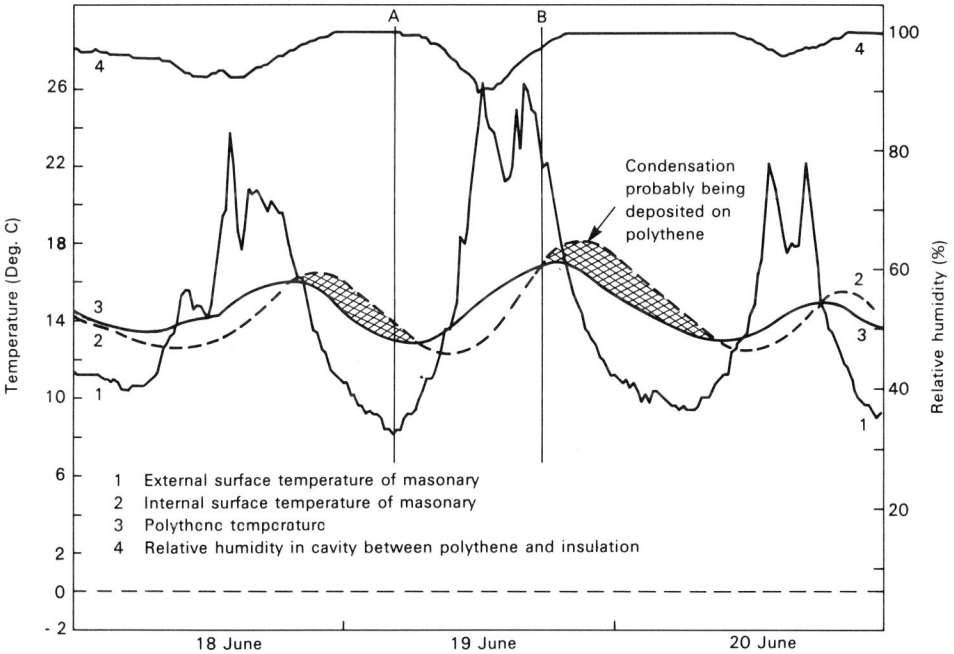

**Figure 2** Temperatures and relative humidity of internally insulated solid walls during 3 days of sunny weather at East Kilbride

| Wall facing | Moisture content % by weight | | |
|---|---|---|---|
| | Stud | Rail | Brickwork |
| ese | 17 | 16 | 0.2 |
| sse | 24 | 48 | 1.8 |
| ssw | 22 | 26 | 0.7 |

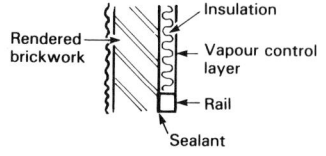

**Figure 3**  Moisture content of drylined walls during summer

W D Biggs, University of Reading, Buro Happold, UK

The author appeared to imply that cavity fill had only a minor effect when compared with cracked rendering. I have grave doubts about cavity fill - the cavity acts as a drying step for moisture penetrating through the outer leaf and, if this is insulated from the heat sink of the building, the outer leaf will get wetter - and stay wetter. Subsequent damage will be the result of freeze - thaw cycling - paper 35 confirms this well.

I find it difficult to accept the phrase 'the small change in temperature and water content of the external leaf' without some evidence of how small is small and what freeze/thaw cycling was in operation. Can the author enlighten me please?

## Author's reply

Insulation to the inside of a wall whether provided by cavity fill, internal insulation, or insulating inner walls will affect the temperature and moisture content of an outer leaf. The main controlling effects when compared with an uninsulated wall are the amount of added insulation, any changes in water vapour permeability and changes in ventilation. In general the outer leaf is likely to be colder in winter and warmer in summer and may be wetter in winter and drier in summer. Measurements on naturally exposed walls of test buildings in the United Kingdom indicate that for fill in cavity walls and internal insulation on solid walls the temperature of the external face is about 1°C lower than insulated walls in cold weather. Moisture contents of individual bricks even in the same wall vary considerably and both higher and lower moisture contents have been measured when comparing insulated and uninsulated walls, but overall the moisture content is somewhat higher.

Considerable site experience has shown that these differences are only a significant cause of damage where the materials in existing buildings already show signs of frost damage, or cracked rendering, which may permit higher than normal moisture contents in the outer leaf. In these instances, insulation of the wall can be expected to increase the rate of deterioration.

# 9 Deterioration of factory units used as metal plating shops

**Brian J Brown,** STATS Scotland Ltd, East Kilbride, Scotland

After 3-4 years of use as metal plating or similar
process industries, two light engineering industrial
units were found to be in a non-lettable condition.
Condition surveys were undertaken to assess the extent
of damage and health and safety risk. Extensive
remedial measures (including the replacement of one
entire floor) were carried out, the total cost of which
was approximately 10 times greater than normal
dilapidation costs.

## INTRODUCTION

Two recently constructed factory units in East Kilbride owned by East
Kilbride Development Corporation were used as electro-plating and metal
finishing works. On vacation of the premises by the tenants severe
deterioration of the internal fabric of both units was found. The
deterioration was so severe as to render the factories as non-lettable
and requiring a considerable amount of remedial measures to bring them
back to a lettable standard.
Condition surveys of the two buildings were carried out to assess the
material properties and life expectancy of the structural components, the
extent of deterioration, and the remedial measures necessary both for
health and safety and structural reasons.

## DETAILS OF THE FACTORY UNITS

The first unit to be investigated (hereafter referred to as Unit 1) was
built in 1981. It is part of a larger construction sub-divided into
separate units and is typical of the type of industrial units let by the
Development Corporation for light engineering purposes. The floor was of
reinforced concrete, 150mm nominal thickness, on a polythene damp-proof
membrane. The walls were of painted concrete masonry blockwork. The
roof was pitched and composed of tapered steel beam trusses with cold
rolled galvanised purlins supporting a double skin of asbestos cement
sheets with glass wool infill.

From October 1981 to November 1984 (a duration of 3 years) the premises
had been used for nickel plating and associated metal treatments. The
building was then vacated and various open vats of chemicals were removed
in February 1985. Subsequently steam cleaning of the internal fabric
took place prior to a condition survey being carried out in March 1985.
The second unit (hereafter known as Unit 2) was investigated in September
and October 1987. This factory is actually a double unit with an
internal partition. The floor and roof construction was the same as in
Unit 1. The construction of the walls differed, however, in being built
of no-fines concrete excepting the toilet walls which were also of
blockwork. The building was constructed in 1973-74 and had for 4 years
prior to September 1987 been used as an electroplating and metal
finishing works.

ELECTROPLATING PROCESSES

Electroplating is generally a 4 stage process involving a) a solvent
clean, followed by b) washing, then c) acid etching (generally
hydrochloric or sulphuric acid) and finally d) immersion in the plating
solution. The type of plating solution used is specific to the type of
plating undertaken and may involve chromic-sulphuric acids for chromium
plating, cadmium and sodium cyanide for cadmium plating, nickel sulphate,
nickel chloride and boric acid for nickel plating and zinc cyanide for
zinc plating (1). Very corrosive and toxic chemicals are thus used.
The internal atmosphere of the units would be classified as very severe
as the internal fabric is exposed to high humidities and acidic fumes.

CONDITION SURVEYS

Both factories were initially visually examined. Destructive and non-
destructive testing was then carried out on the floors, internal walls,
roof system and service components (switchgear, trunking etc). Chemical
analysis for heavy metals (nickel, chromium, cadmium, zinc), chloride,
sulphate and other contaminants such as cyanide was also undertaken.
Corrosion and deterioration products were also assessed by X-ray
diffractometry and electron microscopy.

UNIT 1

Floor

The concrete floor was in a variable condition. No attack or damage was
evident in the office areas and in some parts of the workshop. However,
many areas of the workshop exhibited etched and abraded surfaces with
loose aggregate and/or staining. At its most severe the floor was
stained black, cracked and exhibited white deposits.
4 cores were extracted from the floor through its full thickness and
samples of the hardcore beneath also collected. 3 cores were from
attacked areas and one from an undamaged area. Compressive strength and
density tests were taken from the lower portions of the core from the
undamaged area and one from a damaged part. Results were 21.0 and 28.0
N/mm$^2$ and 2350 and 2330 kg/m$^3$ respectively for strengths and density
(mean values of 24.5 N/mm$^2$ and 2340 kg/m$^3$). Cement content
determinations carried out on the same two cores gave 290 kg/m$^3$ and 330
kg/m$^3$ for the good and damaged areas respectively (mean value 310 kg/m$^3$).

71

The undamaged/unattacked concrete was reasonably well compacted with excess voidage in the 1 to 2% range. The concrete would appear to have been a grade 20 or 25 supply.

Samples of various deposits from the cracked and abraded areas were analysed by X-ray diffractometry and found to be composed of gypsum ($CaSO_4.2H_2O$) and sodium sulphate ($NaSO_4$). The latter is a chemical known to be used in the electroplating industry while gypsum is a reaction product of hydrated cement and sulphate solutions.

Chloride content analysis of the top surfaces of the cores indicated significant contamination by chloride solutions ranging from 0.78 to 3.99 $Cl^-$ % by mass of cement in the attacked areas; the value in the unattacked area was 0.15% $Cl^-$ by mass of cement.

Gradient samples from one core were tested and the results were found to decrease with depth.

Sulphate content results showed a similar trend although the results were not as significant. The top of the core from the unattacked area gave a result of 1.81 $SO_3$ % mass of cement with the middle of core C1 giving a similar result of 1.98 $SO_3$ % mass of cement. All the core top results from the 3 No. damaged cores were higher ranging from 2.49 to 3.15 $SO_3$ % cement. However, as cement may contain up to 3.5% $SO_3$ under BS 12 (2) the results are not as significant as those of the chloride tests. They do, however, suggest the possible ingress of sulphate solution to the concrete surface.

From the physical condition of the floor being abraded, etched, stained, cracked and generally degraded with deposits present together with the chloride, sulphate and XRD results, it is evident that the floor has suffered acid attack together with possible sulphate attack. From core examination the depth of attack is variable ranging from surface only in the areas of surface staining, down to 50mm in abraded, etched and stained areas and down to the entire depth of the core (140mm) in the most severely attacked, stained and cracked area. In this latter area the polythene d.p.m. was also degraded. However, in all the test areas the fill beneath the concrete was found to be uncontaminated with regard to chloride and sulphate giving BRE Digest 250 (3) class 1 conditions.

Blockwork Walls

The majority of the walls did not exhibit visible degradation excepting the presence of crystal growth at the base of the walls adjacent to areas of floor degradation. An exception was the rear wall where areas with obvious splash marks were found to be cratered to up to 10mm depth. Also at this location, a dark deposit towards the base of the wall was noted which on inspection was found to be the damp-proof course in a degraded condition. Although the blocks themselves were found to be in generally undegraded condition this was not the case for the mortar joints which in many cases were cracked, in poor condition and soft. Whereas many of the lower joints were in this condition as may be expected due to their proximity to any floor spillage etc, many of the upper joints were also in poor condition.

Analysis of the joint mortars (3 No. samples) indicated two upper course mortars to be leaner than the expected mix (type III -1:1:5-6) and one indicated appreciable chloride content (0.31 % $Cl^-$ by mass of cement). Sulphate contents were 'normal' in all three cases.

Analysis of yellow and green deposits from wall bases gave appreciable chloride and sulphate contents (0.27-0.39% Cl$^-$, 0.53-3.94% SO$_3$).

Schmidt rebound hammer tests (4) were carried out on 4 No. areas of wall representing visibly good and bad areas. In general every second block course was tested (with exceptions) for a width of three blocks from floor to roof level. 5 No. readings were taken for each block and compared with the visual condition of the blocks and the mortar joints. There was a variation in results between individual blocks (range 16.4 to 35.2) and between individual courses (18.5 to 33.5). Comparison of the results with the visual condition of the mortar joints indicated that lower results were generally obtained in areas of visually poor and soft joints. Although the lower values indicated possible surface degradation, the values were not low enough to give cause for concern. It is probable that the lower courses have been subject to direct contamination and degradation by acidic solutions; however, the upper courses will have been subject to acidic fumes leading to possible lime leach and disintegration of the mortar.

Metal Components

All metal fittings, conduits and trunkings were in poor condition. Light fittings were badly corroded while most conduits and trunking were similarly affected. Switch boxes and fuse boxes had faired only slightly better. Rust from a fuse box recorded excessive chloride and sulphate contents (1.65% Cl$^-$, 6.04% SO$_3$).
The general condition of the tapered beam roof trusses was reasonable; however, the paint system had deteriorated and was in poor condition. Corrosion on the top surfaces of the flanges had been initiated but with little loss of cross-section. Cleats carrying the roof purlins were also corroded as were the purlin bolts. The galvanised purlins were in a very poor condition with most if not all of the purlins showing signs of corrosion. In some cases the corrosion had progressed to the stage where the galvanising had completely disintegrated and corrosion of the steel had commenced with visible rust scaling. The area of the purlins most at risk was the channel formed by the lipped bottom flange. The anti-sag bars between purlins were also corroded.
Chloride and sulphate contents of rust from a beam truss and from the purlins were significant (beam Cl 0.36-0.99%, SO$_3$ 0.36-0.73%; purlin 0.37% Cl$^-$, 0.55% SO$_3$). X-ray diffraction analysis of a rust deposit from the beam indicated the rust to be composed of $\alpha$, $\beta$ and $\gamma$ phases of FeO.OH. The presence of the $\alpha$ (Goethite) and $\gamma$ (Lepidocrocite) phases is to be expected under oxidising conditions; however, the presence of $\beta$ FeO.OH (Akaganeite) is significant in that it only forms in the presence of chloride (5). Scanning electron microscopy of a beam rust sample was also undertaken and the presence of chloride (Cl) and sulphur (S) was noted. Atmospheric corrosion of steel components is rapidly enhanced by even small amounts of either chloride and sulphate but acting together they result in severe corrosion.

Roof

The asbestos cement undersheets were discoloured but in good condition with no signs of any chemical attack. The roof lights although apparently undamaged were severely discoloured with a significant reduction in light infiltration.

## UNIT 2

### Floor

Plastic pipes were buried in the floor leading to a sump containing liquid chemicals. Areas of acid etching and chemical staining were much in evidence with the worst affected at the pipe openings. Despite a relatively high cement content (440 kg/m³) the concrete was found to be variable in quality with variable compaction and honeycombing apparent in one core. This is reflected in the variable strengths and densities recorded (17.0 to 42.0 N/mm², 2310 to 2450 kg/m³).
Chloride contents are high at the surface (0.24 to 1.50 Cl⁻% mass of cement) decreasing downwards with the upper 25mm or more above the 0.20 Cl⁻% by mass of cement threshold value normally accepted for reinforcement corrosion caused by extraneous chloride solutions. Sulphate contents were all normal for OPC despite the fact that gypsum (CaSO₄.2H₂O) was found by X-ray diffraction to be present at the surface of two badly deteriorated areas. Any deleterious effects of sulphates are thus possibly restricted to the surface skin only.
Heavy metal (Ni, Cr, Cd and Zn) and cyanide concentrations are generally high at the surface decreasing downwards, only approaching background levels below 50mm depth. Cadmium (Cd) levels are high throughout.
The crushed rock hardcore below the d.p.m. was found to be generally uncontaminated with respect to heavy metals, cyanide, chloride and sulphate (Class 1 BRE Digest 250 conditions) with the exception of cadmium which again gave higher than normal levels.

### Walls

The walls were of painted and rendered no-fines concrete excepting the toilets which were of plastered blockwork. The walls over the majority of the building were stained brown with the central dividing wall also exibiting yellow and green stains at its base. Contamination was limited to the painted surface only, the rendering being relatively uncontaminated.

### Metal Components

As with Unit 1 most of the metal switchgear, trunking, fuse boxes etc. were corroding externally (internal surfaces were generally corrosion free). X-ray diffraction analysis indicated all 3 No. form of FeO.OH to be present ( $\alpha$, $\beta$, $\gamma$ ); the $\beta$ form (Akaganeite) again signifying chloride enhanced corrosion.
Of the metallic roof components the galvanised purlins were in relatively good condition as were the fixings. The tapered steel beam trusses, however, were exhibiting rust spotting and paint breakdown.

### Roof

The asbestos undersheets appeared relatively undamaged and in good order. Some of the external topsheets especially below the rooflights, however, were found during remedial works to be soft, friable and cracked. A separate investigation was carried out to establish if the chemical process previously carried out in the factory had contributed to this breakdown but all results indicated a negative correlation with the factory process.

## REMEDIAL MEASURES

Both units were in lettable condition before their use as metal plating or similar workshops; Unit 1 as a newly constructed building and Unit 2 nine years old. To bring the buildings back into lettable condition both from structural and health and safety aspects, a considerable amount of remedial work was necessary. In fact almost £46,000 of additional remedial works were carried out by East Kilbride Development Corporation over and above normal dilapidation costs (approximately £5,000) before the units could be re-let.

Due to the amount of remedial action necessary on the floor in Unit 1 to remove all contaminated and damaged concrete, the entire floor was replaced. In Unit 2, all the pipework and the sump were removed and the central area of floor excepting a 1m perimeter strip was essentially replaced and the strip topped by a granolithic screed. The walls in Unit 1 required extensive washing and cleaning together with repointing of many joints, the replacement of the d.p.c. in the rear wall and repainting. In Unit 2 thorough cleaning and re-painting only was required. The roof beam trusses in Units 1 and 2 required thorough cleaning including blast cleaning to remove rust scaling and contamination and were entirely repainted. The galvanised purlins required more attention in Unit 1 than in Unit 2 with the removal of all corrosion and contamination products and the making good with a suitably compatible zinc coating. In both units all electrical metallic components (trunking, switches etc) required replacement.

## CONCLUSIONS

After 3 to 4 years occupancy of standard light engineering industrial units by metal plating or similar process industries, the units were no longer in a lettable condition and required extensive remedial measures (of £51,000 cost) compared with normal dilapidation costs (estimated at approximately £5,000) for the two units. This also resulted in large time gaps between vacation of the premises and re-letting resulting in lost rent revenue when the units were not available for let as the remedial work progressed.

## ACKNOWLEDGEMENTS

Permission to publish the results of the investigations was given by East Kilbride Development Corporation and the assistance of Mr H. Cooper, Estates Department is gratefully acknowledged.

REFERENCES

1. Ross, R. B, 1988.     Handbook of Metal Treatments and Testing. Chapman and Hall, 2nd Edition.

2. BS 12 : 1978     Ordinary and rapid hardening Portland Cement.

3. BRE Digest 250,1981.     Concrete in sulphate bearing soils and groundwaters

4. BS 4408 : Part 4 : 1971.     Non-destructive methods of test for concrete - surface hardness methods.

5 Kassim, J.,T. Baird & J. R. Fryer, 1982.     Electron microscope studies of iron corrosion products in water at room temperature.
Corrosion Science, Vol 22, No. 2, p147-158

# 10 Repair materials and repaired structures in a varying environment

**D R Plum,** The University of Newcastle upon Tyne, UK

Structural life may be extended, in cases where deterioration has occurred, by the use of the correct materials. This paper examines the properties of some of these materials, in particular the polymer concretes and polymer cement concretes. An examination is made of the requirements of a repair, and the appropriate properties by which a material may be assessed as suitable or otherwise. Changes in the environmental conditions such as temperature and humidity are shown to affect the material properties, and the results are compared with the repair requirements.

## INTRODUCTION

The repair of concrete structures has become a burgeoning industry in recent years. Conversion, chloride attack, carbonation and ASR, are words with which structural engineers have become familiar. With the increasing need for repair work to extend structural life has come an increasing demand for new materials having improved properties. Unfortunately it has not always been clear what properties should be improved, or with what they should be compared, or what environments should be specified for them.

Some of the new materials have been based on the use of polymers, whether alone (PC) or as a modifier to cement (PCC). In general in the UK most of the commercial polymer concretes to date have been epoxy resins, while most of the PCC's have been from the acrylic and SBR groups. Special cements and cement replacements are also available, each having a different set of material properties.

The physical testing of these new materials has tended to follow that for concrete (1), although for epoxy resins new test methods have been defined (2). The test methods commonly pay little attention to in-service conditions of temperature and humidity, unless a client so requires. It has become clear that many of the new materials behave in ways different from those of concrete, especially in relation to environmental conditions. It is also clear that when considering a repair patch, screed or coating, the properties of significance need to be addressed, and reliance on the easily measured compressive strength avoided.

REPAIR MATERIALS

Epoxy resins are a common base for the commercial repair materials used in the UK. Polyesters have also been available, but primarily in other parts of the world. The epoxy resin, modified or unmodified, is mixed with a curing agent (polymerising) to give a hardened resin. Sand is usually used as a filler, and the material is further modified by the use of flexibilisers, diluents, pigments, etc. The resulting materials consequently differ from one another, and properties of the hardened PC must be individually determined.

Polymers are also used as modifiers in cementitious materials giving PCC. A number of suitable polymers are available in both the acrylic and SBR groups. The differences are again sufficient for each material to require individual assessment.

REPAIRED STRUCTURES

Repairs to structures are usually required to extend structural life. The function performed by a repair will however vary from one situation to another, and may be required to satisfy several functions at once.

The two principal functions may be defined as
(a) structural, in which a stress carrying function is intended;
(b) cosmetic, in which restoration of structural appearance is a priority.
In both cases other considerations may also be significant, such as wear resistance, skid resistance and permeability.

The properties required of a material to fulfil the requirements of a structural or cosmetic repair are clearly quite different, and may in some matters be completely opposite. The problem of choice of suitable properties is not solved either by seeking those closest to the values for the base concrete, especially when it is remembered that environmental response will probably be different.

The cosmetic repair includes the functions of protection of reinforcement, repair of the concrete, and restoration of the appearance of the structure. Carbonation, chloride induced corrosion, cracking, spalling and surface deterioration may all be problems requiring solution in this cosmetic treatment. The particular requirements for each of these problems is treated elsewhere (3).


REPAIR PERFORMANCE

Structural repairs

Stress carrying repairs to concrete structures will commonly be used for beams (compression zone) and columns. Repairs to walls and slabs are not generally required to carry stress due to the low stress levels in the base concrete. Propping of the structure may take place during repair in extreme cases, but in general dead loads remain in place and hence the repair carries a proportion of the imposed load only.

Assessment of repair performance, in whatever material used, may be carried out at the serviceability limit state (4), but will more usually be required under ultimate conditions.

If F is the force carried excluding the reinforcement contribution
   A is the area of cross-section in compression
   $\sigma$ is the stress
   $\epsilon$ is the strain
and suffices c and r refer to concrete and repair respectively,
then when the concrete reaches ultimate strain $\epsilon_c$ and stress $\sigma_c$,

$$\epsilon_c = \epsilon_r = \sigma_r (1 + \phi_r)/E_r$$

where $E_r$ is the elastic modulus of the repair material
   $\phi_r$ is the creep coefficient of the repair material.

The force ratio $F_r/F$ may be expressed as

$$F_r/F = 1/[\sigma_c (1 + \phi_r)(1/\rho - 1)/\epsilon_c E_r + 1]$$

where $\rho = A_r/(A_c + A_r)$
Typical values of some of the parameters may be taken as

$$\epsilon_c = 0.0035$$
$$\sigma_c = 21 \text{ N/mm}^2$$
$$E_r = 20 \text{ kN/mm}^2 \text{ or } 10 \text{ kN/mm}^2$$

giving $\sigma_c/\epsilon_c E_r$ as 0.3 or 0.6.

The function $F_r/F$ may be plotted against $\rho$ as shown in Figure 1. For these relationships, as derived above, it is assumed that the structure is propped to relieve dead load. As pointed out, the more common case occurs when dead load remains on the structure during repair, and here the strain $\epsilon_r$ will be lower and the force ratio is reduced.

Cosmetic repairs

Patch repairs for protective/cosmetic purposes, together with screeds having similar purposes, are generally not required to carry stress to resist loads. Stress may however arise due to shrinkage or expansion of the material as a result of the initial hardening process, or subsequent thermal and moisture movements. Shrinkage produces effects which are well documented (5) for cementitious screeds, and can produce cracking or debonding/curling. Expansion can produce buckling of the patch/screed accompanied by debonding over large areas.

Theoretical approaches to the problem of screed buckling are possible (6) but have limited value due to the assumptions regarding bond behaviour, the nature of the restraints and assumed imperfections. It is possible however to use a simple approach, sufficient to assess the effects of the principal parameters. Assume that a patch repair has a dimension L in both directions, and is of thickness t. Let an expansion $\epsilon$ be prevented by a force $F_r$ in the repair material, which produces a tendency to a buckling failure. The interface bond stress $p_b$ must be sufficient to prevent such a failure, and may be assumed to be required to resist a force $\beta F_r$. Taking the average bond stress over the patch area to be half the maximum value $p_b$ then,

$$p_b L^2/2 = 2\beta F_r = 2 \beta Lt\epsilon E_r/(1 + \phi_r)$$
$$\text{hence } \epsilon = p_b L(1 + \phi_r)/4\beta t E_r$$

Assuming a small value for L/t of say 10, and a conservative value of $\beta$ of 0.04, then

$$\epsilon = 62.5 \ p_b (1 + \phi_r)/E_r$$

The relationship between $\epsilon$ and $p_b$ is shown in figure 2 for a range of values of $E_r/(1 + \phi_r)$.

## MATERIAL PROPERTIES AND THE ENVIRONMENT

### Compressive strength

Commercially available polymer concretes and polymer cement concretes were tested for strength in accordance with BS 6319 Part 2 (2). Figures 3 and 4 show the result of increasing humidity, which reduces strength in all cases. The reductions vary from one material to another. Removal from the high humidity environments to ambient in general failed to recover the strength loss in PCs, but did achieve some recovery in PCCs.

### Tensile strength

Tensile strength was tested in accordance with BS 6319 part 3 (2). Results showed the same trends as those for compressive strength.

### Elastic modulus

Elastic modulus in compression was tested in accordance with BS 6319 Part 6 (2). Reductions in value were again noted for PC as a result of increasing humidity. For PCC however the opposite effect was noted, but the values were at all environments lower than the control specimens (having no polymer included).

### Compressive creep

Creep tests were carried out on specimens 100 mm long and 15.5 or 17.5 mm diameter. Loading at 10 N/mm$^2$ and 14 N/mm$^2$ was applied after 7 days ambient cure. Creep of PC under different environments is shown in figure 5. Creep of PC appears to be the property most significantly affected by changes of both temperature and humidity. Creep coefficients at 14 days were found to vary from 1 to 30 depending on the material and the environment. PCC was equally responsive with creep coefficients at 28 days varying from 0.1 to 40 depending on the material and the environment.

### Tensile bond

Bond between the repair material and the base concrete was examined using a tension test method (3). In general well prepared and primed concrete was found to give bond strengths close to 3 N/mm$^2$ for PC and produce failure within the top surface layer of the base concrete. For PCC bond strengths were lower, with a greater response to surface preparation. Environmental conditions did not affect the results significantly in either case.

## APPLICATIONS AND CONCLUSIONS

The suggestions given in REPAIR PERFORMANCE above may be applied to particular repair materials on the basis of their measured properties. In a structural repair, the force ratio $F_r/F$ may be estimated for given environmental conditions using the expression given. Table 1 gives the force ratio at $\rho$ = 0.5 for several materials at two environments (ambient and 35°C/90% RH). It may be seen that for all the materials tested, that a rise in temperature and humidity reduces the force ratio, in some cases to a considerable degree. Clearly for this application high modulus and low creep are desirable, coupled with environmental stability.

Table 1. Force ratio for structural repair

| Material | Environment | $E_r$ kN/mm$^2$ | $\rho r$ | $F_r/F$ |
|---|---|---|---|---|
| 1 PC | ambient | 13 | 1 | 0.52 |
| | 35°/90% | 10 | 21 | 0.07 |
| 2 PC | ambient | 26 | 5 | 0.42 |
| | 35°/90% | 18 | 24 | 0.11 |
| 3 PC | ambient | 19 | 1 | 0.61 |
| | 35°/90% | 16 | 3 | 0.40 |
| 4 PC | ambient | 17 | 5 | 0.32 |
| | 35°/90% | 11 | 23 | 0.07 |
| 7 PCC | ambient | 23 | 0.1 | 0.50 |
| (SBR) | 35°/90% | 31 | 41 | 0.11 |
| 8 PCC | ambient | 22 | 0.2 | 0.50 |
| (acrylic) | 35°/90% | 30 | 10 | 0.31 |

In a <u>cosmetic repair</u> the maximum strain $\epsilon$ may be estimated for given environmental conditions. Table 2 gives an estimated strain which may be permitted in several of the materials at two environments. Expansion due to saturation is found to be in the range 0.5 to 2 x 10$^{-3}$ for well formulated materials. It will be noted that this is in general significantly lower than the permissible strain in Table 2. The value of $\epsilon$ is however dependent on the tensile bond stress $p_b$, the value of which can drop to 20% of that given if proper surface preparation is lacking. In addition the value of expansion due to saturation can rise above 20 x 10$^{-3}$ in poorly formulated materials. Clearly for this application low modulus and high creep are desirable, coupled with high tensile bond and low saturation expansion.

The structural life of a building may be extended by the proper use of repair materials. Knowledge of the material properties is needed, and this must be applied on the basis of the purpose of the repair, and the environmental conditions.

Table 2. Permissible strain for cosmetic repair

| Material | Environment | $E_r$ kN/mm$^2$ | $\Phi_r$ | $P_b$ N/mm$^2$ | $\epsilon_{max}$ x $10^{-3}$ |
|---|---|---|---|---|---|
| 1 PC | ambient | 13 | 1 | 2.7 | 26 |
| | 35$^o$/90% | 10 | 21 | 2.7 | 370 |
| 2 PC | ambient | 26 | 5 | 3.5 | 51 |
| | 35$^o$/90% | 18 | 24 | 3.5 | 300 |
| 3 PC | ambient | 19 | 1 | 3.4 | 22 |
| | 35$^o$/90% | 16 | 3 | 3.4 | 53 |
| 4 PC | ambient | 17 | 5 | 3.3 | 74 |
| | 35$^o$/90% | 11 | 23 | 3.3 | 450 |
| 7 PCC | ambient | 23 | 0.1 | 2.5 | 7 |
| (SBR) | 35$^o$/90% | 31 | 41 | 2.5 | 210 |
| 8 PCC | ambient | 22 | 0.2 | 2.5 | 8 |
| (acrylic) | 35$^o$/90% | 30 | 10 | 2.5 | 58 |

Values based on $L/t$ = 10, $\beta$ = 0.04

REFERENCES

1. BS 1881  Methods of testing concrete (1970 on)

2. BS 6319 Testing of Resin Compositions for use in Construction
   British Standards Institution (1983)

3. Plum D.R. 'Epoxy resin repair materials – in the laboratory and insitu'
   Conference – Structural faults and repair – London (July 1987)

4. Fosroc.  Users Manual for Concrete Repair and Protection (1988)

5. Deacon R.C. 'High strength concrete toppings for floors including
   granolithic'.  British Cement Association Advisory Note (April 1976)

6. Boudraa S.E. 'The behaviour of epoxy resins with particular reference
   to the influence of bond strength and curing conditions on the
   performance of screeds'.  MSc thesis, University of Newcastle upon
   Tyne (1985)

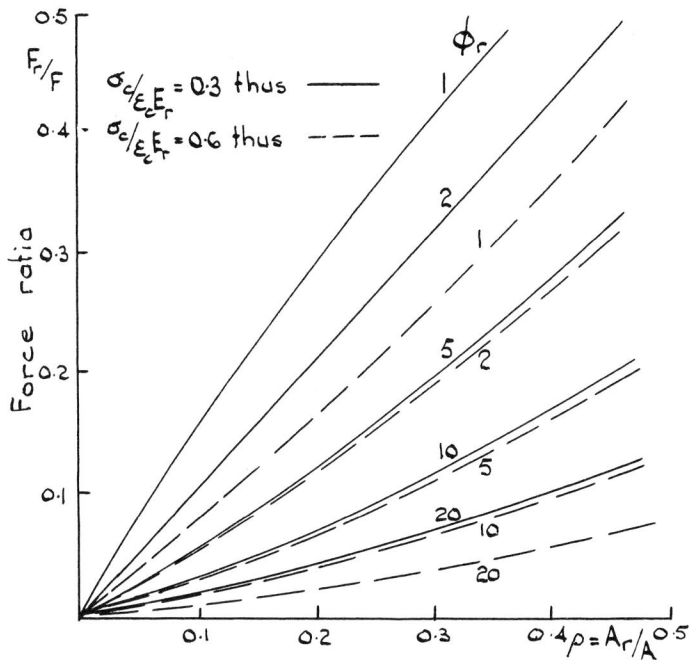

Figure 1. Force ratio - structural repair

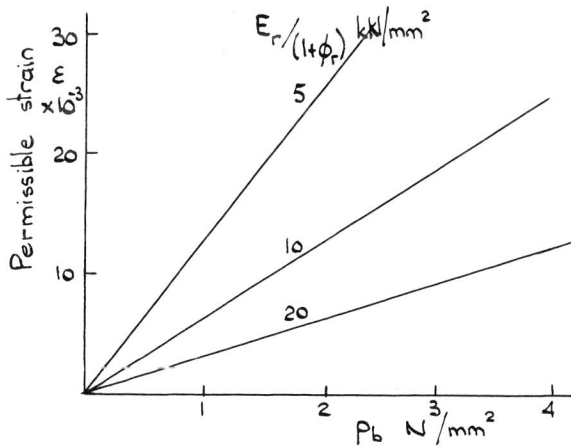

Figure 2. Permissible strain - cosmetic repair

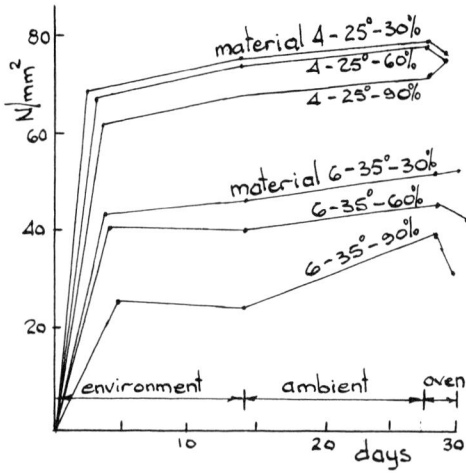

Figure 3.  Compressive Strength, PC

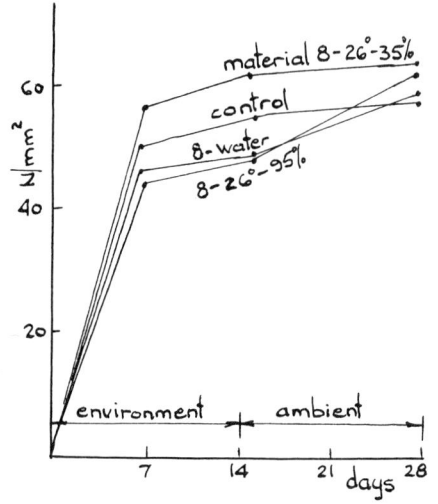

Figure 4.  Compressive Strength, PCC

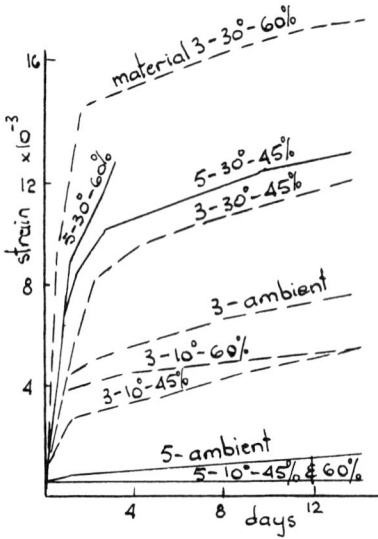

Figure 5.  Compressive creep, PC

84

**Dr G Somerville, C&CA Services, British Cement Association, UK**

In this session  we had a number of Papers on varying topics, but all on the common theme of interaction between aggressive media on the one hand and the fabric or structural elements of the buildings, bridges or whatever on the other.  In my view, work in this are is especially important.  We most certainly do not know enough about the effects of the different types of aggressive media that can attack the structures that we design.  In tackling that, we need to bring together different interests and types of expertise.  For example, in relation to masonry construction, there is the bringing together of expertise represented by Paper N°6 and that represented by Paper N°8.  In this application, and indeed in any other application, there is a need to be aware not only of the technical issues, but also of the financial issues that were typified in this session by Paper N°4.

# 11 Assessing structures through field measured dynamic response

**Bruce J Morgan,** M.ASCE, **Ralph G Oesterle and W G Corley,** Construction Technology Laboratories Inc, Skokie, Illinois, USA

Recent advancements in equipment and procedures utilising the technique known as Experimental Modal Analysis, allows dynamic testing of full-size structures with ambient excitation. This inexpensive and reliable test method for determination of dynamic chacteristics structures provides an excellent way of determining how well actual structure conforms to design assumptions. Also, dynamic characteristics measured periodically can provide a basis for assessing a structure's overall condition after an unusual loading, such as an earthquake or severe wind.

BACKGROUND

Invaluable information concerning the behaviour and condition of a bridge structure can be obtained from knowledge of its dynamic characteristics. Until the mid-1970's, the experimental determination of dynamic characteristics of structures was very cumbersome. The equipment needed to acquire and analyse the data was bulky and heavy. The required procedures were lengthy and expensive.

The situation changed in the mid-1970's due to the development of microcomputers and highly efficient mathematical algorithms for these machines. It became practical to exploit the capabilities of a technique known as experimental modal analysis to determine dynamic characteristics of structures. With this technique, frequency and mode shape information may be obtained in an efficient and straightforward manner. Since 1982, six international conferences on experimental modal analysis have been held. The papers contained in the proceedings (1-6) of the conferences are an excellent source of stae-of-the-art information on the subject.

The dynamic behaviour of all structural systems may be described in terms of a few basic characteristics. Since these characteristics are fundamental in describing and understanding dynamic behaviour, they are briefly reviewed in the first section of this paper. The next section gives examples of the way in which dynamic characteristics may be used to assess the condition of bridge structure. This section is followed by a discussion of the theory and practice of experimental modal analysis including a brief review of the type of equipment used for this purpose. Finally, two examples of structures tested at Construction Technology Laboratories (CTL), using this technique, are given.

## DYNAMIC CHARACTERISTICS

### Degrees of Freedom

A dynamic system has as many degrees of freedom as it has independent coordinates defining its motion. A single mass hanging from a weightless spring only vibrates up and down. One coordinate is all that is required to define the motion, the vertical displacement of the centroid of the mass. Therefore, this is a single degree of freedom system. If a second mass is suspended from the first mass by a second weightless spring, two coordinates, namely, the vertical displacement of the centroids of both masses, are requried to define the motion. This new system, therefore, has two degrees of freedom. A third mass and spring suspended from the second would be defined as having three degrees of freedom, and so forth.

### Normal Modes

A dynamic system has exactly the number of normal modes of vibration as it has degrees of freedom. Associated with each normal mode is a time function referred to as a natural frequency, and a spatial function known as a characteristic shape. An important point to note is that the complete dynamic behaviour of a system in response to any type of excitation or forcing function can be determined by the superposition of behaviour of the individual normal modes.

A single degree of freedom system consisting of a mass hanging from a single weightless spring has one normal mode defined by one natural frequency and a single characteristic shape, that of the mass moving up and down. A two degree of freedom system of two masses and two springs has two normal modes. The first mode has a frequency and characteristic shape associated with both masses moving up and down together. The second normal mode has a characteristic shape of both masses moving up and down always in an opposite sense, or exactly out of phase with one another.

### Damping

A single degree of freedom system consisting of a mass and spring is an idealised one. Once set in motion, it would vibrate indefinitely. A real system in motion vibrates only for a certain duration. The dynamic energy stored in the system is dissipated through friction and the motion eventually dies out. The measure of the energy dissipation characteristics of a system is known as the damping of the system.

### Continuous Systems

Systems that behave like assemblages of springs and masses are known as discrete systems. They have a finite number of degrees of freedom.

A simply supported beam, on the other hand, is a continuous system. Since it may be thought of as being composed of an infinite number of masses and springs, it theoretically has an infinite number of degrees o freedom and normal modes. The dynamic behaviour, however, of a continuous system can very often be adequately described by using a limited number of its normal modes in a dynamic analysis.

Resonance

A structural system responds to a forcing function depending on the relationship between the frequency content of the excitation and the natural frequency of the system. For example, consider a time varying the force defined by Equation (1):

$$F = A \text{ sine } W_O t$$

According to Equation (1), the force is in the form of a sinusoid of frequency $W_O$ which reaches a plus and minus value of A once during each cycle.

If the frequency $W_O$ is much lower than the lowest natural frequency of the system, the structure behaves as if loaded by a static load of magnitude A. If $W_O$ is much higher than the highest natural frequency of the system, the structure essentially does not respond at all. It is as if no load were applied. The interesting case occurs when $W_O$ exactly equals a natural frequency of one of the normal modes of the system. In this instance, vibration, in he characteristic shape associated with the particular normal mode, builds up with each cycle of applied force. If the system had no damping in this mode, the vibration would build up to infinity. The system is said to be in resonance. Even with damping, systems in resonance will exhibit displacements five times or more greater than would be realised by the application of a static force of magnitude A.


BENEFITS FROM DYNAMIC TESTING

One objective of dynamic testing may be to verify that a structure is behaving as predicted by analysis. Assumptions are needed to formulate the analytical models used in the design or evaluation of a structure. Certain assumed properties, such as cross-sectional area and inertia properties, can be obtained by calculation to a good degree of accuracy. Other properties, such as end fixity conditions for beams or support conditions for piers, are known to a lesser degree of accuracy. As an example, structural engineers often assume "fixed" or "pinned" end conditions at the boundaries of structures. similarly, engineers often assume the ends of beams or bottoms of piers are rigidly supported in th vertical direction. In reality, the end conditions can be significantly different than either "fixed", "pinned" or "rigid". Dynamic characteristics can be very sensitive to broundary conditions. Experimental modal analysis demonstrates the actual boundary conditions.

Another objective of dynamic testing is to determine defects or deterioration in a structure from its dynamic characteristics. This can be very useful from a maintenance point of view. Deterioration in a structure always creates a change in its dynamic response. Consequently, with dynamic testing of the structure at regular intervals, comparisons can be made to determine if any change has occurred with time.

Dynamic testing may also be used to determine deterioration at any one point in time by comparing two or more similar structures. If the condition of one structure is known, valuable knowledge can be gained for the other structures by comparison of their dynamic characteristics.

A third use of dynamic testing for discovering defects or deterioration of an individual structure at any one point in time is to compare mesured dynamic behaviour with expected behaviour. Anomalies in measured characteristic shapes can quite often pinpoint specific areas or members of the structure with potential defects. Further detailed testing and inspection are then focused on these specific areas of concern.

All of the above uses of experimental modal analysis can be combined with "tuned" analytical computer modelling to provide a very powerful tool to evaluate problem structures and to design modifications for proper behaviour. As stated previously, once the normal modes of a structure are known, the complete dynamic response to any forcing function can be determined.

In evaluating the condition of a bridge, the following dynamic characteristics can be examined.

## Frequency

Measured frequencies substantially lower than expected would indicate an abnormal loss of stiffness. This may be due to an overall deterioration of the cross section. Lower frequencies measured in one or more spans of a number of similar spans would indicate that a closer examination of the differing spans is called for. Deterioration may be the reason for the reduced frequencies.

## Characteristic Shape

Lack of symmetry in a characteristic shape, when symmetry is expected, is a good indication of a local loss of stiffness and a potential local area of distress or deterioration. A mode shape with a kink in it, or with an otherwise local lack of smoothness, may also indicate a discontinuity in the structure or an area of local failure.

Two normal modes with comparatively close frequencies and only slightly differing characteristic shapes many indicate a condition of distress in a local area. A comprison of the characteristic shapes should indicate the general area of the distress. Two normal modes with very different frequencies, but similar overall characteristic shapes, would indicate that the structure has two load path mechanisms active. Expected behaviour and a close comparison of the mode shapes will indicate if this is normal or abnormal behaviour.

An unusual characteristic shape with only a single support or portion of a support participating excessively in the motion may indicate a soft foundation under the support or a local problem with part of a support. The characteristic shape in question will indicate the location of the problem. Also, a characteristic shape that indicates motion at a support where noe is expected is an indicator of a support or foundation problem.

Frequencies higher than anticipated generally indicate supports stiffer than expected. The slope of the companion characteristic shapes at the supports would indicate actual end conditions. A general seizing of the support bearings, possibly due to corrosion, could be the root cause of this result.

## Damping

Unusually high damping would indicate that more energy dissipative mechanisms are working in the bridge then expected. This may indicate deterioration in the strucuture.

## EXPERIMENTAL MODAL ANALYSIS

The object of experimental modal analysis is to measure the normal modes of a structure. This includes both the natural frequencies and accompanying characteristic shapes.

## Swept Sine Testing

One traditional approach for experimentally determining normal modes is swept sine testing. By this method, a shaker and response transducers are located on the structure. The structure is excited by the shaker, slowly sweeping through a range of frequencies, while the response transducers are being monitored. As the frequency of the shaker approaches the natural frequency of a normal mode, the outputs of the response transducers begin to sharply increase since the structure is approaching a resonant condition. The shaker frequency is tuned until peak responses are obtained from the response transducers. This is the frequency of a normal mode. The characteristic shape is obtained from a comparative analysis of the response transducer outputs at resonance. Damping may be obtained by analysing the decay in the transducer outputs after the shaker is turned off.

This technique can be cumbersome and time consuming. With the advent of the microcomputer and highly efficient mathematical algorithms for these machines in the mid-1970's, it became practical to fully exploit the capabilities of experimental modal analysis to determine normal modes.

## Time and Frequency Domains

Experimental dynamic data is acquired in the time domain, that is, the change in response of a variable with time is measured. To determine normal modes, however, the change in response of a variable over a range of frequencies is desired, that is, data expressed in the frequency domain is desired.

Fast Fourier Transform: The fast fourier transform, of FFT as it is more commonly called, is a mathematical algorithm that very rapidly transfers data from the time to the frequency domains. The FFT algorithm, residing in the memory of an inexpensive, compact microcomputer, has produced a revolutionary effect on experimental dynamics. Experimental data continue to be acquired in the time domain. However, transfer of these data to the frequency domain, where resonant peaks readily indicate normal mode frequencies, becomes rapid and comparatively inexpensive.

## Transfer Functions

Transfer functions re used in experimental modal analysis to determine normal modes. A transfer function is developed as follows. A structureis dynamically excited and responses measured. The excitation record in the time domain is transferred to the frequency domain by FFT processing. The response data in the time domain is also transferred to the frequency domain by FFT. A transfer function is calculated from thee data. A transfer function is a ratio of the response to the excitation displayed as a function of frequency. Resonant peaks indicating normal mode frequencies are generally clearly visible on transfer function plots. Figure 1 is a transfer function plot obtained from the test of a large concrete structure at CTL.

The resonant peaks are clearly visible. Characteristic shapes for each normal mode are determined from a series of transfer functions developed from data taken at a number of points on the structure.

Required Data Acquisition: There is an important point to note concerning the amount of data required for experimental modal analysis. Since normal mode information is developed from transfer functions, only one input and response need to be measured at a time. Either the response transducer is moved from point to point while the excitation remains at one location, or the response transducer is located at one point and the excitation moved. While recording data from a number of points at once is more efficient, it is not necessary. Only two channels of data acquisition, one input and one output, are actually required.

Damping: Damping is also derivable from transfer function relationships.

## EXPERIMENTAL TECHNIQUE

## Response Measurement

Transfe functions can be derived from a wide variety of response measurements. Accelerometers, velocity, or displacement transducers are all acceptable response measurement instruments. Transfer functions have even been developed from strain gage data dynamically recorded in a number of locations on a structure.

## Excitation

As discussed previously, the excitation force variation with time is transferred into the frequency domain by FFT. Therefore, it is not necessary to excite the structure with any paticular type of forcing function or device. The only requirement of the excitation is that it have frequency content at the frequencies of the normal modes, so that resonant peaks will appear in the transfer functions.

Basically, three types of excitation are used.

Shakers:  Shakers are still commonly used as excitation sources. They need not be used in swept sine mode, however.  Random excitation with a frequency band wide enough to encompass all the expected modal frequencies of interest is all that is required.  One advantage of shaker excitation is the ability to concentrate energy into a comparatively narrow frequency band to improve the signal to noise ratio of the measurement system, in other words, to make the resonant peaks more clearly discernable.

Impact:  Impact may be used to excite a structure.  A properly designed impact device can be tailored to have the desired frequency content to resonate the normal modes.  Impactors have the decided advantage of being simple and inexpensive excitation sources.  A number of specially designed impact hammers with heads instrumented with load cells are commercially available.  Figure 2 shows an impactor developed at CTL to excite large structures.  It consists of a weight dropped into a sand bed that, in turn, is mounted on a load cell.  This device has been used successfully to excite a number of both steel and concrete structures.

Ambient Excitation:  One of the most powerful improvements in dynamic testing techniques brought about by new technology is the ability to develop transfer functions from structural response measurements made under ambient loading conditions.  Wind or traffic loadings are sufficient to excite a bridge for this type of measurement.  Since a measure of the excitation is not available, the processing of the response transducer data is performed differently than for the case where excitation force data are present.  Since the forcing function is generally at a very low level, seismic quality instruments are required to acquire the response data.

Data Acquisition

A number of dynamic signal analysers are commercially available to convert the analog time domain data into digital form, and then process the data into the frequency domanin via FFT algorithms.

Data Analysis

There are two directions to take once the data have been acquired and transferred into the frequency domain.

Self-Contained Analysers:  There are commercially available units that combine the data acquisition and analysis functions.  The same unit that contains the analog to digital converter and FFT hardware also contains the computer softwarenecessary to extract frequency, characteristic shape and damping from the frequency domain data.

Modal Analysis Software:  There is commercially available modal analysis
software for a number of the most popular desk top microcomputers.  This
software accepts frequency domain data from the dynamic signal analyser
and performs the necessary manipulations to produce the normal mode
information.

Animated Characteristic Shape:  All analysis systems produce
characteristic shapes that can be animated on either a self-contained
screen or on the screen of the desk top microcomputer.  The shapes in
animation can be viewed from any angle, at any speed of vibration and at
any amplitude.  They can also be superimposed on each other for
comparative study.  This very flexible animation feature is extremely
useful in diagnostic work involving tests of existing structures like
bridges.


EXAMPLES

Tollway Girder

Figure 3 shows a bridge girder removed from the Illinois Tollway.  The
girder was taken from a 25-year-old bridge that wa demolished as part of
a change in right-of-way.  A number of tests of this girder performed at
CTL primarily focused on determining the condition, durability, and
serviceability of the girder after 25 years of service.  A modal analysis
of the girder was also performed.  The sand drop impactor, shown in
Figure 2, was used for excitation.  A three by seven grid of points was
laid out on the upper surface of the girder.  An accelerometer was
mounted at one point and the impactor moved from point to point.
Twenty-one transfer functions were developed and the first series of
normal modes analysed.  In this instance, two very similar characteristic
shapes existed forwidely differing frequencies.  From the superposition
of these characteristic shapes in an animated display a separation along
one side of the girder was noted.  There was, in fact, a delamination
along one side of the girder in exactly the area indicated.  The
delamination, noted in Figure 3, was visible because the girder had been
removed from the bridge.  It would not have been visible in the actual
bridge in service.

Hollow Core Slab

Figure 4 shows a hollow core slab tested under ambient wind conditions.

The slab was held at an angle to the wind by the two wooden supports
noted in Figure 4.  For this test, wind of less than ten miles per hour
was sufficient to excite the slab.  Two seismic accelerometers were used
to acquire the response data.  Excellent normal mode data were obtained
with two very interesting results.

The wooden supports were found to be considerably softer than
anticipated.  The slab moved on the supports essentially as a rigid body
in one of the normal modes of the system.  Also, a kink in one of the
characteristic shapes occurred in an area where a crack existed in the
slab.  The crack had not been observed previously.

CONCLUSION

Experimental Modal Analysis is an extremely efficient way of developing
the dynamic characteristics of actual bridge structures in the field.
These characteristics can be used to verify analytical models. More
importantly, however, the dynamic characteristics provide valuable
insight into the condition of a bridge, or changes in condition of a
bridge, broadening the use of modal analysis into the area of bridge
evaluation, inspection, and maintenance.

REFERENCE

1) "Proceedings of the 1st International Modal Analysis Conference and
   Exhibit", November 8-10, 1982, Orlando, Florida.

2) "Proceedings of the 2nd International Modal Analysis Conference and
   Exhibit", February 6-9, 1984, Orlando, Florida.

3) "Proceedings of the 3rd Internatinal Modal Analysis Conference and
   Exhibit", January 28-31, 1985, Orlando, Florida.

4) "Proceedings of the 4th International Modal Analysis Conference and
   Exhibit", February 3-6, 1986, Los Angeles, California.

5) "Proceedings of the 5th International Modal Analysis Conference and
   Exhibit", April 6-9, 1987, Imperial College of Science and Technology,
   London, England.

6) "Proceedings of the 6th International Modal Analysis Conference and
   Exhibit", February 1-4, 1988, Orlando, Florida.

NOTE: Above proceedings available from IMAC, Union College, Graduate and
      Continuing Studies, Schenectady, New York, 12308.

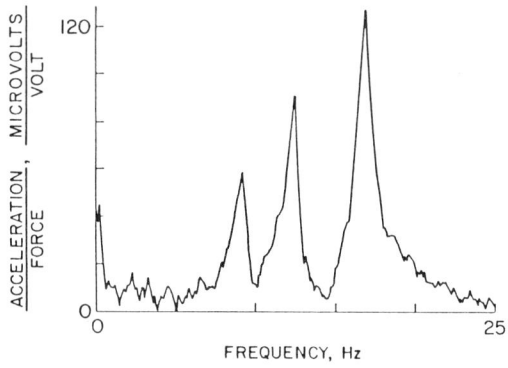

Figure 1    Transfer Function

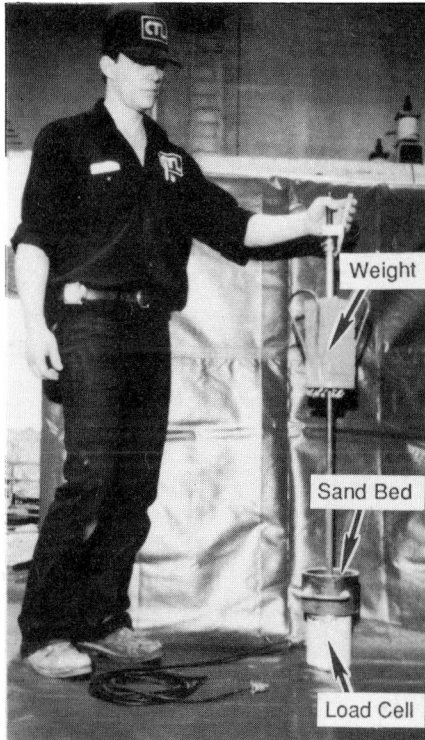

Figure 2    Sand Drop Impactor

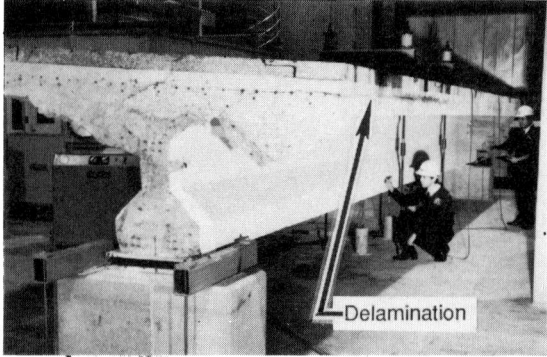

Figure 3   Tollway Girder

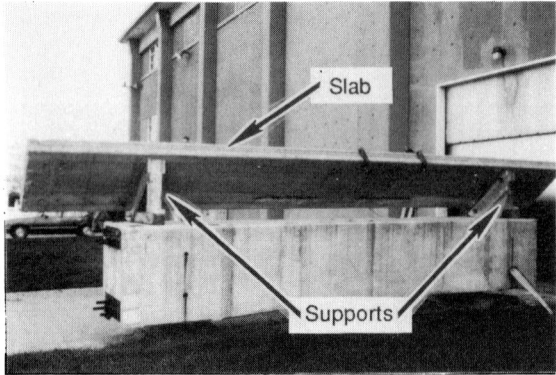

Figure 4   Hollow Core Slab

# 12 Monitoring of service reservoir foundation condition by dynamic testing techniques

**S L Matthews,** Technical Director, Structure Testing Services (UK) Ltd, UK
**R T Heslop,** Formerly Technical Services Manager, Sunderland and South Shields Water Company, UK
**P S White,** Formerly Structural Engineer, Structure Testing Services (UK) Ltd, UK

**Summary** - Dynamic response measurements can be employed to monitor the condition of the foundation of a reinforced concrete service reservoir without taking the installation out of service. The procedure was developed in response to the serious problems experienced by the Sunderland and South Shields Water Company with the formation of voids below and around some of its service reservoirs. Measurements are undertaken on a periodic basis by means of an electro-dynamic vibrator located on top of the columns supporting the reservoir roof. Comparison is made with 'baseline' data, enabling changes with time to be identified in the condition of the reservoir foundation and concrete structure. Interpretation is aided by finite element modelling.

## Introduction

1. The geological succession in the Sunderland area, coupled with a history of coal mining, poses particular problems for the Sunderland and South Shields Water Company. Gradual collapse of the drift overburden deposits into the natural fissures within the underlying Magnesian Limestone, widened during ground disturbance caused by mining activities, creates sub-surface voiding. This process, exacerbated by leakage from service reservoirs, allows voids to migrate upwards through the overburden.

2. Such an incident led in 1979 to the partial collapse of a service reservoir. The resulting loss of some 70 megalitres of water, (approximately half a day's supply for the 550,000 population served by the Company), is estimated to have caused erosion of some 500 m$^3$ of glacial till. The loss of foundation support over an area of around 1000 m$^2$ caused differential settlements of up to 1.2 metres. This rendered three-quarters of the complete reservoir complex unusable.

3. The cost of replacing the lost storage was estimated to be about £3 million at 1981 prices.

4. In response to the need to detect the presence of voids before structural damage occurs to a reservoir STS developed a technique for dynamically exciting the floor slab by applying vibration forcing to the column heads.

The technique is designed to detect voiding under the floor slab by monitoring the dynamic response of the column head. By this means the column is used as a 'stinger' or push-rod to couple the vibrator and response transducer to the floor slab, as shown in Figure 1. Hence, the reservoir can be tested without being decommissioned.

## The Concept of Detecting Structural Change by Monitoring Dynamic Response

5. The principles of dynamic monitoring have been known for a long time, however, it is only fairly recently that they have been applied to civil engineering structures in order to detect structural change.

6. The principle as applied to service reservoirs can perhaps more easily be explained in relation to a simple structure such as a beam.

7. A beam has natural frequencies which are known for different support conditions from established theory. Thus a beam can be tested dynamically and the experimental and theoretical results can be compared, as shown in Figures 2 and 3. This comparison enables an assessment to be made of the condition of the structure as found. In addition the suitability of the test technique can be reviewed together with the adequacy of the theoretical model. Monitoring of the structure can then continue in order to detect any changes in its dynamic characteristics.

8. By incorporating structural defects within the theoretical model an assessment can be made of the sensitivity of the dynamic properties to varying defect types and severity.

9. The theoretical behaviour of a damaged beam structure can be seen in Figure 3, whilst Figure 2 shows the corresponding experimental response.

10. Thus, it can be seen that with appropriate testing and modelling an assessment can be made of structural changes. This principle forms the basis for the dynamic monitoring of service reservoir foundations. Although considerably more complex than a simple beam structure a service reservoir will exhibit its own dynamic characteristics. A structure does not have to be considered particularly flexible or 'lively' before dynamic techniques can be used. Provided sufficiently sensitive sensors are used, dynamic behaviour can be monitored for most structures.

11. The purpose of monitoring the service reservoirs is to detect evidence of any voids which may develop. The presence of a void below a slab has two principal effects on the dynamic system. It leads to a local reduction in stiffness and also a localised loss of mass. The dynamic properties of the system will change in a manner dependent upon the relative influence of these two factors on the overall system. In the case of a beam structure with a crack-like defect there is a significant loss of local stiffness with a negligible loss of mass. This leads to a reduction of resonant frequencies with little change in modal mobility, as shown in Figure 2.

12. Figure 2 also illustrates another important feature relating to structural changes, namely that each mode of vibration will respond differently depending upon the relationship between the mode shape and the location of the structural change. In the case of the beam structure the modes which exhibit greatest frequency shift are those which have antinodes near the defect location.

13. The relationship between void location and reservoir mode shapes implies that the results obtained by exciting a column will be most sensitive to voiding directly under that column or deterioration of the column itself. A void which occurs at an intermediate position between columns will affect all the adjacent columns to a differing degree. By analysing the column responses an estimate can be made of the void location.

14. Having detected a significant change in the response at a column the reservoir could then be drained for detailed testing directly through the floor slab.

## Reservoir Testing Procedure

15. The purpose of the test technique is to excite the reservoir structure together with its foundations in order to assess its condition from its resonances and its frequency response. The method used to do this involves applying forcing inputs to the reservoir roof at each column head in turn.

16. The vibrator is mounted vertically above the longitudinal axis of the column in the manner illustrated in Figure 1. Coupling to the column head is by means of an expanding metal insert fitted into a drilled hole. A similar fixing is used to couple the accelerometer to the column head. A permanent surface box and cover was installed over each column head.

17. During testing a randomly varying force is applied to the column head by the vibrator and the frequency of the applied forcing varying between 0 and 400 Hz. The magnitude of the forcing applied to the structure is limited, but is sufficiently large for the natural frequencies of vibration to be identified by the high gain transducers employed to monitor response. Data obtained during testing comprising the vertical accelerations monitored at the column head and the applied forcing, are analysed on site using a dual channel dynamic signal analyser. The processed data is stored onto digital cassette for further examination and future reference.

## Development Studies

18. The development and application of dynamic testing methods to service reservoirs was supported by a programme of experimental studies and finite element analyses. The type of dynamic testing employed has much in common with modal testing, the theory and practice of this field have been discussed in detail by Ewins (1).

19. The experimental work employed a number of physical models of service reservoir and related structures which ranged from approximately one-half to one-tenth scale, culminating in field trials at a number of reservoirs. This work was used to:

   - demonstrate comparative responses for solid and voided slabs
   - confirm the ability of the system to detect voids at depth
   - investigate the relative merits of different procedures and different items of equipment
   - determine typical effects and values for damping, added water mass, ground stiffness and effective ground mass in order to calibrate the analytical model
   - investigate the excitation power level requirements for field testing

20. The finite element method was used to develop the analytical model, initial work being directed to establishing the validity of the model and for providing guidance to the experimental work. This process culminated in an analytical sensitivity study in order to:

- estimate the void size which could be detected directly under a column base
- estimate the depth to which voiding could be detected directly under a column base
- estimate the ability to detect voids offset from the column base under a continuous floor

and to generate guidelines regarding the characteristic changes in frequency response that could be expected for different:

- foundation types (pad or continuous footing)
- ground stiffness (rock or clay)
- void location

In addition analyses were performed to investigate the sensitivity of the response to changes in column modulus.

21. Damping was considered in these studies, with hysteretic damping values ranging from 10% to 40% for the various materials forming the model. The damping values chosen enabled the relative participation of the different materials involved to be reflected in the frequency response. This was achieved by the calculation of modal damping values, based on the relative material strain energies for the various modes of vibration. Thus a mode which involved significant ground motion therefore had a higher damping value than a mode which involved only structural motion. Damping had the effect of merging resonances in the frequency response, where closely spaced resonances existed. Thus a heavily damped response was much harder to interpret than a lightly damped response.

22. Comparison of the analytical results with experimental results indicated that actual changes in dynamic responses could be signficantly more pronounced than those predicted analytically. For example, the analytical model suggested that voids could not be detected at depth in rock. However, experimental work demonstrated that practically this was possible. The analytical studies suggested the following conclusions:

**Pad Footing on a Soft Foundation (Clay)**

- Voids directly under a pad footing are detectable from a plan size of $1m^2$ upwards
- Voids of around 3m square in plan are detectable up to a depth of between 2-3 metres

**Pad Footing on a Stiff Foundation (Rock)**

- Voids are not detectable directly under a pad footing, allowing for a maximum void size of 1.5m square in plan before the footing becomes unsupported. (An unsupported footing can be detected!)
- Voids at around 3m square in plan are not readily detectable at depth

## Continuous Slab on a Soft Foundation (Clay)

- Voids directly under a continuous slab are detectable from a plan size of 2-3 metres square upwards
- Voids of around 3m square in plan are detectable within a depth of about 0.5 metre
- Voids of around 3m square in plan directly under a continuous slab are detectable up to eccentricities of about 1 metre

## Continuous Slab on a Stiff Foundation (Rock)

- Voids directly under a continuous slab are detectable from a plan size of 2-3 metres square upwards
- Voids of around 3 metres square in plan are not readily detectable at depth

## Column Stiffness

- It is estimated that a drop of around 20% or greater in the column stiffness can be detected.

The sensitivities are improved where thin base slabs are employed.

## Monitoring of a Service Reservoir

23. The Sunderland and South Shields Water Company currently uses dynamic testing techniques to monitor the foundation condition of three of its service reservoirs. Figure 4 illustrates a typical frequency response function obtained from site measurements. Good quality data is obtained to below 10 Hz, this being judged by the coherence function (not shown) available on the dynamic signal analyser.

24. Figure 5 presents an analytical assessment of the influence of voiding upon the frequency response of a typical column. In this case it will be noted that changes occur in amplitude and frequency of response. The mode of vibration at 46 Hz exhibits the largest frequency shift because the mode shape at this frequency involves a greater participation of the ground mass. This can be appreciated by inspection of the mode shapes at 16 Hz an 46 Hz shown in Figures 6 and 7, respectively. For the 'ground' mode at 46 Hz the column is almost rigid, since the base displacement follows the roof displacement. The implied finite element analytical model is shown in Figure 8.

25. The dynamic measurements carried out have been used in two ways:

- to make comparison against baseline measurements, to enable changes with time to be identified
- to make relative comparisons between individual column positions

This has enabled comment to be made in respect of effective ground moduli and variations in the concrete structure.

26. Should a void be suspected the reservoir would be decommissioned and other investigations conducted using a number of specialist procedures including ground probing radar.

## Conclusions

27. A method of dynamic response testing has been developed which enables the foundation and structural condition of service reservoirs to be monitored during their life without the need to take them out of service.

## Acknowledgement

Acknowledgement is due to the support and encouragement given by the Sunderland and South Shields Water Company in the development and advancement of this technique.

## References

(1) D J Ewins, Modal Testing: Theory and Practice, Research Studies Press Limited, 1986.

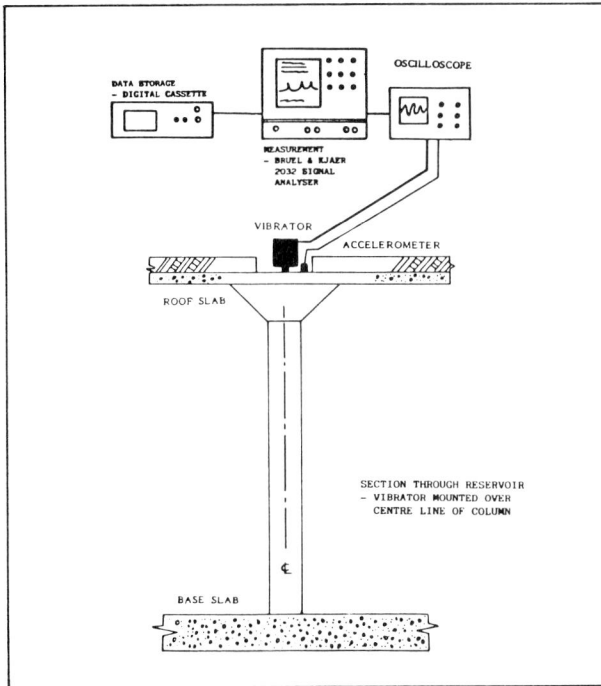

**FIGURE 1    SCHEMATIC ARRANGEMENT FOR VIBRATION TESTING TO RESRVOIRS**

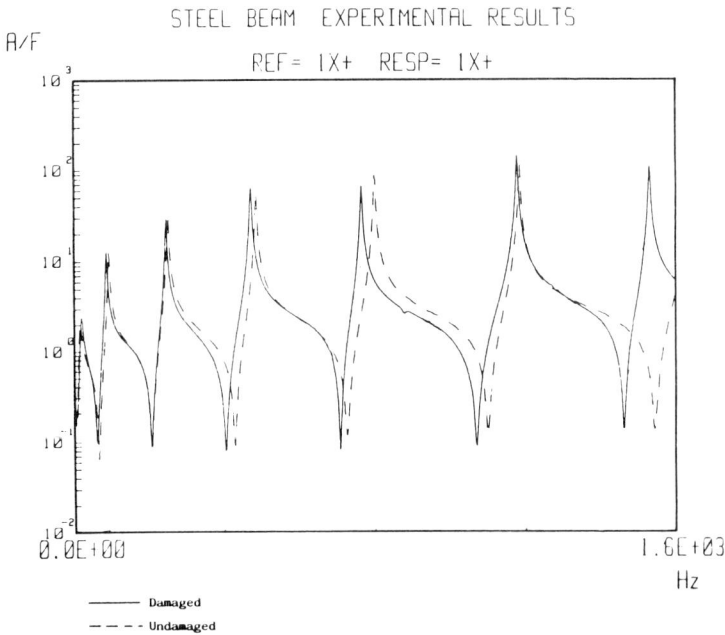

**FIGURE 2    EXPERIMENTAL RESULTS FOR STEEL BEAM**

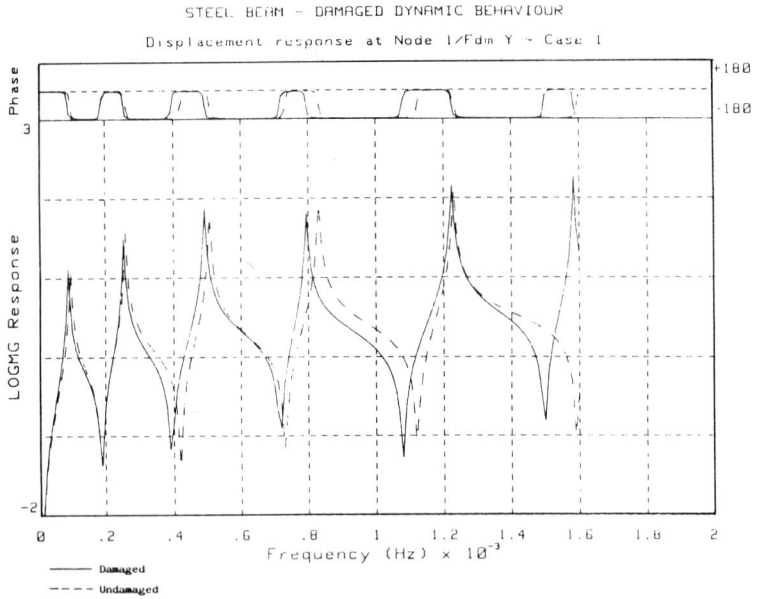

FIGURE 3   THEORETICAL BEHAVIOUR OF STEEL BEAM

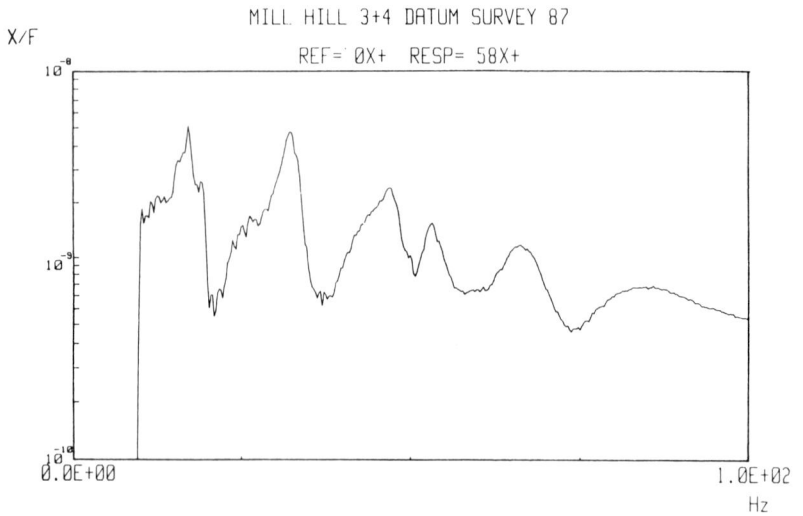

FIGURE 4   TYPICAL MEASURED DYNAMIC RESPONSE FOR RESERVOIR COLUMN

104

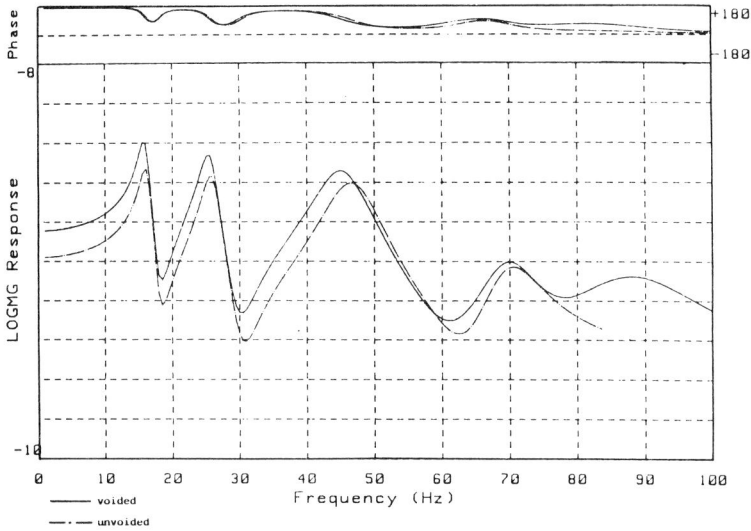

SUNDERLAND MILL HILL 3+4 5m pad  2.5m void  clay/rock  C960 22.1.88

Displacement response at Node 166/Fdm Z ~ Case 1

voided
unvoided

**FIGURE 5   THEORETICAL DYNAMIC RESPONSE OF RESERVOIR COLUMN**

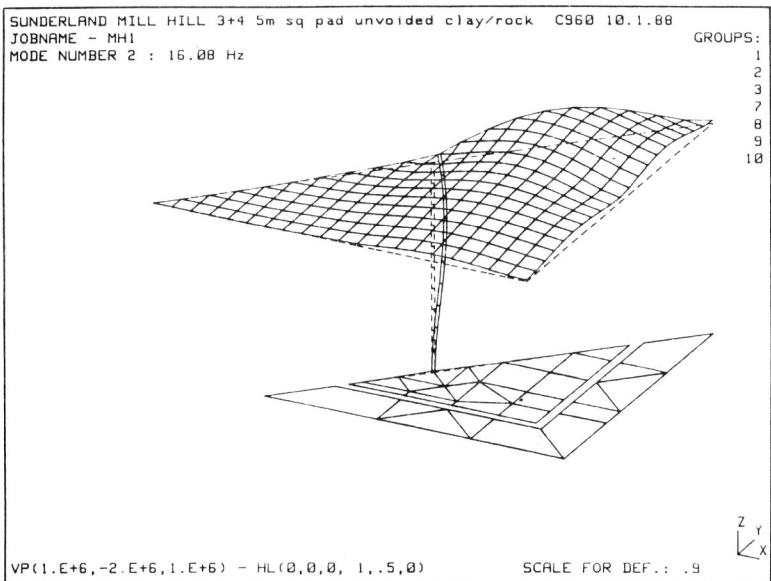

SUNDERLAND MILL HILL 3+4 5m sq pad unvoided clay/rock   C960 10.1.88
JOBNAME - MH1                                                    GROUPS:
MODE NUMBER 2 : 16.08 Hz                                              1
                                                                     2
                                                                     3
                                                                     7
                                                                     8
                                                                     9
                                                                    10

VP(1.E+6,-2.E+6,1.E+6) - HL(0,0,0, 1,.5,0)        SCALE FOR DEF.: .9

**FIGURE 6   MODE SHAPE OF VIBRATION AT SUPERSTRUCTURE FREQUENCY**

105

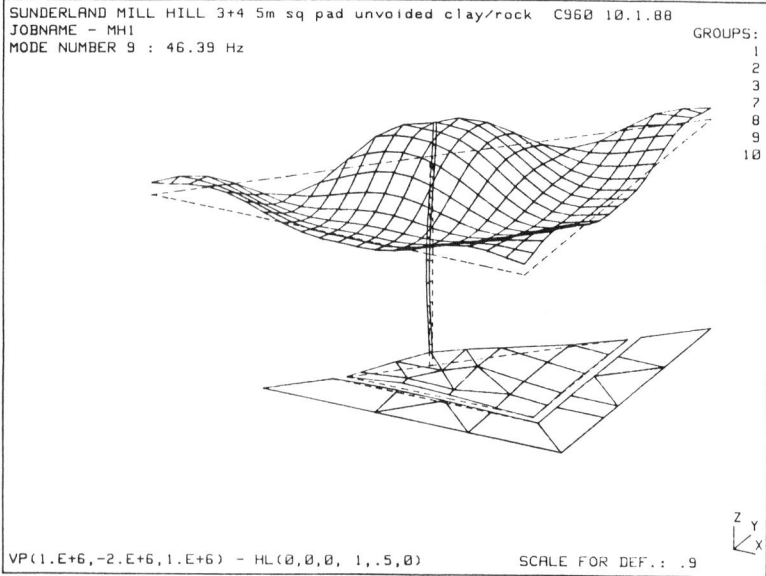

FIGURE 7   MODE SHAPE OF VIBRATION AT 'FOUNDATION' FREQUENCY

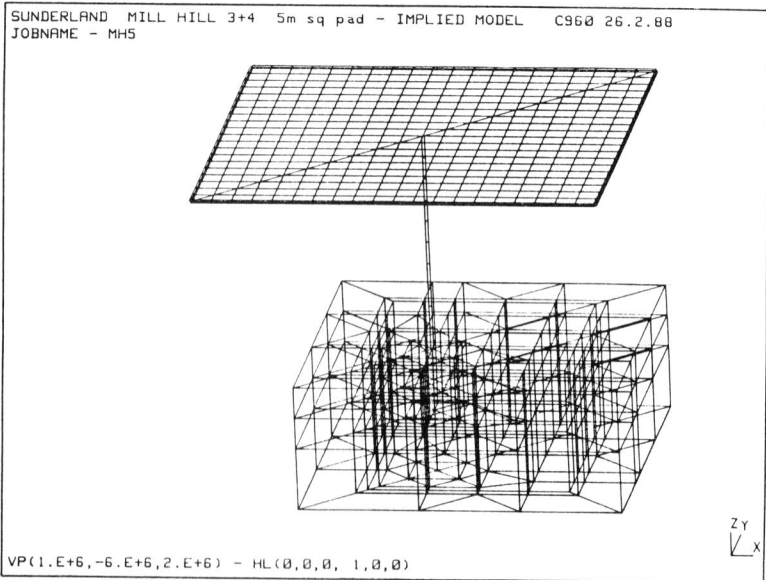

FIGURE 8   IMPLIED ANALYTICAL MODEL

# 13 Physical properties of structures investigated by dynamic methods

**Werner J Ruecker,** Federal Institute for Materials Research and Testing (BAM)

A methodology for the surveillance and inspection of structures is described. It is based on vibration measurements and their assessment with regard to time-dependent and spectral properties such as damping, amplitude maxima, natural frequencies and eigenmodes. The state of stiffness of the structure is deduced from the changes in measurement values. The procedure is demonstrated by measurement results.

## INTRODUCTION

The load bearing capability of structures changes with increasing service time. The causes for such changes can be ageing processes of the material but also hidden faults or external influences such as dynamic loads or static overloading. If the occurrence of damage can already be detected by visual inspection it can be expected that the extent of the interior damage of the structure is already rather critical, resulting in sudden failure with catastrophic consequences, especially for people. But also unexpected reduction of the serviceability of the structure can cause extensive economic losses. In case of damage an effective reconstruction is not possible as the exact location and extent of the damage are frequently not known. If reconstruction work is carried out it is possible that the work is unnecessarily extended or existing damage is not recognized. Therefore, non-destructive testing methods used for early damage detection and for the assessment of the structural integrity are a basic precondition for assuring the safety of the structural system. A very effective method to achieve these goals·is the inspection of the actual state of the structure by vibration measurement methods.

## DEFINITIONS

If methods based on vibration measurements are conceptionally developed one has to distinguish clearly between **permanent** building surveillance and **periodic** building inspection. Building surveillance is most likely applied in cases where a sudden deterioration of the structural integrity is to be expected with dangerous effects on the surrounding environment and, especially, for people. The aim of the method is to concentrate only on a few physical parameters, namely on the values known to be critical, when selecting the measurement and assessment values. These selected values are permanently measured and immediately evaluated according to different methods adapted to the relevant purpose. In many cases alterations of the condition of the structure are recognized by characteristic values evaluated directly from time signals or by comparison with Fourier spectra or parts thereof.

Building inspection is carried out by extensive measurements of the dynamic behaviour of the investigated system and by subsequent subjection of the system to a detailed analysis of early damage detection. Depending on the intensity of the analysis following the measurement, a very early detection of damage including a failure diagnosis is possible. Especially suitable for this kind of analysis are identification methods which determine the stiffness and damping behaviour of a system based on the measured eigenvalues and eigenmodes. Therefore, building inspection is the essential precondition for the application of building surveillance methods. Furthermore, is is also a necessary precondition for planing, execution and control of the improvements achieved by reconstruction measures.


## BASIC EQUATIONS FOR SYSTEM IDENTIFICATION

The statement where and to which extent a structure is damaged can only be achieved by analysing the mathematical model of the structure. A dynamically excited system without damping is decribed by the equation of motion

$$\mathbf{M} \; \mathbf{u}(t) \;\; + \;\; \mathbf{K} \; \mathbf{u}(t) \;\; = \;\; \mathbf{p}(t).$$

Taking the homogeneous solution of this equation

$$- \; \mathbf{M} \; \mathbf{U_0} \mathbf{\Lambda} \;\; + \;\; \mathbf{K} \; \mathbf{U_0} \;\; = \;\; 0$$

one gets the real matrix of the eigenvectors $\mathbf{U_0}$ and the diagonal matrix of the eigenvalues $\mathbf{\Lambda}$. Damped systems, which are most common, have complex eigenvectors with non-proportional real and imaginary parts. In order to receive a clear illustration, equations for systems with damping are not included in the paper. It is, however, referred to the fact, that the consideration of damping is in general of great importance for system identification purposes. The basic idea of system identification consists of the determination of the system parameters, namely stiffness matrix $\mathbf{K}$, damping matrix $\mathbf{D}$, and mass matrix $\mathbf{M}$, by measurement of the natural frequencies and eigenmodes. The relations between measurable modal parameters and the system parameters to be identified are summarized in Table 1 for undamped as well as for damped systems.

If it were possible to exactly measure and evaluate the eigenvalues and eigenmodes, the system parameters could be identified by means of the equations given in Table 1. In this case it is necessary to make sure that every natural frequency in the frequency range of interest is measured. The same applies to the computed sequence of the eigenvalues of the mathematical model. Both preconditions can generally be accomplished if the work is carried out with accuracy and experience. Finally, methods must be developed which allow exact measurements of the modal values and their evaluation as well as necessary updating processes.

Table 1: Relations between modal values and system parameters (after (1))

| | no damping | viscous damping | hysteretic damping |
|---|---|---|---|
| mass-matrix $M^{-1}$ | $M^{-1} = \dot{U}_s \dot{U}_s^T = \sum_{i=1}^{n} \phi_{si} \phi_{si}^T$ | $M^{-1} = \dot{U}_B \Lambda_B \dot{U}_B^T = \sum_{i=1}^{2n} \lambda_{Bi} \phi_{Bi} \phi_{Bi}^T$ | $M^{-1} = \dot{U}_D \dot{U}_D^T = \sum_{i=1}^{n} \phi_{Di} \phi_{Di}^T$ |
| damping matrix $B$ | | $B = -M \dot{U}_B \Lambda_B^2 \dot{U}_B^T M$ $= -\sum_{i=1}^{2n} \lambda_{Bi}^2 M \phi_{Bi} \phi_{Bi}^T M$ | $D = Im [(\dot{U}_D^T)^{-1} \Lambda_D \dot{U}_D^{-1}]$ $= Im (M \dot{U}_D \Lambda_D \dot{U}_D^T M)$ |
| stiffness matrix $K$ | $K = (\dot{U}_s^T)^{-1} \Lambda_s \dot{U}_s^{-1}$ $= M \dot{U}_s \Lambda_s \dot{U}_s^T M$ $= \sum_{i=1}^{n} \lambda_{si} M \phi_{si} \phi_{si}^T M$ | $K = -M \dot{U}_B \Lambda_B^3 (I - \dot{U}_B^T M \dot{U}_B \Lambda_B) \Lambda_B \dot{U}_B^T M$ | $K = Re [(\dot{U}_D^T)^{-1} \Lambda_D \dot{U}_D^{-1}]$ $= Re (M \dot{U}_D \Lambda_D \dot{U}_D^T M)$ |

If dynamic methods are applied it is of special interest to know in which way and to what extent occurring damage influences the dynamic measurement values. Mainly the following changes will occur: 1. The damping behaviour will generally increase. For concrete this leads to an unsymmetric hysteresis. Due to this effect unsymmetric positive and negative vibrations occur in the time signal. In the frequency domain the resonance peaks widen. 2. With increasing damage more and more new degrees-of-freedom and non-linearities will appear partly leading to stronger vibrations in the time domain and to additional and subharmonic frequencies in the frequency domain. 3. Due to modifications of the degrees-of-freedom of the system the kinetic and potential energies of the system change, too. Therefore both, the strains and the vibration velocities, are values which possibly indicate occurring damage. It should, however, be preferred to measure the vibration velocity as this kind of measurement is easier to perform. It has to be mentioned here that in this case the results can only be correctly interpreted with reference to the static system. 4. Any modification of the system always causes a finite change of the stiffness $K$, the damping $D$, and the mass $M$ which subsequently is the reason for alterations of the eigenvalues and eigenmodes. It can be shown that changes in stiffness result in additive changes of the eigenvalues and changes in the modal mass lead to multiplicative changes of the eigenvectors (1). Therefore, both effects are important for the identification of damage. As for the normally occuring extent of damage the modal mass changes are expected to be rather small, the deviation of the natural frequencies seem to be a preferable indicator.

EXPERIMENTAL RESULTS

For the developement of dynamic methods used for early damage detection and analysis of the bearing capability it is very important that an unmistakable, clearly defined relation exists between typical damage and its projection in the time and spectral domain. In order to gain experience and to test the basic suitability of the method laboratory test were carried out on 2 prestressed concrete beams. The beam shown in Fig. 1 was damaged by a 2.5 cm deep notch in the quarter-point and, in addition, about 1/3 of the prestressed steel cables were cut. The beam was excited by impulses and the responses were measured by velocity transducers. Fig. 2a shows the Fourier spectrum of a measurement point in the centre of the beam excited in the beam centre and Fig. 2b shows the spectrum at the quarter-point with the beam excited near the quarter-point. The Figs. show, that by chosing the right position for excitation and pick-up (here close to the location of the damage) the damage can be seen very well in the spectrum. In addition, it is important to note that the damage can only be made visible in certain natural frequencies and modes, depending on the location of damage.

The 14 m long beam shown in Fig. 3 was loaded statically in several steps (dead load, 4 times and 7 times the service load) followed by dynamic tests after each step. Fig. 4 shows clearly the decrease of the eigenvalues with increasing load. If the loss of stiffness is calculated from the eigenvalues by forming the square of the frequency ratios a stiffness reduction of 11 % is obtained at an increasing factor of 4 and of about 40 % at a factor of 7 of the service load. These values correspond exactly to the ones determined by the static tests and gained by the measurements.

It should be noted here, although it is not included in this paper, that changes in the values of local and global damping allow correspondingly good damage predictions.

After the laboratory tests on simple beams had been carried out the experiments were extented to bridge constructions in situ. The structure shown in Fig. 5 was damaged by cracks in its main beams and in the cross-sectional members. The bridge was repaired by closing the cracks and by enforcement of the main girders by prestressed steel cables. Dynamic tests were conducted before as well as after the reconstruction work. Fig. 6 shows the Fourier spectrum of a measuring point in the middle of the bridge. It can be seen that the natural frequencies show higher values now. From these values the increase of stiffness can be calculated which amounts to approximately 15%. Besides the spectra, which give only information on the global stiffness distribution and its alteration, the eigenmodes of the structure must be measured in order to receive the local stiffness, damping and possibly mass distribution. The values of the local distribution of those physical parameters are obtained by using the equations given in Table 1. As an example Fig. 7 shows the first two eigenmodes evaluated by measurements compared to the ones evaluated by a finite element model. Their agreement in shape as well their correspondence of the eigenvalues is relatively good, however, for the higher eigenmodes, which indicate damage in a more suitable way, this is not always the case.

This is based on the fact that for the beam bridge investigated here, the eigenmodes are extremely complex and the conventional measurement technique with discrete pick-ups is not a suitable tool for this type of bridge (2).

The steel-timber-arc-bridge, shown in Fig. 8, tends to have extremely large dynamic amplitudes under pedestrian traffic. The stiffness characteristics of the bridge had to be examined by dynamic inspection. If a satisfying comparison between the measured results and the results from a finite element computation of the bridge could be achieved, a proposal for the reconstruction based on the finite element analysis should be developed. During the measurements the bridge was dynamically excited by a cable which was first stressed and than suddenly cut. The vibrations were measured simultaneously at all measurement points (approx. 30). In Fig. 9 the spectra for horizontal and vertical motion are given. As can be seen, there are a lot of natural frequencies relatively close together and there is also only little damping in the system. In Fig. 10 the eigenmodes are compared which show excellent agreement between measurement and computation. The reasons for this agreement are suitable excitation, extremely precise measurements and detailed modeling of the structure. The example demonstrates that - due to the good agreement - dynamic stiffness analysis can be carried out even for extremely complicated structures if suitable measurement and computation techniques are used.

CONCLUSION

A method for the surveillance and inspection of buildings based on dynamic measurements has been presented. The experiments carried out on simple beams in the laboratory and on different types of bridge constructions in situ demonstrate that changes in the natural frequencies indicate changes in stiffness. After the combination of the measured eigenvalues and eigenmodes with the results of a model of the structure the local stiffness variation could be computed. However, care must be taken that the measurement technique is extremely precise, the evaluation techniques are suitable and the modeling of the structure is most accurate. In the near future, non-linear effects, which may also turn out to be a good indicator for damage detection, will be analysed by means of the so-called cepstrum analysis, which may also prove to be a suitable tool and might give new perspectives in the field of early damage detection.

REFERENCES

(1) Natke, H.G.: Einführung in Theorie und Praxis der Zeit-reihen- und Modalanalyse, Vieweg-Verlag (1983)

(2) Ruecker, W.J.: Zustandsprüfung von Bauwerken mit dynami-schen Methoden, BAM-Bericht (1989), (under preparation)

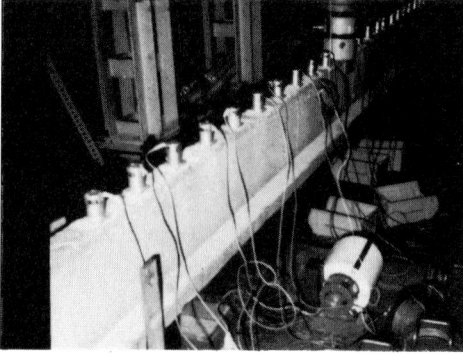

Figure 1:
Beam with attached velocity transducers and exciter

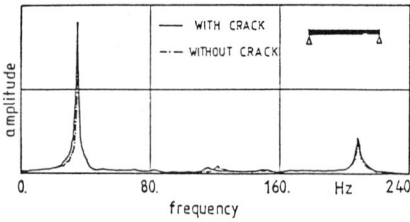

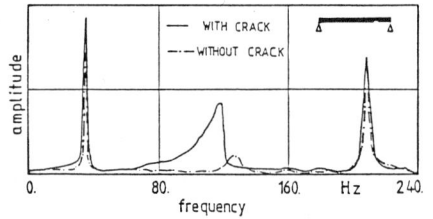

Figure 2: Fourier spectra of measurement points in the beam centre (left) and quarter-point (right)

Figure 3:
Beam with attached velocity transducer and marked cracks after application of 7 times service load

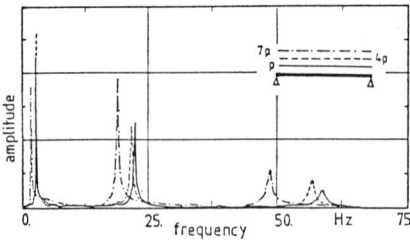

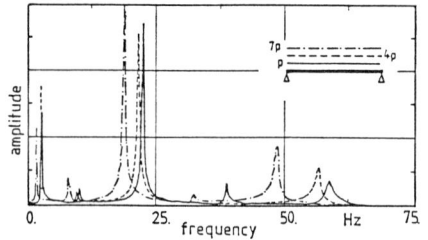

Figure 4: Fourier spectra of measurement points in the beam centre (left) and quarter-point (right)

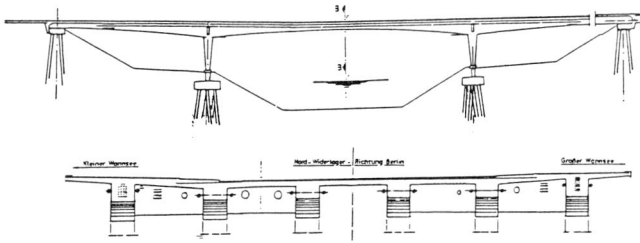

Figure 5: View of the investigated "Wannsee"-bridge

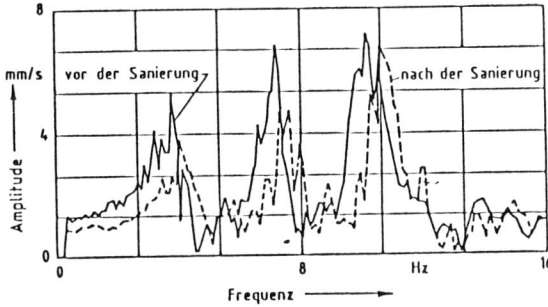

Figure 6: Fourier spectrum of a measurement point before and after reconstruction

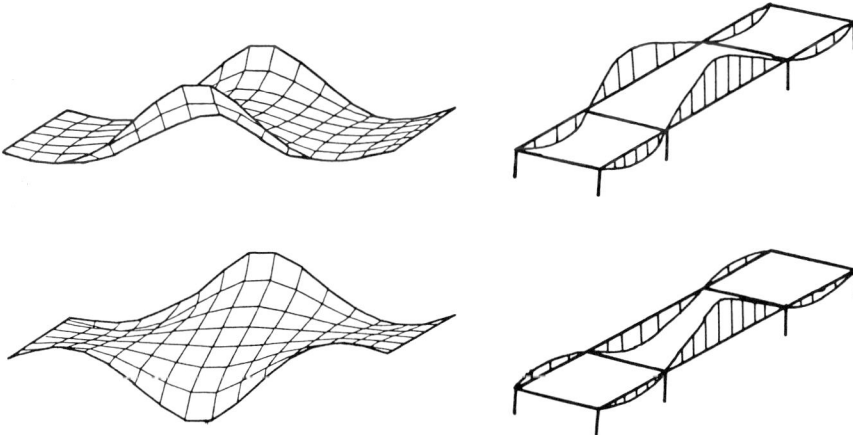

Figure 7: Comparison of the first two eigenmodes

113

Figure 8:
View of the investigated
"Sechser"-bridge

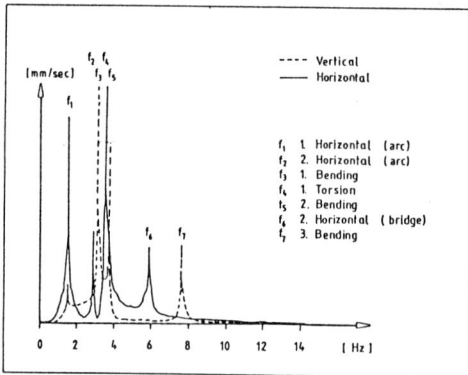

Figure 9:
Fourier spectra showing
the natural frequencies

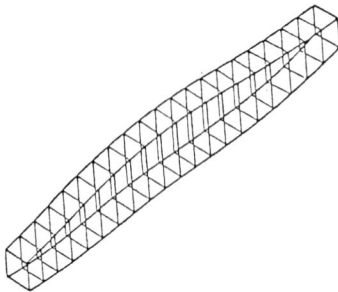
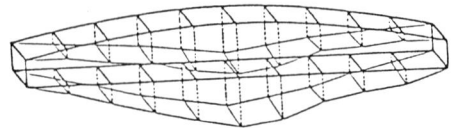

$f_3$  1. Bending

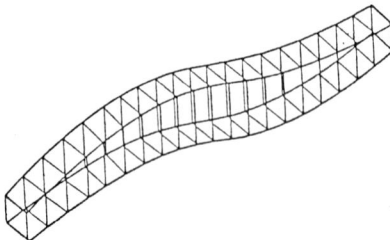
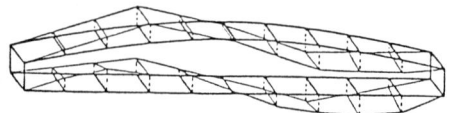

$t_5$  2. Bending

Figure 10:  Comparison of two eigenmodes

J G M Wood, Mott MacDonald, UK

The three papers on dynamic response of structures all show damage or deterioration producing a relatively small shift in the frequency diagram. The frequency response of structures depends partly on the support conditions and the wave mode in the structure, but predominantly on the Youngs Modulus on the concrete. The literature on the Youngs Modulus of concrete shows a wide scatter. Generally measured stiffnesses of structures in proof loading tests show stiffnesses significantly different from those calculated using the normal published values of E design, even when change in stiffness with age is considered. Could the authors of the papers explain how the problem of the uncertainty of E in structures and hence the uncertainty in the expected range of frequency response can be differentiated from changes due to faults in the structure.

# 14 The effectiveness of radar for the investigation of complex LPS joints

**R C de Vekey,** The Building Research Establishment, Watford, UK
**G Ballard,** GB Geotechnics Ltd, Cambridge, UK
**B W Adderson,** The Building Research Establishment, Watford, UK

Blind trials of a typical RADAR system have been carried
out on laboratory prepared full-size models of sections of
the 'H2' joint used in TWA large panel system buildings.
The findings have been analysed to try to give an objective
assessment of the capability of RADAR investigative
techniques to assist the engineer to make decisions about
the structural condition of the buildings.

## INTRODUCTION

Effective non-destructive techniques for investigating building
structures have long been sought and much research has been devoted to
the application of either transmitted or reflected electromagnetic
radiation. Much work was done on transmitted microwave radiation at BRE
in the 1960s by Watson et al (1,2,3). Recently the hardware for RADAR
(RAdio Direction And Ranging) has improved and equipment became
available which was designed to allow geological / geotechnical
investigations eg of substrata, water tables, discontinuities such as
mines,caves etc. with a scale of tens or hundreds of metres. The
background, theory and application are covered in a comprehensive book
by Ulriksen (4). With further improvements, particularly at shorter
wavelengths, it became possible to investigate discontinuities. on a
finer scale of metres and fractions of metres and RADAR was applied to
the location of pipes, eg by Blears and Daniels (5) and Daniels (6,7),
the investigation of soil moisture, by Annan and Davis (8), of cracks
and fractures in the ground, by Glover and Rees (9) and the quality of
foundations and roadbases eg. by Daniels (10), Holt and Eales (11)
Clemena. (12) and Rosetta (13). Users of the shorter wavelength
equipment then sought to apply the same method to the investigation of
the quality of walls, floors and joints of buildings which have
discontinuities on the same scale. Examples are given by Botros et al
(14), Transbarger (15) and Carr and Cuthbert (16).
Particularly, in view of the intense interest in the potential
performance of Taylor-Woodrow Anglian (TWA) buildings, in the light of
progressive collapse in the Ronan Point gas explosion, an attempt has
been made to utilise radar, in conjuction with more conventional
techniques, to assess the quality of the components of the key
horizontal joints. Since the early work on Ronan Point itself in 1984
(17), the technique has been used on a range of TWA and other system
buildings of the 60s and 70s. There is, however, very little specific
previous open literature on this particular application.

For the past two years the BRE has been carrying out a series of tests on a range of commercial and developmental radar systems to try to arrive at an objective assessment of the usefulness of the technique for the investigation of concrete system buildings. This paper covers one part of the work; an attempt to make an objective semi-quantified laboratory assessment of the use of one radar system by blind trials of the radar on full scale model joints made to carefully controlled specifications.

## EXPERIMENTAL PROCEDURE AND RESULTS

Concrete elements have been cast which are typical of the quality and composition observed in TWA buildings and can be assembled into models of the H2 joint. Table 1 gives the concrete properties and Figure 1 shows the assembled form in section. All the units are 700mm long. Several blind trials have been carried out all using a Geophysical Survey Systems Inc. model 4800 pulsed radar device with a transducer tuned to 1GHz.

## Phase 1 Preliminary scans of individual model joints.

Before the first full trial was carried out a preliminary survey was carried out on the joint specimens assembled but with no in-situ concrete or dry-pack mortar. Figure 2 shows a typical scan in progress and Figure 3 the transmitter/receiver, recorder and battery power supply. This trial was to check that the 12 model specimens (a) were sufficiently dried out, (b) gave a reproduceable response from joint to joint (ie were reasonably identical) and (c) were reasonably similar in response to real joints.

This showed that the specimens were satisfactory in respect of points (a) and (b) but there were some differences from real TWA components. The first main problem is the size, imposed by the requirements of transportability etc. meant that the edge effects complicated the signal for either end of the scan length. Additionally, the reinforcing mesh in the upper wall unit came right to the surface whereas in TWA units it is buried in 10-20mm of concrete. This caused additional radar response and made interpretation of the data from the frontal scans more difficult. In real structures a power cable is sometimes present along the inner edge of the drypack joint which acts as a useful reference point. The metal angle provided did not behave quite in the same way. It was felt, however, that these variations could be allowed for in the subsequent trials.

## Phase 2 Scans of filled individual joints.

The in-situ concrete was installed in the gap between the end of the floor slab and the lower panel upstand. This was either of good quality well compacted concrete or of poorly compacted gap-graded concrete as depicted in Figure 4. The compositions of the concretes are given in Table 1. Additionally, reinforcement was incorporated in some units and not in others. Dry-pack mortar was then placed in a 50mm thick layer over the joint, and the upper wall unit was installed. The dry-pack joint was either fully filled except for a 25mm strip of fibreglass insulation or had a pattern of cavities as depicted in Figure 5. The

Table 1 Concrete used for model joints

| ELEMENT | COMPOSITION WEIGHT% | | | | | THEORETICAL DENSITY OPC | | FRESH CONCRETE DENSITY OPC | | HARDENED DENSITY | CONCRETE STRENGTH | No |
|---|---|---|---|---|---|---|---|---|---|---|---|---|
| | OPC | AGGREGATE size | | | WATER | | | | | | | |
| | | 20-10 | 10-5 | 5 mm | | $kg/m^3$ | $kg/m^3$ | $kg/m^3$ | $kg/m^3$ | $kg/m^3$ | $N/mm^2$ | |
| Walls (estimated 2% air content) | | | | | | | | | | | | |
| VU/L | 13.2 | 35.5 | 15.8 | 27 6 | 7.9 | 2387 | 314 | 2345 | 309 | 2308 | 39.2 | 12 |
| Floors (estimated 6% air content) | | | | | | | | | | | | |
| F | 15.3 | 0 0 | 30.5 | 45.8 | 8.4 | 2376 | 363 | 2232 | 341 | 2213 | 35.9 | 12 |
| Cover panel (estimated 5% air content) | | | | | | | | | | | | |
| C | 13.1 | 0.0 | 39 5 | 39.5 | 7.9 | 2387 | 314 | 2261 | 298 | 2231 | 36.5 | 12 |
| Good quality infill | | | | | | | | | | | | |
| | 13.2 | 0.0 | 31 7 | 47.5 | 7.6 | - | - | - | - | 2234 | 38.3 | 3 |
| Poor quality infill | | | | | | | | | | | | |
| | 11.8 | 82.8 | 0.0 | 0 0 | 5.4 | - | - | - | - | 1636 | 4.0 | 3 |
| Dry-pack mortar | | | | | | | | | | | | |
| | 11.8 | 0.0 | 0.0 | 82.8 | 5.4 | - | - | - | - | 1890 | 13.9 | 3 |

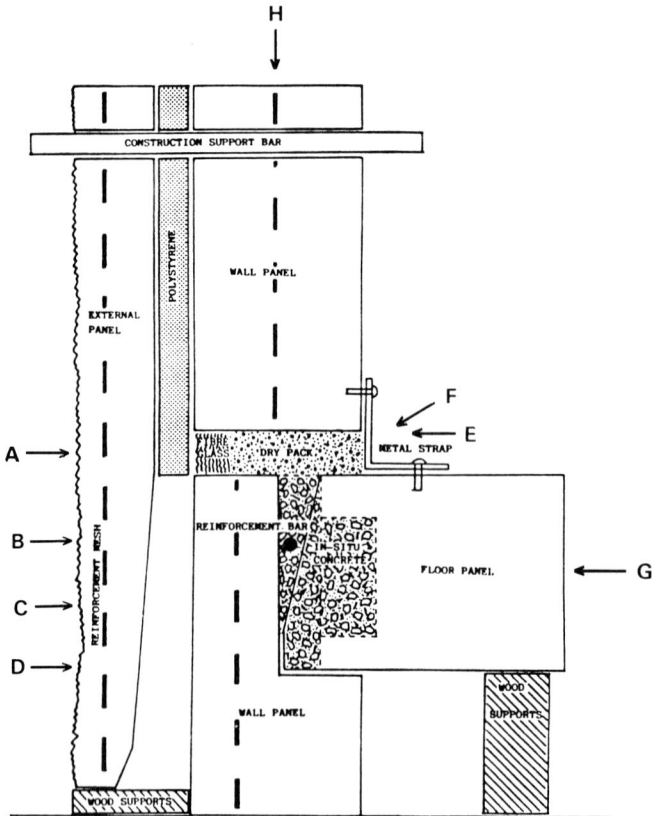

Fig. 1 Assembled form of a model joint
(Horizontal RADAR scan lines lettered A-H)

Fig. 2 **Scanning in progress**

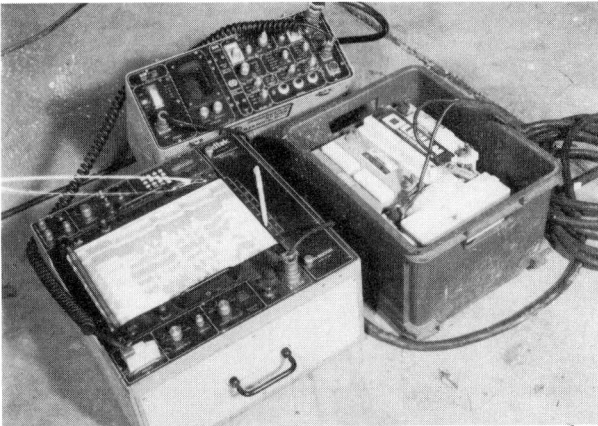

Fig. 3 **Tranmitter/receiver**, recorder and power supply

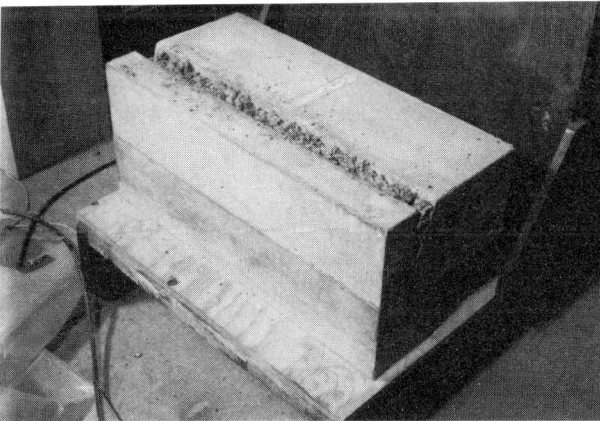

Fig. 4 **Typical** poor infill using gap-graded concrete

plan layout of the cavities is given in Figure 6a  Finally the
insulation board and front cover panel were installed and the ends were
covered over to prevent inspection. Table 2 lists the state of the 12
panels.

Table 2  Results of Phase 2 radar survey of isolated model joints from
exterior face (A,B)

| JOINT No. | IN-SITU CONCRETE AS MADE | RADAR REPORT | DRY-PACK ZONE AS MADE | | | % VOIDS RADAR | | |
|---|---|---|---|---|---|---|---|---|
| 1 | good + rebar | some voids + rebar | 12 | 12 | 12 | 5, | ?, | 7 |
| 2 | voided no-rebar | voided, no-rebar | 0 | 0 | 0 | 0, | 6. | 2 |
| 3 | good + rebar | good / small voids + rebar | 0 | 0 | 0 | 10, | ?, | 10 |
| 4 | voided no-rebar | voided, no-rebar | 12 | 12 | 12 | 17, | 5, | 10 |
| 5 | good, no rebar | good / small voids no rebar | 0 | 0 | 0 | 12, | 0, | 11 |
| 6 | good, no rebar | good / small voids no rebar | 0 | 0 | 0 | 12, | 0, | 7 |
| 7 | good, no rebar | good / small voids + rebar | 12 | 12 | 12 | 21, | 5, | 12 |
| 8 | voided + rebar | voided  no-rebar | 0 | 0 | 0 | 11, | 4, | 12 |
| 9 | voided + rebar | voided. no-rebar | 0 | 0 | 0 | 14, | 0, | 8 |
| 10 | good + rebar | good / small voids + rebar | 12 | 12 | 12 | 16, | 7, | 16 |
| 11 | voided no-rebar | voided + rebar | 12 | 12 | 12 | 26, | 4, | 12 |
| 12 | voided + rebar | voided + rebar | 12 | 12 | 12 | 8, | 0, | 4 |

Analysis:  Good in-situ 5/6. voided in-situ 6/6, rebar 4/6, no rebar 4/6
Dry pack 15/36 (Within ±5%)    (Probability=0.41)
Dry pack 26/36 (Within ±10%)   (Probability=0.72)
Dry pack 34/36 (Within ±20%)   (Probability=0.94)
Dry pack 36/36 (using 50% critical method)(Probability=1.0)

After curing and drying out of the concrete and mortar, horizontal radar
scans were carried out at various levels on the front face of the
specimens at positions A-D on Figure 1. Additionally, some vertical
scans were tried but were not found to be very helpful. The scans were
then analysed and a report was submitted giving the findings in terms of
a qualitative estimation of the quality of the in-situ concrete and
whether a reinforcing bar was present or absent and a semi-quantitative
estimate of the void level of the dry-pack. These are given in Table 2.
One problem encountered was the edge effects which make it more
difficult to interpret the data at each end of the scan.

## Phase 3 Scans of joints in joined rows of six

In this trial the units were mortared together in two rows of six
samples to try to reduce the problems of edge effects and the dry-pack
was reformulated with much larger voids shown in Figure 6b for group B
and 6c for group A. The distribution of the good/poor infill, rebar, and
dry-pack are given in Tables 3 and 4. In this run scans were made in
positions A-C and additionally at E and F equivalent to skirting level
in a real building and G which is only possible in this experimental
set-up. The radar predictions from scans A-C are given in Table 3 and
from E-G in Table 4.

120

Fig. 5 **Drypack in phase 2 tests**

6a. Voided and unvoided dry-pack used in Phase 2

6b. Dry-pack layout of panel B phase 3 (other half
    mirror image)

6c. Dry-pack layout of panel A phase 3 (other half
    mirror image)

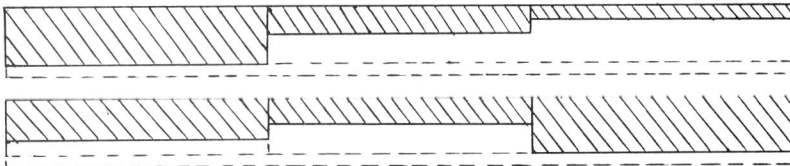

6d. Dry-pack layout of both panels in phase 4
    (upper left hand end, lower right-hand end)

Fig. 6 **Plans of dry-pack**

Table 3  Results of phase 3 radar survey of aggregated model joints from exterior face (A,B)

| JOINT No. | IN-SITU CONCRETE AS MADE | RADAR REPORT | DRY-PACK ZONE - % VOIDS AS MADE | RADAR |
|---|---|---|---|---|
| 1 | voided + rebar | coarse voids + rebar | 0, 0, 0 | |
| 2 | voided + rebar | coarse voids + rebar | 85 75 70 | |
| 3 | good no rebar | fine voids + rebar | 50 30 5 | |
| 4 | good no rebar | fine voids + rebar | 5 30 50 | |
| 5 | voided + rebar | coarse voids + rebar | 70 75 85 | |
| 6 | voided + rebar | coarse voids + rebar | 0 0 0 | |
| 7 | good + rebar | Good no rebarebar | 0, 0, 0 | 15, 15, 10 |
| 8 | good + rebar | Good no rebar | 0, 20, 50 | 15, 15, 40 |
| 9 | voided + rebar | coarse voids no rebar | 70, 60, 50 | 60, 40, 30 |
| 10 | voided + rebar | coarse voids no rebar | 50, 60, 70 | 25, 25. 60 |
| 11 | good + rebar | fine voids no rebar | 50, 20, 0 | 40, 0, 0 |
| 12 | good + rebar | fine voids no rebar | 0, 0, 0 | 0, 10, 20 |

Analysis: Good in-situ  4/6 (if fine voids=50% good), voided in-situ 6/6
        Rebar present 4/6, rebar absent 4/6
        Dry pack 3/18 (Within +5%)        (Probability=0.17)
        Dry pack 9/18 (Within +10%)       (Probability=>.50)
        Dry pack 16/18 ( ithin +20%)      (Probability=0.89)
        Dry pack 14/18 (using 50% critical method)(Probability=0.78)


Table 4  Results of phase 3 radar survey of aggregated model joints from interior faces (E,F)

| JOINT No. | IN-SITU CONCRETE AS MADE | RADAR REPORT | DRY-PACK ZONE - % VOIDS AS MADE | RADAR |
|---|---|---|---|---|
| 1 | voided + rebar | coarse voids | 0, 0, 0 | G, G, G |
| 2 | voided + rebar | coarse voids | 85 75 70 | V, V. V |
| 3 | good no rebar | Good | 50 30 5 | V, PV, G |
| 4 | good no rebar | Good | 5 30 50 | G, PV, V |
| 5 | voided + rebar | coarse voids | 70 75 85 | V V, V |
| 6 | voided + rebar | coarse voids | 0 0 0 | PV, PV, PV |
| 7 | good + rebar | Good | 0, 0, 0 | G, G, G |
| 8 | good + rebar | Good | 0. 20, 50 | G, G, V |
| 9 | voided + rebar | coarse voids | 70, 60, 50 | V, V, PV |
| 10 | voided + rebar | coarse voids | 50, 60, 70 | PV V, V |
| 11 | good + rebar | Good | 50, 20, 0 | V, G G |
| 12 | good + rebar | Good | 0. 0, 0 | G, G, G |

Coding of the dry pack radar observations:  G=good unvoided (0-19% voids), PV=part voided (20-49% voids) V=voided (50-100% voids)

Analysis: Good in-situ  6/6, voided in-situ  6/6 , Rebar - no data
         Dry pack 29/36 (using 50% critical method) (Probability=0.81)

122

# Phase 4 Scans of joints in reformatted joined rows of six

In this trial of the two rows of six samples, the in-situ concrete was rearranged and the dry-pack was reformulated with a new voids pattern shown in Figure 6d. In this case, as an experiment, the normally impossible scans from point H and G were tried. The results are given in Tables 5 and 6 and typical radargrams are shown in Figure 7.

**Table 5  Results of phase 4 radar survey of aggregated model joints from rear face (E,G)**

| JOINT No. | IN-SITU CONCRETE AS MADE | RADAR REPORT | DRY-PACK ZONE - % VOIDS AS MADE | | | RADAR | | |
|---|---|---|---|---|---|---|---|---|
| 1 | voided + rebar | medium voids no rebar | 25, | 25, | 25 | G, | G, | G |
| 2 | voided + rebar | coarse voids no rebar | 75 | 75 | 75 | V, | V, | V |
| 3 | good + rebar | good no rebar | 75 | 75 | 75 | V, | V, | V |
| 4 | voided no rebar | coarse voids no rebar | 50 | 50 | 50 | PV, | PV, | PV |
| 5 | good no rebar | mostly good no rebar | 50 | 50 | 50 | PV, | PV, | PV |
| 6 | good + rebar | good no rebar | 25 | 25 | 25 | G, | G, | G |
| 7 | good no rebar | good no rebar | 0, | 0, | 0 | PV, | PV, | PV |
| 8 | voided no rebar | coarse voids no rebar | 0, | 0, | 0 | G, | G, | G |
| 9 | good no rebar | good no rebar | 0, | 0, | 0 | PV, | PV, | PV |
| 10 | good no rebar | good no rebar | 50 | 50, | 50 | G, | G, | G |
| 11 | voided no rebar | coarse voids no rebar | 50, | 50 | 50 | PV, | PV, | PV |
| 12 | voided no rebar | coarse voids no rebar | 0, | 0, | 0 | G, | G, | G |

Coding of the dry pack radar observations: G=good unvoided (0-19% voids), PV=part voided (20-49% voids) V=voided (50-100% voids)

Analysis: Good in-situ 5.5/6, voided in-situ 6/6,
        Rebar 0/4, no rebar 8/8
        Dry pack 27/36 (using 50% critical method) (Probability=0.75)

**Table 6  Results of phase 4 radar survey of aggregated model joints from rear face angled (E,G angled)**

| JOINT No. | IN-SITU CONCRETE AS MADE | RADAR REPORT | DRY-PACK ZONE - % VOIDS AS MADE | | | RADAR | | |
|---|---|---|---|---|---|---|---|---|
| 1 | voided + rebar | coarse voids + rebar | 25, | 25, | 25 | G, | G | G |
| 2 | voided + rebar | coarse voids + rebar | 75 | 75 | 75 | V, | V, | V |
| 3 | good + rebar | Good + rebar | 75 | 75 | 75 | V, | V, | V |
| 4 | voided no rebar | coarse voids no rebar | 50 | 50 | 50 | PV, | PV, | PV |
| 5 | good no rebar | good + rebar | 50 | 50 | 50 | PV, | PV, | PV |
| 6 | good + rebar | good + rebar | 25 | 25 | 25 | G. | G, | G |
| 7 | good no rebar | good no rebar | 0 | 0, | 0 | G, | G, | G |
| 8 | voided no rebar | coarse voids no rebar | 0. | 0, | 0 | G, | G, | G |
| 9 | good no rebar | good no rebar | 0. | 0, | 0 | G, | G, | G |
| 10 | good no rebar | good no rebar | 50, | 50, | 50 | PV, | PV | PV |
| 11 | voided no rebar | coarse voids no rebar | 50, | 50, | 50 | PV, | PV, | PV |
| 12 | voided no rebar | coarse voids no rebar | 0, | 0, | 0 | G | G, | G |

Analysis: Good in-situ 6/6, voided in-situ 6/6 ,
        Rebar 4/4, no rebar 7/8
        Dry pack 30/36 (using 50% critical method) (Probability=0.83)

## ANALYSIS AND DISCUSSION

The physics of the analysis of radar traces is still in it's infancy and much of the interpretation is still on the basis of experience and 'fingerprinting' of the return signal. Much of the difficulty in making objective quantitative predictions of behaviour is because the objects being investigated are very close to the antenna and coupling can occur particularly to metallic components. However using basic data on the dielectric constants of the constituent materials, eg Von Hippel (18), and the corresponding pulse velocity. information about the position of interfaces can be derived from the measured timings in relation to known 'landmarks' such as the front of the specimen, the back face or the metal angle at the wall/floor junction. Additional information about the void content of well characterised materials can be derived from the absolute velocity if it can be deduced.

Analysis of the in-situ concrete and the rebar is on the basis of a simple bi-state statistic; ie good/not good or present/absent.

Where the approximate estimate of the dry pack/void ratio was made, analysis is on the basis of the accuracy of prediction. The requirements for accuracy are obviously dictated by the need to make a sound engineering decision but probably need to be higher for semi-filled joints than for nearly full or nearly empty joints. eg if the method indicates 90% fill even an error $+$of -20% could give an acceptable engineering conclusion whereas 50-20% might not. The easiest way of deriving a probability is to set a critical void level and take as correct all predictions that would lead to a correct decision in relation to the critical level. A value of 50% seems reasonable in that it could lead to a halving of the presumed factor of safety. In practice the critical void level may differ from 50% depending on the magnitude of the load, ie near the top or bottom of the building, and whether the load was a vertical (axial or eccentric) or shear load. An additional consideration is whether it is satisfactory to transfer all the load via the in-situ concrete. This method automatically leads to a form of parabolic response which is more sensitive near the critical value. To avoid making subjective decisions samples with exactly 50% void have been taken as 50% correct. On the above basis the accuracy of prediction is given at the foot of each of the relevant tables (2-6). Percentage accuracies are also given in cases where a finite figure has been assigned by the radar analyst.

The radar was good overall at judging infill concrete quality (on average 0.94 probability) and a bit unreliable at detecting rebar (on average 0.73 probability). Dry-pack quality, using the critical 50% method gave an average $+$0.83 probability but a poorer figure if overall accuracies better than -20% were required. Problems were experienced because of edge effects and because the dry-pack void sizes were on the borderline of radar discrimination in phase 2. Experimentation with the survey positions indicated that the dry-pack could be measured most effectively by surveying from points E and F (internal skirting). This would of course require the removal of any strengthening angle but obviates scaffolding. The in-situ concrete is tackled most effectively from the front in real situations although the impractical points H and G gave very good results as well. A position angled up from the internal ceiling might give a similar response to position G (but was not practical in the experimental set-up).

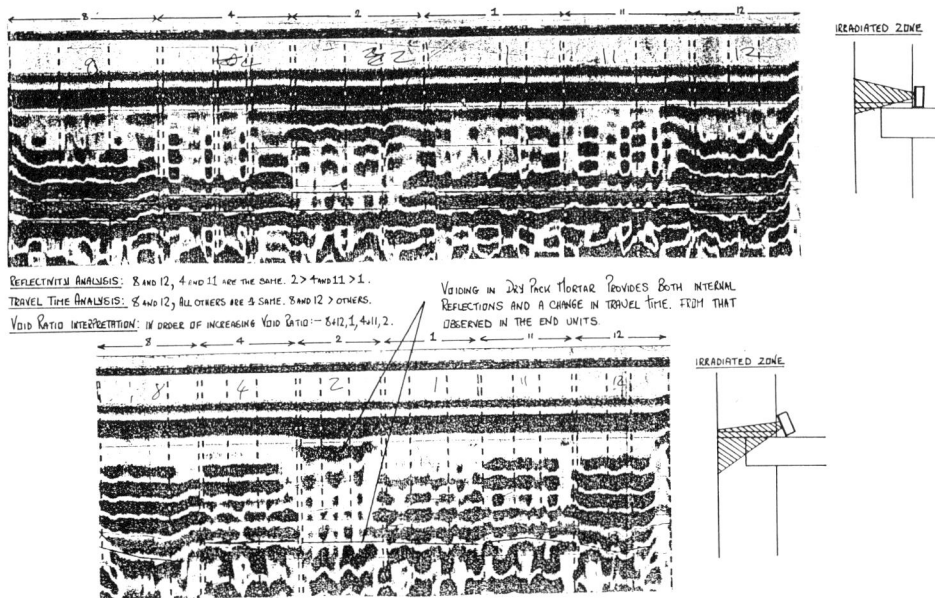

REFLECTIVITY ANALYSIS: 8 AND 12, 4 AND 11 ARE THE SAME. 2 > 4 AND 11 > 1.

TRAVEL TIME ANALYSIS: 8 AND 12, ALL OTHERS ARE A SAME. 8 AND 12 > OTHERS.

VOID RATIO INTERPRETATION: IN ORDER OF INCREASING VOID RATIO:— 8+12, 1, 4+11, 2.

VOIDING IN DRY PACK MORTAR PROVIDES BOTH INTERNAL REFLECTIONS AND A CHANGE IN TRAVEL TIME. FROM THAT OBSERVED IN THE END UNITS.

Fig. 7 Typical radargrams with position diagrams

## CONCLUSIONS

With experience information can be derived about the quality of in-situ concrete and drypack in typical LPR joint systems. In a joint as complicated as the H2 joint because of the diversity of interfaces present and thus the complexity of the derived signal, the probability of making a correct engineering judgement from RADAR data seems to be around the 0.7-0.9 level depending on the item of interest and the scan position. This should give a reasonable if approximate guide to whether the technique is viable on grounds of safety and cost-effectiveness.

## ACKNOWLEDGEMENTS

The work described has been carried out as part of the research programme of the Building Research Establishment of the Department of the Environment and this paper is published by permission of the Director.

## REFERENCES

(1) A.Watson, 'The non-destructive measurement of water content by the microwave absorption method', CIB Bull.1960,3 pp15-16.

(2) Boot,A.R. and Watson.A., 'Application of centrimetric radio waves in non-destructive testing', ASTM/RILEM symposium on application of advanced and nuclear physics to testing materials, ASTM SPT no.373, (1964).

(3) Watson,A., 'Measurement of water content in some structures and materials by microwave absorption', Proc.RILEM/CIB symposium on moisture problems in buildings, 2, pp6-8, (1965)

(4) P.Ulriksen, Application of Impulse Radar to Civil Engineering, 1981

(5) Blears,A.S.,and Daniels D.J., 'Gascopact: A new approach to pipe and cable location', IGE 49th Autumn meeting, (1983).

(6) Daniels,D.J., 'Gascopact', Pipetech 84, ppm1-m5 (1984).

(7) Daniels,D.J., 'Location of underground services by ground probing radar', NODIG 85, pp5 1.1-5.1.6, (1985).

(8) Annan,A.P. and Davis,J.L. 'Electromagnetic detection of soil moisture', Canadian Journal of remote sensing, V3, 1, pp76-86, (1977)

(9) Glover,J.M. and Rees,H.V., 'Crevasse detection using ground probing radar', Geological Survey of Canada and Fedeeal Panel of Energy Research & Development. Seminar apapers May (1988).

(10) Daniels,D.J., 'Short pulse radar for stratified lossy dielectric layer measurements', IEE Proc -F, V127, Part F, No.5, Oct (1980).

(11) Holt F B., Eales,J.W., 'Nondestructive evaluation of pavements', Concrete International, June, pp41-45, (1987).

(12) Clemena,G.G., 'Evaluation of overlaid bridge decks with ground penetrating radar', Virginia Dept.of Highways and Transportation, Richmond,PB,82 221839, (1982).

(13) Rosetta,J.V., 'Feasibility study on the measurement of Bridge deck Bituminous overlay thickness by pulse radar'. Geophysical survey systems inc. W.O. #507-0 Hudson, New Hampshire,USA

(14) Botros,A.Z.,Olver,A.D.,Cuthbert L.G.,Farmer,G.A., 'Microwave detection of hidden objects in walls', Electronics Letters, V20, 20, pp824-825, (1984).

(15) Transbarger,O., 'FM radar for inspecting brick and concrete tunnels', Materials evaluation, 1985

(16) Carr,A.G., Cuthbert,L.G.,Liau,T-F., 'Signal processing techniques for short-range radars applied to the detection of hidden objects'. Proc.7th European Conf. on Electrotechnics EURCON, Paris, (1986).

(17) The structural adequacy and durability of large panel system dwellings, Building Research Establishment Report, 1987

(18) Von Hippel,A.R., Dielectric materials and applications, Chapman and Hall, (1954) and Handbook of the Am. Inst. of Physics 2nd ed. Intnl. critical tables, McGraw Hill Book Co. (1963).

# 15 The stiffness damage test – a quantitative method of assessing damaged concrete

**T M Chrisp, R S Crouch and J G M Wood,** Special Services Division, Mott, Hay & Anderson Consulting Engineers, 20–26 Wellesley Road, Croydon, England, UK

To assist in the development of cost effective management strategies for deteriorating reinforced concrete structures, Mott, Hay & Anderson have been developing tests to quantify the amount of damage within the concrete. This paper reports one such test, the Stiffness Damage Test (SDT), developed to assess the change in the stress/strain response of concrete damaged by alkali aggregate reaction (AAR) or overstress from impact, thermal or other loadings.

## INTRODUCTION

The structural behaviour of reinforced concrete involves the interaction of many nonlinear physical phenomena. This makes the formulation of rational analytical procedures difficult and has lead to the current reliance upon very simplified empirical approaches in design and analysis codes. These empirical methods have generally been developed for sound concrete from the study of large bodies of experimental data, assembled over many years. However as the stock of bridges and buildings deteriorate the original test data and codes cease to provide a valid basis for strength evaluation.

One possible approach to the problem would be to undertake a large scale empirically based programme of load testing existing and created structures with advanced stages of deterioration in order to record their actual response. However this approach would be prohibitively costly. A more cost.effective technique is to monitor real structures as deterioration progresses and to record any changes in material behaviour (eg stress/strain/time response, strength and ductility). An analysis of structural behaviour up to the limit state may then be carried out using finite element techniques with modified properties. This can then be calibrated against the results of a more limited set of proof load and ultimate load tests.

To aid integrity assessment of AAR damaged concrete structures, Mott, Hay & Anderson [1] have been evolving both insitu structural monitoring (such as proof loading, expansion and crack monitoring, and pulse velocity surveys) and physical testing techniques on concrete cores.

In order to make decisions on the management of structures with AAR it is necessary to estimate both the 'expansion and damage to date' and the 'potential for further expansion and damage'. The recommendations of the Doran Committee [2] provide a framework for this and give some interim guidance on methods. While petrography, uspv and summation of crack widths provides indications for the estimation of 'expansion and damage to date' they tend to give a high scatter of results and are difficult to relate to physical properties used in design appraisal. Nixon and Bollinghaus [3], Swamy and Al–Asali [4], Hobbs [5] and Akashi et al. [6] have evaluated changes in tensile and compressive strength, uspv and Youngs' modulus (E) with expansion resulting from the progressive development of AAR. This shows a reduction in E as being a sensitive measure of developing degradation. Walsh [7] highlighted the link between crack density and cyclic stress/strain response in rocks. Crouch [8] considered these in suggesting a new test for concrete – The Stiffness Damage Test (SDT).

At Mott, Hay & Anderson, we set out to develop the SDT as a non–destructive stress/strain measuring technique to provide a measure of concrete material damage which can be related to compressive and tensile strength changes, free expansion to date and petrographical changes. This paper reports its development on cores damaged by alkali silica reaction (ASR).

## THE STIFFNESS DAMAGE TEST – ASSESSING DAMAGE IN STRUCTURAL CONCRETE

The SDT has been developed in Mott, Hay & Anderson as a simplified method of measuring damage within concretes from deteriorating structures. The early results were reported by Wood and Crouch [9] to the ASR Forum in 1987. Chrisp [10] reports the details and the refinement of the test, and compares the results of cores taken from a range of structures with the results from cylinder compression, torsional shear strength [11] and uspv tests.

The test involves the measurement of the strain response of concrete cores under low cyclic load. In order to avoid reinforcement whilst coring structures, the diameter of the core has been chosen as 70mm. Normally the length is fixed between 175mm and 200mm long in order to maintain an aspect ratio between 2.5 and 2.75 to 1. Some longer cores have been tested to detect the variation in damage along the specimen. To minimise damage imparted to the core by testing, a maximum stress of 5.5 $N/mm^2$ is applied at a rate of 0.1 $N/mm^2/s$, then unloaded at the same rate. The load cycle is servo controlled, through a microprocessor and is repeated a further four times. Longitudinal strain is recorded over the central 70mm of the core in order to avoid platen end effects and bedding in, but also to maintain a gauge length of over 3 times the maximum aggregate size. The displacement of three linear transducers is recorded by the microprocessor every 0.5 $N/mm^2$.

In 18 months over 400 such tests have been undertaken at The University of Bristol. The concrete tested came from both ASR damaged concretes and controls. Some cores have been tested once after coring and then again following a period in a 100% relative humidity environment to accelerate the ASR expansion. Initial state and any subsequent changes in stiffness characteristics are recorded and the findings incorporated into the structure's management strategy.

## RESULTS EMERGING FROM THE STIFFNESS DAMAGE TEST

The following parameters, calculated for each SDT, are averaged over the last four of the five load/unload cycles, (refer to Figure 1):

1.  Chord Loading Stiffness (Ec) – Slope of the loading response (a–b).

2.  Unloading Stiffness (Eu) – Slope of the response immediately after unloading (b–c).

3. Damage Index (DI) – Area of the hysteresis loops normalised over the stress range.

4. Plastic Strain (EP) – Non recovered strain at the end of the unload cycles (d–e).

5. Non Linearity Index (NLI) – The slope of the loading response to half the maximum load (a–f), divided by the Ec, and represents the degree of convexity or concavity of the loading curve.

The data and calculated parameters are stored on a microcomputer database program allowing easy access and manipulation of the results. Table 1 shows an extract from the database which includes the results from the SDT, core details, and uspv and strength results.

SUMMARY RESULTS OF STIFFNESS DAMAGE TEST

STRUCTURE – ANON

| Core Number | Mix Design Strength (N/mm^2) | Observed Crack Code (0–9) | Loading Stiffness Ec (kN/mm^2) | Unloading Stiffness Eu (kN/mm^2) | Damage Index | Plastic Strain | Non-Linearity Index | USPV (km/s) | Cylinder Strength (N/mm^2) | Torsional Strength (N/mm^2) |
|---|---|---|---|---|---|---|---|---|---|---|
| 923 | 30 | 6 | 14.3 | 47.6 | 31.56 | 24.1 | 0.95 | 4.4 | 0.0 | 1.64 |
| 916 | 30 | 6 | 22.5 | 45.2 | 15.45 | 13.1 | 1.00 | 4.6 | 34.3 | 0.00 |
| 854 | 30 | 0 | 38.5 | 52.0 | 3.98 | 2.1 | 1.02 | 4.7 | 0.0 | 0.00 |

TABLE 1 – Summary of Results from the Stiffness Damage Test

Figure 1 shows the cyclic stress strain response for two cores from the same nominal mix, but from different structural elements; one with a low and one with a high degree of ASR surface cracking. Not only is the Ec significantly lower for the core from the fractured element, but also the degree of stress/strain hysteresis is markedly larger when compared to that of the core from the undamaged element. The area under the last four hysteresis loops (DI) is likely to be related to the crack opening, closing and sliding activity, and thus represents a measure of the damaged state of the specimen in that direction. The average value of this dissipated energy, normalised over the stress range, offers a useful new measure of concrete quality.

Figure 2 shows a plot of Ec and Eu versus DI. The cores were taken from a concrete of the same mix but from different areas in an ASR affected structure. Cores from uncracked elements show a Ec and have a low DI, while cores from cracked elements show a low Ec and a have high DI.

Figure 2 also shows a near constant Eu. It is suggested that Eu provides a measure of the Ec of undamaged concrete. The small drop in the Eu, as concrete becomes more affected by ASR, may be the result of softening of the reactive particles or some fractures remaining open during unloading. The Eu of a concrete therefore, may provide the control value with which to compare the Ec and thus gives an alternative measure of damage. This could be of particular value in discriminating between low stiffness through fracturing and low stiffness due to a high water–cement ratio. Use of the loading modulus alone loses this capability.

Typical stress/strain curves for concretes with preorientated fractures, and normal and high/water cement ratios are shown in Figure 3; [10] discusses these in detail.

## CONSIDERATIONS OF CRACK ORIENTATION IN CORE SAMPLES

A core recovery programme is often expensive and potentially structurally damaging. This limits the number that can be extracted from a structure. Cores generally can only be taken from lightly stressed areas. In ASR affected concretes this may bias the sample. Lowly stressed volumes could exhibit an isotropic expansion/fracturing due to the ASR, whereas more highly compressed regions may exhibit either very little fracturing under a hydrostatic field or highly orientated fracturing under deviatoric fields.

Many load carrying reinforced concrete elements are in a state of low triaxial, or near biaxial, compression. Biaxial elements such as walls and slabs will be less willing to expand in-plane because of the restraint provided by the reinforcement and the over-riding compressive stress state. Thus it may be expected that cracking due to ASR could occur preferentially, parallel to the free surfaces (i.e. in-plane). A compression test on a core, extracted normal to the wall might not reveal any strength loss since the cracks will not contribute to any weakening. Nevertheless, one would expect a significant reduction in the effective stiffness of the core until crack closure takes place. Taking cores along the plane of the wall might be a more difficult operation but it is seen to be much more relevant since fractures in the direction of loading will significantly weaken the core. Thus relying only on cores taken out-of-plane could seriously over-estimate the strength of the element.

It is therefore desirable to relate the results of the SDT to orientation and the insitu stress state. By building up data on the relationship between the SDT parameters, expansion history, insitu stress state and changes in uniaxial compressive strength and tensile strength, it may be possible to study triaxial behaviour in finite element analysis of actual damaged elements in structures.

## CONCLUSIONS

The Stiffness Damage Test, carried out upon core samples, provides a distinctive signature of damaged concrete. Both energy dissipated in the compression cycling and the loading stiffness (normalised with respect to initial unloading modulus) represent useful new measures of damage in deteriorating concrete.

The orientation of a core relative to the principal stress directions is a crucial parameter in fractured core strength assessments.

The use of computers to control the loading rig, capture the data and analyse the results has produced a quick, reliable, cost effective method for assessing the amount of damage in concrete.

Updating of the Stiffness Damage Test is continually being undertaken as the results from a wide range of investigations by Mott, Hay & Anderson become available. Its relationship to expansion, strength and petrographic changes has a high priority in our current programme of work.

## ACKNOWLEDGEMENTS

The authors would like to thank Dr.P.Waldron of The University of Bristol and Dr.S.Cullimore for their skill in undertaking the core testing programme.

**REFERENCES**

[1]     Wood, J.G.M., Johnson, R.A., Norris, P.: 'Management strategies for buildings and bridges subject to degradation from alkali aggregate reaction', 7th int. con. on alkali-aggregate reaction, Copenhagen, (June, 1983)

[2]     Doran Committee, 'Structural effects of alkali silica reaction', Inst. Struc. Engrs., (December 1988)

[3]     Nixon, P.J., Bollighouse, R.: 'The effect of alkali-aggregate reaction on the tensile and compressive strength of concrete', Durability of Building Materials, vol.2, p.243-248, (1985)

[4]     Swamy, R.N., Al-Asali, M.M.: 'Engineering properties of concrete affected by alkali silica reaction', ACI Mat. Jour., p.367-374, (September-October 1988)

[5]     Hobbs, D.W.: 'Alkali-silica reaction in concrete', Thomas Telford Ltd, (1988)

[6]     Akashi, T., Amasaki, S., Takagi, N., Tomita, M.: 'The estimation for deterioration due to alkali aggregate reaction by ultrasonic methods', University of Ritsumeikan, Hanshin Expressway, 7th Int. Conf. on AAR, Ottawa, Canada, (August 1986)

[7]     Walsh, J.B.: 'The effects of cracks on the uniaxial elastic compression of rocks', J.Geophys. Res., vol.70, p.399-411,(1965)

[8]     Crouch R.S.: 'Specification for the determination of stiffness damage parameters from the low cyclic uniaxial compression of plain concrete cores. Revision A', Mott, Hay & Anderson, Special Services Division, Internal Technical Note, (1987).

[9]     Wood J.G.M., Crouch, R.S.: 'Recent developments in the structural assessment of alkali aggregate reaction', Mott, Hay & Anderson Technical Note to the ASR Forum, (9th October 1987)

[10]    Chrisp, T.M.: 'The cyclic response of damaged concretes under low uniaxial compression', research thesis, The University of Bristol, (forthcoming).

[11]    Norris, P., Wood, J.G.M., Barr, B., 'A method of strength determination using concrete cores subjected to torsion loading',  mag. of concrete research, (1989).

[12]    Kotsosvos, M.D.: 'Effect of testing techniques on the post-ultimate behaviour of concrete in compression', Materiaux et Constructions, Vol.16, no 91, p.3-12, (1983)

CORE FROM UNCRACKED ELEMENT

| Loading Stiffness | = | 38.5 kN/mm^2 |
| Unloading Stiffness | = | 52.0 kN/mm^2 |
| Damage Index | = | 4.0 |

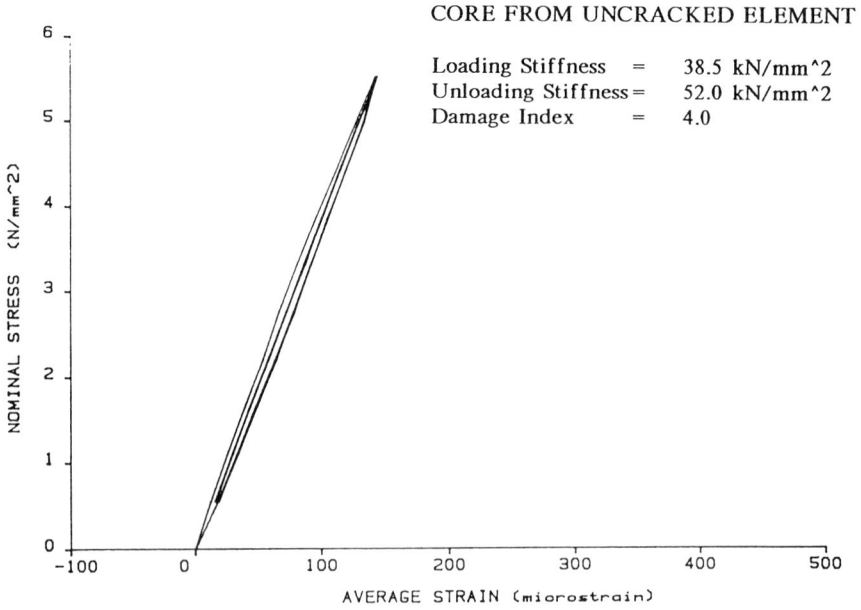

Special Services Division,
Mott, Hay & Anderson,
Croydon, England.
tel: 01-686-5041

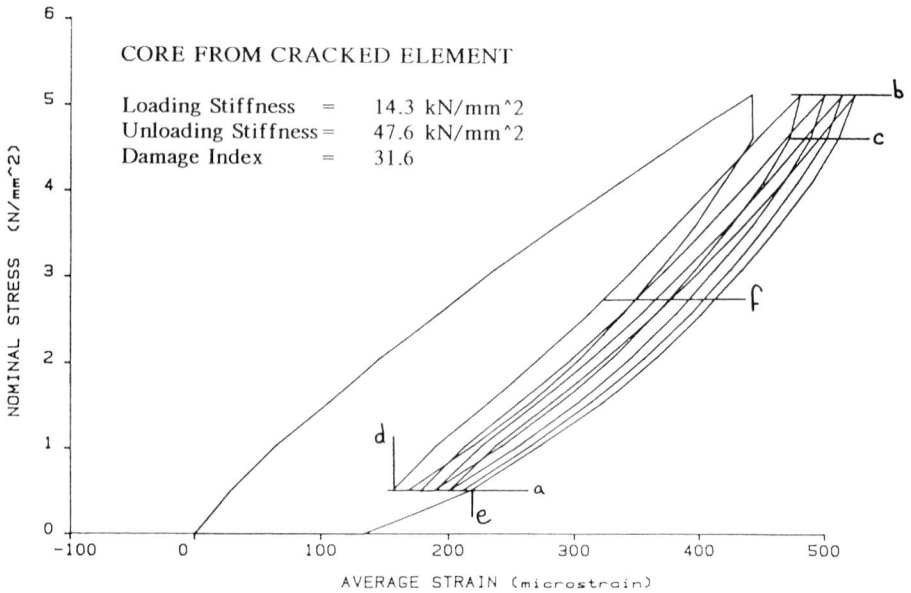

CORE FROM CRACKED ELEMENT

| Loading Stiffness | = | 14.3 kN/mm^2 |
| Unloading Stiffness | = | 47.6 kN/mm^2 |
| Damage Index | = | 31.6 |

FIGURE 1

# Core Response Under Low Cyclic Load
## based upon 120 cores of same mix

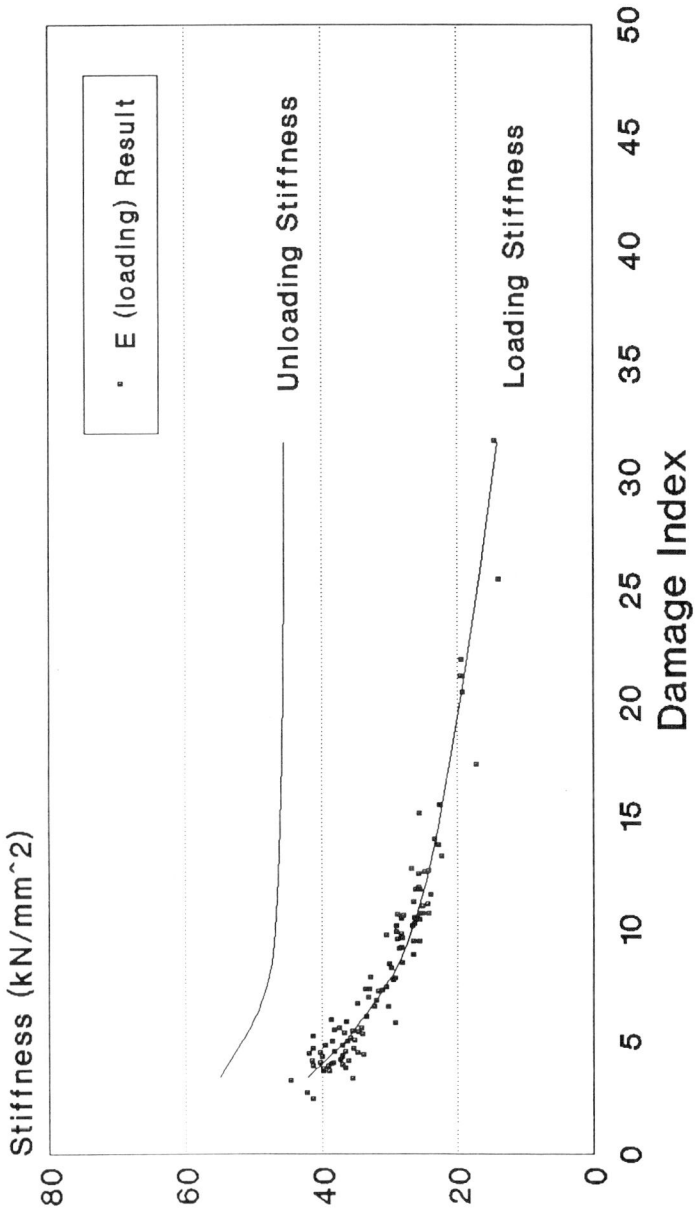

FIGURE 2

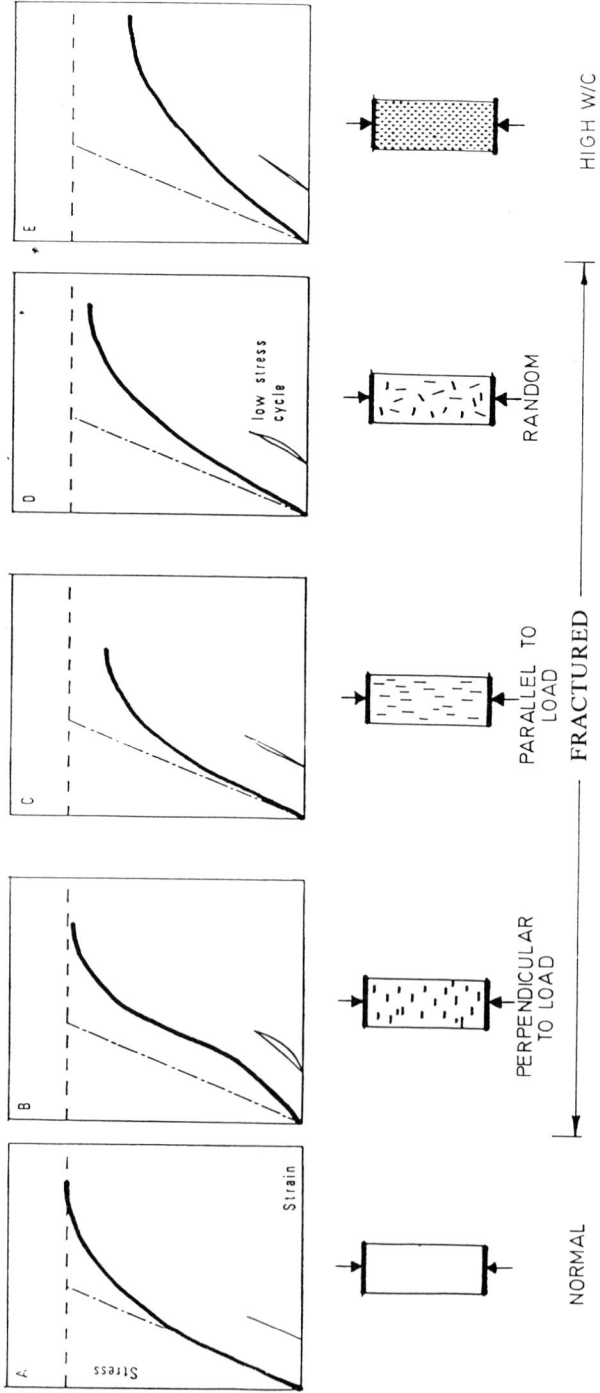

STRESS-STRAIN RESPONSE OF NORMAL & DAMAGED CONCRETES UNDER UNIAXIAL COMPRESSION

FIGURE 3

**P Jackson, Gifford & Partners, UK**


The stiffness damage test provides a means of quantifying the material damage to concrete. However, like the authors, we are involved in the management of structures affected by AAR and it is not clear to me how the results of the test will be used in this area. To us as structural engineers, the value of material tests lies in the information they give us about the likely behaviour or life of our structures. Research has shown that standard material tests are not a good guide to the behaviour of AAR affected structures; the stiffness and strength of the structures is less affected by the deterioration than the tests imply. Since Dr Chrisp has shown us that his test correlates well with standard tests, it must suffer from the same problem so how do you apply the results?

Some studies have found that the stiffness of structures which are suffering from AAR is much as would be predicted using the undamaged material properties. If so, one might even suggest that the value of this new test lies in its claimed ability to estimate the undamaged E value from tests on damaged specimens.

Author's reply

In putting his first question Mr Jackson has not fully appreciated our aims in relating the Stiffness Damage Test (SDT) to traditional strength tests. The British Standard 8110 'Structural use of concrete' is based on experimentally determined relationships between stiffness and compressive and tensile strength, for 'normal undamaged, undeteriorated concrete elements'. These values are often used by Engineers carrying out assessments on damaged structures thus assuming that the relationships still hold. For concrete with AAR we know the relationships break down. At Mott MacDonald, by comparing the SDT parameters to the range of traditional strength tests, we are progressively establishing the relationships between strength and stiffness for deteriorated concrete. This provides a substantially improved basis for appraisal. This approach is detailed in our paper "Towards quantification of microstructural damage in AAR deteriorated concrete", which is to be presented to the International Conference on the Fracture of Concrete and Rock, The University of Wales, Cardiff, September 1989.

With regard to Mr Jackson's comments about structure stiffness, we agree that the stiffness measured from the SDT is not the same as the structure stiffness. However the structural response must relate to the aggregate of the properties of the elements within it. We mention in our introduction that in order to provide structure response data, a large programme of proof load testing would need to be carried out on deteriorating structures. Unfortunately, this would be a prohibitively costly exercise. By developing the damage parameters of the SDT we have borne in mind our Clients criteria of providing a cost effective appraisal technique. Used in conjunction with other methods of appraisal we feel that the SDT can provide useful information that can be used in subsequent analysis and to produce a coherent structural management strategy.

Mr Jackson incorrectly likens the unloading stiffness (Eu) measured by the SDT to the structure stiffness and has misunderstood the fundamental theory of Eu. We put forward that Eu is a measure of the concrete stiffness before any deterioration has taken place. This is based upon the fracture mechanics principal that cracks that exist in the mesostructure of the concrete matrix are considered to be fully closed when the peak stress applied by the test is reached. Immediately upon unloading only the concrete is considered to respond thus giving a stiffness of concrete minus deterioration. This is distinct from Mr Jackson's theory that it is a function of structure stiffness i.e. reinforced concrete section and arrangement.

# 16 Inspection practices for concrete structures in the UK

**S R Rigden, E Burley, A Poole,** Concrete Research Group, Queen Mary College, University of London, UK
**A Christer,** Department of Mathematics, University of Salford, UK

This paper describes the Authors recent research into current practises used by large U.K. public organisations when carrying out the regular inspection of their concrete structures. The quality of the inspection and repair records kept by various inspecting bodies is also discussed. The authors proposals for the development and application of operations research models to the inspection and repair process is described.

## INTRODUCTION

In April 1987, following the award of a SERC research grant, the Queen Mary College Concrete Research Group in conjunction with Tony Christer Professor of Operation Research at Salford University began a one year pilot study into the technical and financial systems used by public bodies to inspect and maintain their concrete structures. The purpose of this study was to consider the viability of developing models that would predict rates of deterioration of defects in concrete structures and describe the relationship between degeneration and inspection frequency, inspection method and inspection and repair cost. The long term aim of the authors is the development of condition based inspection models for major concrete structures which allows the inspection and maintenance activity to be accurately modelled. This would allow inspections to be more economically phased and the timing of repairs more accurately determined. As well as leading to improved asset management, such models should also lead to gains in terms of extended life, serviceability and structural integrity.

## METHOD OF RESEARCH

To obtain information on the management of concrete structures, a large number of structured but informal interviews took place with senior or principal engineers, architects and surveyors from county councils, water authorities and London Boroughs. The files some of these organisations kept in relation to their inspection and repair activity were examined and a number of inspections attended. Meetings were also held with the Water Research Council (WRC) and the Department of Transport (DTp). Less formal contact was established and maintained with commercial organisations specialising in the inspection, repair and maintenance of structures.

The objective of these discussions and also of the ensuing examination of records and inspection visits was to obtain an understanding of the following:

(a) the organisation systems and methods used by each body to inspect its structures

(b) the inspection and repair records produced by each organisation and how these might be modified to provide data for a predictive model

• (c) the perception each organisation has as to the problems associated with the repair and inspection process.

By achieving these objectives it was hoped that the viability of developing predictive models would be proved and the need for further research into the degeneration process in real structures firmly established.

To supplement the information obtained from the participating bodies and to obtain a wider perspective of the problem associated with the repair, maintenance and operation of concrete structures, a workshop was held at Queen Mary College in December 1987. Delegates from over twenty Government, Local Authorities, Water Authorities and Commercial companies attended the workshop. The programme reflected the objectives of the pilot study whereby speakers, representing owning bodies, first presented their approach to inspection and were then followed by members of the research team who outlined the benefits of their proposed research. The ensuing discussion assisted in confirming the authors view that much work needs to be directed at determining the rate of deterioration of defects in real concrete structures and accurately determining the inspection needs of individual structures. The enthusiasm with which the practising engineers at the workshop were prepared to consider and debate the application of advanced operations research techniques to their inspection problems was very striking and encouraging.

The results of the study are considered under the following headings:

(a)  Inspection and Assessment
(b)  Records
(c)  Costs

Inspection and Assessment

Inspections are generally undertaken by inspectors who can be graduate engineers, technicians or clerk of works. Most inspectors have other duties as continuous inspecting is considered to be tedious and therefore self-defeating, as realiability declines. By using graduate engineers who are also active in design, the intention is to create feedback and therefore reduce the incidence of faulty details being unthinkingly repeated. Inspectors who are also clerks of works are provided with continuity as they can supervise the remedial work they identified, and perhaps recommended. The use of consultants and other outside bodies is kept to a minimum as they are considered to be more expensive than in-house staff normally. Detailed cost comparisons between the use of in-house staff and outside consultants was not though undertaken by the authors so the validity of the premise could not be verified.

All inspection reports are reviewed by the engineer/architect/surveyor responsible for the structure who will authorise whatever action is necessary. Where there is cause for concern, much use is made of photographs which often remove the need for a site visit. Most defects are monitored before any remedial action is taken which should ensure that remedial measures are only initiated after making use of considered engineering judgement. This point was emphasised at all meetings with the participating organisations and at the workshop.

Most structure owning organisations said their greatest concern was the correct timing of a repair. Research organisations and repair companies also indicated that they are frequently asked about the progress of defects and for advice on the best time to initiate repairs. They also indicated that they found difficulty in giving authoritative answers to these questions. Due to the difficulty in deciding the timing of repair, little effort is made to establish an order of repair in terms of severity. Most repairs were costed and added to the bottom of the list to be undertaken in turn. A number of organisations, especially the county councils, had started a priority rating system but had found it to be impractical and had either abandoned it or were letting it fall into disuse.

Regular routine inspections are not yet carried out by all organisations. County councils have the most developed system, carrying out general inspections every two years and principal inspections every six years on bridges belonging to themselves and the DTp. These frequencies were recommended by DTp and are in accordance with recommendations in the EEC publication 'Bridge Inspection'. It was indicated though that these frequencies represented a starting point and in some instances the return period of a principal inspection has been extended to nine years.

Water authorities do not have the same highly developed system, (except when dealing with structures coming within the Reservoirs Act) though many, but not all, do have a routine inspection programme. Those organisations participating in the project tried to inspect their structures every two to four years, though for some structures the return period could be as high as seven years. The inspection frequency is a function of the resources available and is not generally related to material performance. Defects that are being monitored tend to be inspected more frequently and at a rate that reflects the severity of the problem.

Only one of the London Boroughs visited had in place an organised regular inspection programme. This inspection programme probably exists because a major structural failure which resulted in loss of life had previously occurred in the Borough. The others carried out investigations only in response to complaints received from tenants or reports from housing officers. These complaint initiated investigations were often more extensive than the routine visual inspections carried out by other bodies. However, all investigations of particular faults tend to be more extensive due to the need to identify the source and cause of the defect. The one borough that has an inspection programme derived its five year cycle from the resources available within the department.

All routine inspections are visual examinations of the external envelope of a structure. Little equipment is used and one organisation also believed that cover meters provided little useful information. However evidence was established that clearly showed that close visual inspection with a minimum of simple hammer testing was essential in order to avoid completely missing areas of defective concrete that would be hidden from binocular surveys and other remote viewing methods.

The main function of these inspections is to provide an early warning of defects such that their development can be monitored. Consequently these inspections are generally very quick and use only the simplest of tools. The introduction therefore of more sophisticated and time-consuming methods will not happen until they are proved to be effective and cost-saving.

Records

Inspections are a relatively new phenomenon and consequently few historical records exist. It was said by more than one organisation that until they began an inspection programme they did not know how many structures they had. This applied to both county councils and London Boroughs. A number of London Boroughs devoted their initial energies to their high rise stock and have still to fully assess their low rise stock.

Current records are generally not suitable for developing a predictive model as where they existed they are generally descriptive in nature. All organisations possessed archive material but it was often difficult to obain, unreliable and inconsistent in the breadth & depth of coverage. The unreliability often started with the failure of contractors to provide accurate 'as built' drawings at the conclusion of construction. The extent of this problem was well illustrated by one London Borough who on starting to investigate a multi-storey residential block found, contrary to the insitu concrete construction shown on the 'as built' drawing, that they were dealing with a precast load bearing panel construction. Incidents of tower blocks being several storeys higher than shown in the records were also reported.

The most developed system of records belonged to the bridge sections of county councils. Each bridge has its own card index and file and is cross referenced by name, number and location. The card index gives the bridge name, location, function and construction material. The file contains previous inspection reports, load assessments where appro-priate, a history of work carried out on the structure and a summary of work requiring action, together sometimes with a cost estimate. There is rarely any record of what the repair actually cost.

These records, however, in general do not contain a sufficiently detailed breakdown of the faults diagnosed, to enable them to be used in their present form, as data for a predictive model.

One commercial organisation was found to possess detailed quantitative records that would be of immediate use in developing and using a predictive model. The company concerned, CAN U.K. Ltd., offer an inspection service to structure owners. Since their inception in 1983 they have accummulated records on well over one thousand major structures. These records are a computer listing that is consistent in detail and presentation and is noticeably superior to the inspection records generally kept. Because CAN are not the structure owners and might not themselves carry out the repairs initiated by their inspections, their cost records are not as extensive as their records of the faults encountered. CAN is able to supply, as part of their reporting service, a computer package that enables the structure owner to maintain a log of all defects found on the structure. The system enables analysis of the defects to be carried out by selection and sort criteria and repeat surveys to be logged for comparative purposes. The system is also being expanded to enable repair information to be kept alongside defect records. It is to be hoped that where appropriate, owners will then expand the inspection records supplied by CAN (UK) to include details of repair costs.

Little effective use was made of computers by most of the other organisations investigated. The most advanced system was one that provided a three monthly prompt with regard to inspections due. This system also gives the results of the previous inspection. Every organisation but one intended to introduce some form of computer orientated information system but, significantly, either no positive steps towards such a goal had been taken, or the objectives achieved were far below those outlined in discussions with the researchers.

Costs

Cost of inspection vary considerably from £59 for a general inspection of a small concrete bridge to over £10,000 for a multi-span motorway viaduct. A high rise structure can be inspected and reported on for £1,000. Reservoirs have to be emptied and the cost of a one day inspection can be £3,000. It is therefore clear that on large structures the inspection frequency should be selected to match the behaviour of the construction material in order to obtain the maximum cost benefit from both the inspection and the structure.

One major problem is the difficulty of accurately estimating the cost of repairs. Previously on high rise buildings a binocular survey identified the need for repair and provided the information on which estimates were based. These estimates were consistently too low and caused budgeting problems due to the need to divert funds from other sources to pay for the increased costs of repairs discovered once work commenced. Detailed inspections were reported to have reduced this error to around 15%.

Problems of estimating still exist on bridges where repair work has often to be undertaken in confined or restricted areas. One organisation stated that further research into the estimating of repairs would be most welcome as control of budgets was a critical factor in the management of inspection and maintenance. This may be one reason why estimated costs are easily available but actual costs are lost in the accounting system!

CONCLUSIONS

It was found that even between similar organisations which owned a similar cross-section of structures that there was a great variation in the intensity and frequency of their inspection programme. In most cases the size of the budget available for inspection was the major factor in determining the inspection programme, this budget perhaps increases in size by a few percentage points each year. A move towards 'optimum' inspection programmes would then, it seems, lead to a substantial change in the level of activity of a large number of organisations. As an 'optimum' programme of inspection should lead to savings in the aggregate cost of repairing and inspecting and it is seen that substantial savings to the economy as a whole would result fromm such changes; though it is expected that the actual cost of inspection alone would rise.

With a few exceptions, the records currently being kept by concrete structure owning bodies are inadequate for the purposes of developing and using inspections and maintenance models as envisaged by the Authors. Where high quality records are kept by commercial inspection and repair companies they might be better suited for this purpose as they generally have large volumes of data on a variety of structure types over a greater geographical area. This provides greater potential for comparative performance evaluation.

Even without the availability of effective inspection and repair models a substantial improvement in the quality of inspection and repair records has the potential to dramatically improve the management of these operations. Easily availabe data on the progress of defects would help identify both the need for and the optimum timing of repairs. The potential for feed-back to design engineers should also not be underestimated. A clear correlation between structure type, structure and age to defect, defect frequency, environmental conditions, and repair costs could result in a substantial improvement in both the design and detailing. of structures. Such studies have in isolated cases been made in regard to building structures, to good effect. It is understood that the Department of Transport is seeking to improve the quality and effectiveness of the records it keeps on its many buildings. The great potential that this vast quantity of engineering data potentially holds for the handful of designers is impressive to contemplate.

Many of the people responsible for the inspection and repair of concrete structures are aware of both the shortfall in their recording system and the need to identify the optimum for repairing concrete defects. This is reflected in their intention to install computer management systems to handle the information generated by inspections. However, many are unsure and indeed confused about how to achieve the best system and this is reflected in the fact that non have a fully effective system in operation.

Inspections are generally a visual appraisal of the external envelope of a structure and little use is made of modern non-destructive techniques as it is thought little information is available to suggest they can be used cost effectively. A number of collaborating organisations expressed the view thay more guidance on the use and effectiveness of such techniques in regular inspections was required.

The belief that better value for money and greater control of budgets could be achieved if the rates of degeneration of concrete defects and the optimum time for repair could be established was strongly expressed by many of the participants.

Inspection frequencies are based on what is considered to be good engineering practice and on the resources available to carry out the work. No attempt is made to correlate quantitatively inspection frequency and the behaviour of concrete in individual structures.

## Delay Time Maintenance Modelling

Delay time maintenance models have been applied to the modelling of the inspection and repair of mechanical plant. The authors are now seeking to develop, validate and apply such models to the maintenance problems of concrete structures. The extended time span controlling the development of defects in concrete structures and the often life threatening consequences of failure, when combined with the random nature of the initial development of such defects makes the modelling process complex but critically important.

In using delay time analysis it is assumed that defects arise at some initial point when they are first detectable then deteriorate over time (the delay time) until a stage is reached when repair is considered essential. Essential in this context can be determined by either the need to minimise the life time cost of repair or to ensure the service-ability of the structure.

The length of the delay time is seen to be a function of:-
a) The type of defect
b) The material used in the structure
c) Environmental factors
d) The sophistication and efficiency of the inspection techniques used
e) The "essential" time of repair criteria

The principal objective of the inspection process is then seen to be to identify a fault and make a repair within the delay time. Excessive repetitions of inspections within the delay time are not cost effective and inspections at a greater interval than the delay time result in increased repair cost and possibly the loss from service of the structure or worse.

Engineering judgement and adequate historical records will be required to enable subjective estimates of delay times to be made, taking into account the factors listed in a) to e) above. Once these delay times can be determined, the models can be calibrated and tested. Once available they should result in a substantial improvement in the management of the inspection and repair process.

## RECOMMENDATIONS

From the study it is quite clear that further work investigating the rate of degeneration of defects in concrete structures is necessary in order to assist owners in economically maintaining the serviceability and more accurately determining the appropriate time of repair. Such an investigation should include:-

1. a study of the behaviour of particular defects, to provide input to models to determine quantitatively the consequences of alternative inspection policies which would then lead to determining the appropriate inspection technique and frequency.

2. the development of a predictive model, based on the findings of 1 above, that will relate inspection frequency and technique to rates of deterioration, serviceability and repair.

3. a financial assessment of the techniques of inspection including an extensive examination of the NDT methods not often used at the regular inspection stage.

A substantial improvement in the breadth, depth, consistency and above all availability of inspection and repair records kept by structures owners bodies is essential. Once this data is available it must then be fully exploited.

S F Ray, Bingham Cotterall, U.K.

The authors reported that 'a substantial improvement was required in the breadth, depth, consistency and above all availability of inspection and repair records kept by structures owners'. Please would the authors confirm, based on their pilot study, what standard information should be collected and how it should be recorded.

It would also be useful to know key dates when aspects of their work are likely to be completed and if their findings will be published.

## Author's reply

The work reported on at the Conference relates to a low cost 12 month long pilot study. As such we would not wish at this stage to make firm recommendations in regard to the information to be kept in relation to the inspection and repair process. The following comments should then be seen as preliminary only.

The records kept must start with accurate as built drawings. This problem of inaccurate on grossly inaccurate as built drawings has been around as long as we have had buildings, paper and writing equipment. It seems to me though that some improvement could be made to the accuracy of as built drawings if a large sum of money is in the bill of quantities to cover the cost of producing them, and that the drawings are produced and submitted where possible in stages as the work progresses. The drawings would then be checked by staff familiar with the work actually carried out. A major problem will as built records has also been caused in the public sector by continued reorganisation of the authorities concerned. Much greater importance should be given to ensuring that existing records are maintained and transferred when such reorganisations take place.

Once an inspection cycle commences the records kept should relate to:

a)  Current and special use of structure, including loadings.

b)  Frequency and intensity of inspection, including details of equipment used. The system would be expected to act as a point as to when the next inspection is to be carried out and the type of inspection required. Any special limitations or restrictions on the inspection should be noted as part of the point.

c)  Details of faults found. These details need to be consistent in presentation and detail so variations with time can be clearly identified. We often find that it is difficult to identify change with certainty due to the use of vague descriptive terms on the use of dimensions which have been estimated by size. Crack widths cannot be accurately measured by eye or by ruler. Even the extent of cracking (number of column forces cracked) can be a function of the observer and weather conditions. Simple Demec strain gauges should be used to measure strains as a minimum solution. The

# 17 Monitoring the stress-strain behaviour of prestressed concrete structures

**R Wolff and H J Miesseler,** Strabag Bau–AG, Cologne, West Germany

Today great emphasis is placed on the early detection of defects in load bearing structures. By integrating of optical fiber sensors into composite materials used as prestressing tendons, the composite material become controllable and one speaks of so-called "intelligent load bearing structures". If optical fiber sensors are integrated into the concrete or glued on the concrete surface it is possible to monitor the event of cracks and where they occur. The structural application of this permanent monitoring process and is initial cases of application are shown in this paper.

## 1. INTRODUCTORY REMARKS

The development of optical fiber sensors is based on the current proven state of the art and quality standards of optical sensors which have already been in use for almost a decade in the telecommunications cable sector. In telecommunications engineering the optical sensor is first and foremost a signal transmitter. In contrast with conventional cable signal transmission, the optical sensor has an immense efficiency potential, the reason for its high significance being the excellent light permeability attained.

Factors of influence on the attenuation of light utilised in telecommunications engineering have been researched and, with regard to an attenuation-stable transfer of light, reduced to the lowest possible value. Light attenuation which is undesirable in telecommunications engineering can be used as a sensor effect in dependence with the mechanical demands of optical sensors.
The aim in the development of optical fiber sensors by Felten&Guilleaume is to achieve the highest possible test signal as a result of mechanical alterations in this sensor.

## 2. THE OPTICAL FIBER SENSOR

If a ray of light is conducted through an optical fiber sensor, losses occur in areas of microbending. When measured, the resultant loss of light is shown as a change of attenuation and expressed in decibels. In order to achieve extent of faults found needs extent and duration of the inspection, to be cross reference with the equipment and tests used and the access available.

d) Most records do not include details of the repairs actually carried out and the costs incurred. Estimates of the cost to be included are often included. This problem is seen to be very similar to the as built drawing problem, though in this case it is the maintenance engineer compounding his own problems rather than the construction engineer doing it for him!

The authors are awaiting funding from SERC to carry out a further three year study in this area. Subject to the successful conclusion to this work detailed guidance on record keeping will be produced.

an attenuation change when longitudinal tension occurs, use is made of the knowledge that microbending can also be produced as a result of radial pressure on the optical fiber sensor.
This effect can be obtained when the optical fiber sensor is provided with a fine wire strand (Fig. 1).

From a specific lay ratio, when tension occurs the diameter of the circle enclosed by the coil shows a greater reduction than that of the optical fiber sensor. Consequently the wire strand radial exerts pressure on the optical fiber sensor, producing microbending which in turn causes a corresponding change of attenuation, thereby changing the optical fiber sensor into an optical fiber extension sensor. To provide mechanical protection the optical fiber sensor is surrounded by a bonded coating of, for example, longitudinal glassfibers embedded in resin matrix. The sensor then has a diameter of approx. 2 mm.

The optical fiber sensor shows changes of attenuation as a function of the applied strain, (Fig. 2). When optical extension sensors are integrated into the construction component to be monitored, it is possible to determine changes in stresses from measured changes of strain.

The stress-strain behaviour of concrete may occur evenly but also be characterised by local irregularities such as cracks, therefore in addition to the integral light attenuation monitoring process, the so-called transit measurements assist in the recognition of local cracks.

Moreover, the intensified attenuation signal at the position of the crack can be registered by using the back- scatter measuring technique originating in telecommunications technology.

By superposition of the attenuation curves of the loaded and unloaded sensor, both the location of the crack and local changes in stress can be detected.

The current equipment available enables the locations of three dimensional stress behaviour to be determined with decimetric accuracy, even when the sensors are installed kilometers apart. For monitoring purposes, infrared light rays are transmitted through the optical sensor and changes in the passage of light integral recorded with the aid of electronic measuring equipment.

To a large degree, the monitoring intervals can be determined as wished, making permanent monitoring over a long period possible. Should the integral monitoring process show a change in the light attenuation, the back-scatter measurement will show the location and extent of the change.

Today, two measuring principles can be used for integral monitoring; the process where both ends of the sensor are acces-

sible  is shown in Fig. 3. The reference optical fiber sensor
serves to compensate temperatures and correct possible variations in transmitter  efficiency  and  receiver  sensitivity.
The transmitter transforms the electrical signals of the control  device  into optical signals and couples these into the
sensor and reference optical fiber sensor. The light of  both
the  optical  sensor  and  the reference fiber sensor will be
transformed into electric signals in  the  receiver  and  the
two  signals  separated by mechanical switching. The electric
signals of the receiver are processed in the monitoring  computer and saved on suitable data storage equipment.

The  second  integrated monitoring system is shown in Fig. 4.
It is characterised by ray splitting and connection  of  both
transmitter  and  receiver  to  the  optical  fiber extension
sensor and the reference optical sensor at the same end.  The
optical  fibers  of both sensors are equipped with reflectors
at the opposite end.  Light  received  from  the  transmitter
will  thereby  be  reflected. It then returns to the point of
origin and from there is again diverted over the  ray  split-
ter to the level receiver and evaluated.
This  system  is used in the case where one end of the sensor
is no longer accessible after installation.


3. MODEL TESTS

To investigate the suitability of the  optical  fiber  extension sensor for the monitoring of engineering structures, investigations  under  practical  conditions  have been carried
out at the Institut für Baustoffe, Massivbau und  Brandschutz
of  Braunschweig Technical University, one trial having been
made on a reinforced concrete slab,  18  cm  thick  and  with
3.62  m  span. For  this  purpose,  the  sensor was arranged
within a longitudinal phase on the tensile face of the  slab.
The  results  of these tests are shown in Fig. 5, wherein the
attenuation of the optical fiber extension sensors  is  shown
as a function of the force bending the concrete slab.

The  curve shows points of discontinuity caused by the development of cracks.

4. INTELLIGENT PRESTRESSING TENDONS

The optical fiber sensors demonstrated can, of  course,  also
be  integrated into fiber composite materials for utilisation
as prestressing reinforcement. The optical sensors  installed
during  production  of  the  individual  fiber composite bars
enable an insight into the stress-strain  behaviour  characteristics  of  the  structural elements prestressed with these
bars, thereby allowing conclusions to be drawn regarding alterations in the state of stress and location of same.

Prestressed structures can be monitored by means of integrated sensors, and could  then  be  described  as  "intelligent"
load  bearing  structures.  Work  in  this  field is promoted

under an EC research and development programme (Project BRITE 1353).

Glass fiber composite materials for use in reinforced concrete structures, developed and employed for the first time by the STRABAG / BAYER Joint Venture 'HLV' (Heavy Duty Composite Materials) in a research programme covering many years, are a corrosion resistant alternative to conventional prestressing steel. The prestressing tendons produced with this new material comprise 19 glass fiber bars (HLV - heavy duty composite bars) with a nominal diameter of 7.5 m and working load of 600 kN (Fig. 6).
The cross-section of the individual glass fiber bars comprises 68 % glass fibers and 32 % unsaturated polyester resin. The 1670 N/mm² longitudinal tensile strength of the material is a result of the high glass fiber content with rigid uni-directional orientation. The tensile diagram is shown in Fig. 7.


5. ULENBERGSTRASSE BRIDGE IN DÜSSELDORF (FRG)

The 'Ulenbergstrasse' bridge in Düsseldorf (FRG) -Fig.8- was a world premiere for a reinforced concrete bridge designed for heavy road traffic (Load Class 60/30) where prestressing tendons comprising glass fiber bars with a high tensile strength were used instead of prestressing steel tendons. The bridge was commissioned by the Düsseldorf Municipality and construction executed by the 'HLV - Elements' JV between 1985 and 1986 as a demonstration project of the Federal Ministry of Research and Technology. The Ulenbergstrasse bridge is a two-span, solid slab bridge with clear spans of 21.30 and 25.60 m.

This bridge is prestressed with 59 heavy duty composite bar tendons each comprising 19 glass fiber bars with a nominal diameter of 7.5 mm. The cross-section of the bridge will be permanently monitored by optical fiber sensors manufactured by Felten and Guilleaume Energietechnik AG, two sensors being integrated into the concrete of the deck slab and another two into one of the 19 glass fiber bars of two of the HLV tendons (see Fig. 9). This unique, latest state of the art, monitoring of a structure has been in effect on the Ulenbergstrasse bridge since its inauguration in July 1986.

Results obtained during a week in August 1987 are given in Fig. 10. The temperatures in the interior of the bridge, as constantly monitored, and also the bridge elongations measured at the bridge ends by means of sensor attenuation, are shown. From the records it can be seen that the temperature variations affect bridge strains and thereby the attenuation curve of the sensor, i.e. the optical fiber extension sensor duly reproduces the strain of the bridge. No discontinuity can be read from the graph. This indicates that no relevant cracks have occurred in the concrete during the monitoring period. This conclusion is applicable to the whole of the almost two years observation period. Sensor monitoring of the Ulenbergstrasse bridge at Düsseldorf thereby confirms a perfectly normal behaviour pattern for the structure, when evaluated according to the aforementioned aspects.


## 6. PEDESTRIAN BRIDGE MARIENFELDE IN BERLIN (FRG)

A pedestrian bridge in the Marienfelde district of Berlin (FRG) has been constructed in 1988 as a two-span, TT-beam bridge providing a pedestrian traffic area and a bridle path. The Senat für Bau- und Wohnungswesen awarded the contract for the bridge, which has clear spans of 27.58 m and 22.90 m and a cross-section width of 5.00 m (Fig 11). The prestressing of the bridge is designed as external prestressing without bond and will be achieved by means of seven, 19 bar glass fiber tendons. The nominal diameter of the glass fiber bars is 7.5 mm and the working load per tendon 600 kN. All seven glass fiber prestressing tendons are permanently monitored by integrated optical fiber sensors (Fig. 12). In addition, four optical fiber sensors are embedded longitudinally into the concrete to provide for permanent monitoring of eventual cracks within the concrete cross-section.


## 7. CONCLUSION

Stress and strain with optical fiber sensors along the total length of prestressing tendons are made possible.
Additional fiber sensors inside the concrete section of bridges monitor the occurence of cracks and where they occur.

The optical fiber sensor's function may be compared with sensible nerve fibers, reacting upon changes in stress and strain of composite materials or concrete. Electronic equipment enables permanent survey.

Higly sensible remote controls of concrete structures - this is, of greatest value in the future.


## REFERENCE

1. Noack, G.
   Lichtdämpfung als Kontrollmaß, Überwachung von Dehnungen und Spannungen mit Lichtwellenleitern. Industrie Anzeiger, Ausgabe 46/47 (1987) Seite 10 - 12

2. König, G., Ötes, A., Giererich, G., Miesseler, H.J.
   Monitoring of the Structural Integrity of Bridge Ulenbergstrasse in Düsseldorf - presented at the IABSE Colloquium Bergamo 1987 - Monitoring of Large Structures and Assessment of their safety"

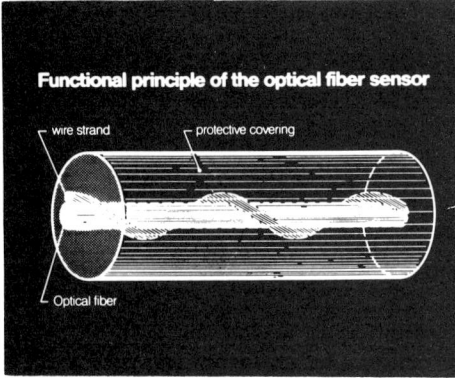

**Figure 1** Functional principle of the optical fiber sensor

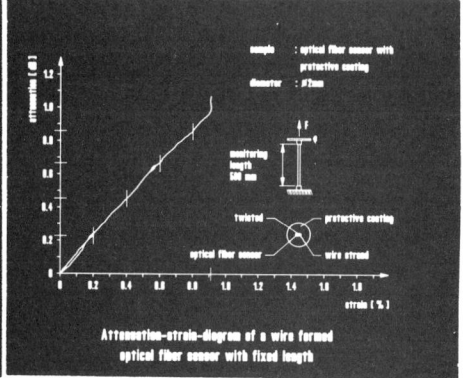

**Figure 2** Attenuation-strain-diagramm

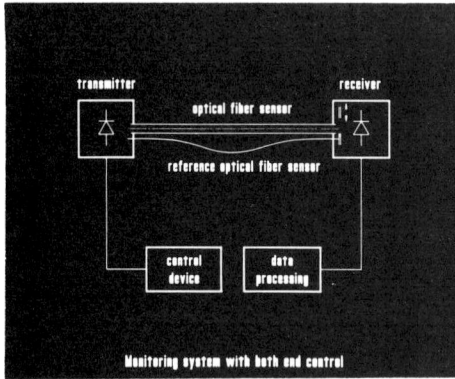

**Figure 3** Monitoring System with both end controll

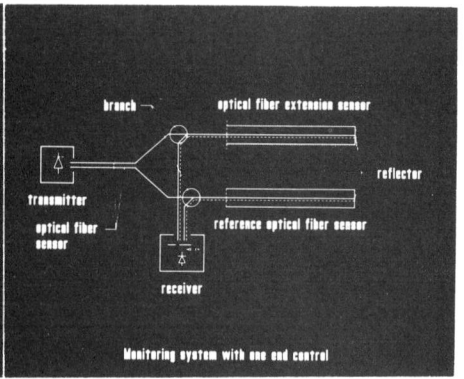

**Figure 4** Monitoring System with one end control

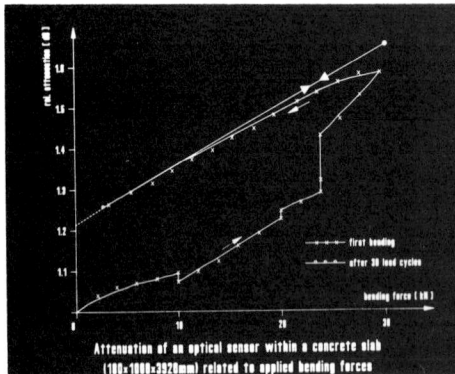

**Figure 5** Attenuation of an optical fiber sensor within a concrete slab

**Figure 6** "HLV" prestressing tendon

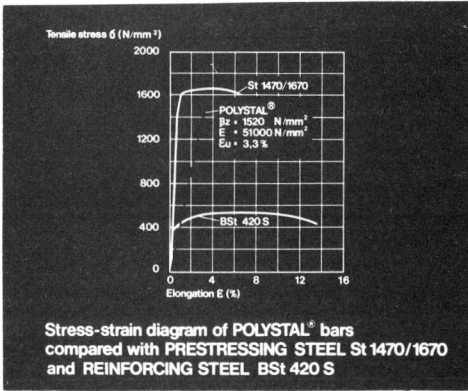

Figure 7   Stress-strain
diagramm

Figure 8   Ulenbergstraße-
bridge in
Düsseldorf

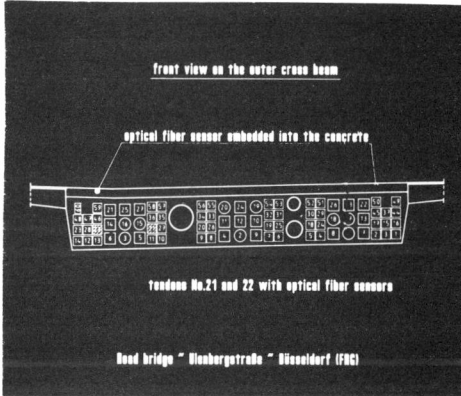

Figure 9   Arrangement of the
optical fiber
sensors

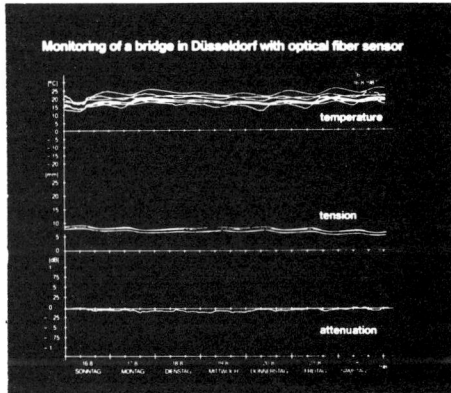

Figure 10   Results of a
monitoring period

Figure 11   Pedestrian Bridge
Marienfelde

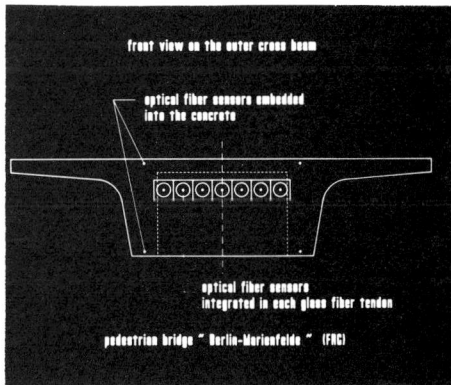

Figure 12   Cross section of
the bridge

Further Information:

STRABAG BAU-AG
Head Office
Siegburgerstrasse 241
5000 Cologne 21
Federal Republic of Germany
Tel.:     0221/824-2418 (Dipl.-Ing. Miesseler)
Telex:    8871050
Telefax: 0221/824-2936


FELTEN & GUILLEAUME ENERGIETECHNIK AG
Schanzenstrasse 24 - 30
5000 Cologne 80
Federal Republic of Germany
Tel.:     0221/676-2754 (Dr. Levacher)
Telex:    887787-0 fgd
Telefax: 0221/676-2646

Dr G Somerville, C&CA Services, British Cement Association, UK

This Session fell naturally into two parts, with the first three Papers being concerned with measuring the response of the total structure to some form of dynamic loading. It became clear in the presentations that the stage of development of this technique at the moment is such that we are effectively comparing "finger-prints", i.e. we need to repeat the test on the structure at intervals and make comparisons of the prints so obtained and then evaluate the significance of any differences between them. In the generality of things, I think this technique has a number of advantages in that it does relate to the actual structure. However, it also has a number of disadvantages in that it is difficult to isolate out the contributions from materials and from components which may be structural or non-structural; nor is it possible to identify the particular effect of boundary conditions, separate from the contribution from individual elements.

The second half of the Session presented techniques that fall into two sub-categories. The first of these is the identification of deficiencies in the structure, i.e. the presence or otherwise of reinforcement and its location if present or the location of voids. It seems to me that the techniques in solving those problems have still some way to go to be truly effective.

The second sub-group is concerned with measurements taken on small elements extracted from the structure itself, usually in the form of cores which may be used to measure the strength or to prepare sections for chemical analysis, etc. This undoubtedly provides useful information (there are many other techniques available additional to those described in this Session). However, the real problem with these techniques is assessing the data obtained in relation to the performance of the structure itself. We all know, for example, that the results from a standard cube, tested in accordance with BS1881, bear little relation to the strength of the concrete in different types of structural elements; so too with be the taking of cores or other types of samples.

Taking the different types of test together, we have a dilemma : the techniques for assessing the structure overall are limited, in that we cannot actually isolate out the cause of any effects that we may measure, and the sampling of small parts of the structure has serious limitations in how the data are used to assess the performance of the structure itself.

I believe that we need to develop both techniques much more in the future. I further believe at this instant in time that it would be quite wrong to put too much dependence on either technique in isolation.

# 18 To better reactor containments

**Prof Dr -Ing J Eibl,** University of Karlsruhe, West Germany

The lecture on the above-mentioned subjects will report on endeavours in Germany to build more effective
concrete containments for pressure water reactors

The usual risk-philosophy where risk is estimated by the product of
probability of occurence times the caused damage in case of a failure
is more and more rejected by the public if applied to nuclear power
plants. To the author's opinion people have a sound feeling that the
first factor of this product tends to zero, while the second to an
undefined height in case of a heavy core melt accident. Acceptance of
a technique which is unavoidable at least for the near future may
probably only be gained if the damage in case of a failure can be limited to a well determined final magnitude. From here also for civil
engineers a big challenge arises as probably new type of containments
could help to reach this goal.

Starting from this basis of reasoning the author has made some first
preliminary investigations. A containment proposal is presented which
probably could help to control the following effects resulting from a
heavy core melt accident:

- resistance against a helium explosion
- a nearly passive removal of afterheat
- a core catcher with a passive cooling system.

First strategies how to govern rupture of a pressure vessel are also under discussion.

Reporting of this work the author intends to stimulate the imagination of creative civil engineers for this field of work.

# 19 Twenty years of surveillance experience with prestressed concrete pressure vessels at Wylfa nuclear power station

**P Dawson, CEng, MICE,** Taywood Engineering Ltd, UK
**R A Vevers, CEng, MICE,** Central Electricity Generating Board, UK

SUMMARY

The paper describes in-service surveillance experience obtained at
Wylfa Nuclear Power Station which has been operated by the Central
Electricity Generating Board (CEGB) for nearly twenty years. Each of
the two nuclear reactors at the station are contained by a prestressed
concrete pressure vessel (PCPV) which provides the main pressure
boundary and biological shield.

The Nuclear Site Licence Condition under which the CEGB are the
Licencees, requires that the PCPVs are inspected on a regular basis.
These requirements are detailed in the Station Maintenance Schedule.

This paper gives details of results obtained from the statutory
in-service inspection programme of the PCPVs. The programme includes a
detailed examination of the concrete surface and a selection of
prestressing anchorages; in addition a one per cent sample of tendons
are selected for anchorage load checks using a 'lift-off' technique and
the results compared with design predictions. A number of tendons are
removed for corrosion inspection of the prestressing strand and samples
selected for testing to determine their mechanical properties. The
vessels and their foundations are surveyed by optical means to confirm
that settlement and tilt remain with design limits.

The results obtained from in-service inspections have shown that the
PCPVs are still operating well within their design limits and their
overall performance has been excellent.

INTRODUCTION

Each of the two prestressed concrete pressure vessels (PCPVs) at Wylfa
Nuclear Power Station is a concrete sphere, 96 ft (30 m) internal
diameter, with an average wall thickness of about 12 ft (4 m). Each
vessel is prestressed with over 1200 tendons with an ultimate tensile
capacity of 820 tons. The average length of a tendon is about 120 ft
and their combined total weight is about 2250 tons. These vessels are
believed to be the most heavily prestressed structures yet built.

Under operating conditions, the pressure of the reactor coolant gas
within the vessel is about 380 psi gauge (27 bar) and its temperature
is about $400^{\circ}$C. Foil insulation on the inside of the vessel liner,
and cooling water, circulating in pipes welded to the back of the

liner, control the inner face concrete temperatures to about 40°C. The outside face concrete temperature is typically about 20°C.

Construction of the first vessel commenced in 1964. It was stressed during 1968 and proof pressure tested at the end of that year. The second vessel was stressed during 1968 and proof tested in 1969. The reactors were effectively brought into service in 1969 and 1970 respectively.

A cut-away perspective drawing of one of the vessels is shown in Figure 1. Detailed information on the design and construction of the Wylfa vessels is given in Refs. 1 & 2.


DETAILS OF PRESTRESSING SYSTEM

Each of the vessel tendons is made up of 36 strands with an external diameter of 0.6 ins (15 mm) and a guaranteed ultimate tensile strength of 51,000 lbs (227 kN). The tendons are anchored at each end by three Freyssinet 12/0.6" anchorages seated on a concrete anchor block as shown in Figure 2. This approach was adopted in order to minimise the total number of tendons, and thereby the potential congestion of ducts, in conjunction with using the largest prestressing anchorage (and jack) then available.

Each tendon was stressed to a total load of 615 tons (75% GUTS) at the anchorages at each end before lock-off. Because almost all the tendons are heavily curved, friction between a tendon and its duct resulted in the load at the mid-point of the tendon being considerably less than the applied jacking load at the anchorages. In addition, as the tendon load was transferred from the prestressing jack to the corresponding anchorage during lock-off, a reduction of tendon load occurred at the anchorage as a result of the inward movement of the male cone as it wedged within the female cone. The extent of this reduction depended on the tendon curvature near the anchorage and the particular extent of pull-in, but was generally about 200 tons. Typical tendon load profiles on completion of stressing are shown in Figure 3.

At an early stage in the design of the Wylfa vessels it was decided that, contrary to conventional prestressing practice, the tendons would not be grouted into their ducts on completion of stressing. This approach was adopted to retain the facility to be able to remove and replace tendons during the life of the vessel, should this ever be deemed necessary. Three consequences arose from this decision:

a)  The anchorages are required to carry the full tendon load throughout the operational life of the vessels. This includes the hypothetical 'ultimate load' situation; in consequence it had to be demonstrated that the anchorage design was capable of sustaining 90% of the guaranteed ultimate tensile strength of the tendons.

b)  An alternative method of protecting the tendons against corrosion had to be developed and implemented. This was achieved by a combination of wax and greases.

c)  The anchorages were available for subsequent load measurement.

159

SURVEILLANCE PROGRAMME

The Nuclear Site Licence Conditions, under which the CEGB is the
Licensee, requires that PCPVs are inspected on a regular basis. These
requirements are detailed in the Station Maintenance Schedule. Each
reactor is shut down once every two years so that each year there is an
inspection of one of the two vessels. This normally coincides with the
scheduled shut-down for reactor maintenance when the reactor and the
internal parts of the vessel are inspected and serviced.

In order to fulfil the licensing requirements, the CEGB nominates a
member of its Civil Engineering Branch staff to act as the "Appointed
Examiner". The Appointed Examiner is responsible for ensuring that
inspections of the civil engineering features of the vessel are carried
out prior to, and during, the shut-down period. A report is submitted
to HM Nuclear Installations Inspectorate of the Health and Safety
Executive summarising the results of the examinations carried out on
the vessel. The Appointed Examiner will, if satisfied, recommend that
the vessel is put back into service. The Appointed Examiner's report
is standardised and covers the following items:

a)  Concrete Surface Examination

    A visual survey of the vessel surface is undertaken and any
    increase in the size of pre-identified cracks, or the presence of
    new cracks since the previous survey are reported. Cracks which
    are considered to be propagating are measured and the rate of
    growth is established.

b)  Tendon Anchorage Inspection

    A sample consisting of a minimum of 1% of all anchorages is
    inspected for signs of damage, corrosion, slippage or other
    deterioration.

c)  Tendon Anchorage Load Checks

    Measurements of the load carried by tendon anchorages are carried
    out by "lifting-off" a number of tendons in a sample of at least 1%
    of all tendons. These load checks are tabulated and presented in
    graphical form so that current and past trends can be compared with
    design expectations and the maximum permissible limits of loss of
    load.

d)  Tendon Corrosion Examination

    A number of strands are removed from selected tendons. These are
    inspected for signs of deterioration in their protective coating
    and for underlying corrosion or mechanical damage. Lengths of
    strand are tested to determine their mechanical properties, such as
    strength and ductility. Small samples of the protective grease or
    wax are analysed to determine moisture content and chemical content
    e.g. chlorides. Cumulative records of any corrosion found are kept

in a statistical format and these records are compared with permissible levels.

e) Foundation Settlement

A precise survey of levels is undertaken and the results tabulated and presented in graphical form so that current and extrapolated settlements can be compared to design expectations and maximum permissible changes in level.

f) Vibrating Wire Strain Gauges

Vibrating wire strain gauges (VWSGs) embedded in the vessel concrete are read at monthly intervals. At approximately 5 yearly intervals, a report is presented comparing VWSG readings with a theoretical strain analysis, which takes account of time-temperature dependent tendon relaxation and concrete creep, as further evidence of continued operation within expected limits.

g) Vessel Concrete Temperatures

Thermocouples measuring concrete temperatures are read regularly and checked for conformity with the operating rules for the vessels. Particular attention is paid to hot-spots, areas around steam pipework penetrations and the stand-pipe zone.

h) Main Reactor Coolant

The Station Staff regularly monitor losses from the primary circuit and ensure these do not exceed statutory limits. A summary of daily losses of gas from the coolant circuit is presented in a report.

i) Vessel Cooling Water System

Routine visual inspections of tendon anchorages and pipework penetrations are carried out in order to detect evidence of water leaks from the liner cooling system.

Further information on the programme of in-service surveillance of the Wylfa vessels is contained in Ref. 3.


RESULTS OBTAINED FROM INSPECTIONS

A detailed description of the methods and equipment that are employed in surveillance programme at Wylfa is given below together with an assessment of the results which have been obtained. The equipment used is standard civil engineering equipment with a number of small refinements.

a)  Concrete Surface Examination

All surface cracks, their lengths, widths and changes in dimensions
are recorded.  The widths of cracks are monitored for changes
between inspections by means of demountable "Demec" strain gauges.
All the cracks found in these surveys are narrow (typically less
than 0.2 mm).

An assessment has been made of the crack measurements taken during
the service life of the vessels and a comparison made between crack
widths when the PCPV was pressurised and unpressurised.  The
results of the readings were plotted and the worst cracks
photographed.  The results of this work confirmed that a small
number of cracks were propagating but that the growth rate was so
small (less than 25 microns per year) that the growth was of no
structural significance.  The change in crack width with pressure
was insignificant.

b)  Tendon Anchorage Inspection

The condition of anchorages has remained good.  Minor cracking of
the concrete surface has been observed.  Such cracks are monitored
for possible propagation by the means described in paragraph a)
above.

c)  Tendon Anchorage Load Checks

When a tendon is initially stressed, the reaction force from the
prestressing jack is taken via a stressing foot onto the female
cone of the anchorage as shown in Figure 2.  During stressing, the
male cone floats free within the cage of strands at the nose of the
jack.  During lock-off the male cone is pushed firmly into the
female cone and is subsequently pulled in yet further as load is
released from the jack and transferred to the anchorage.

For a load measurement exercise, the stressing foot is replaced by
a shimming foot which encircles the female cone.  The prestressing
jack is then connected to the projecting strand ends.  As the jack
load is increased, the load in these strands increases but the
anchorage does not move until the applied jacking load exceeds the
load applied to the anchorage by the anchored tendon.  In
consequence, the jack load at the instant the anchorage starts to
move is a measure of the load in the tendon which was being carried
by the anchorage beforehand.

The detailed techniques for this 'lift-off' load measurement
exercise have been developed and refined during the operational
life of the reactors.  The current approach is to fix electrical
deflection transducers to the shimming foot such that they register
on small projections fixed to the anchorage.  A graphical record is
then obtained of anchorage movement and jack pressure and the jack
load at the instant of lift-off is obtained by interpreting this
graph.

At the time the Wylfa vessels were designed, allowances were made for the expected extent of time-dependent load loss in tendons following their initial stressing. This loss has two components:

a)  A loss due to tendon relaxation, which is defined as the reduction in load of a stressed tendon held at constant length (change in stress at constant strain),

b). A loss due to concrete creep and shrinkage. Creep is defined as a reduction in length of a (concrete) element held under constant load (change in strain a constant stress).

The essential purpose of the periodic measurements of tendon anchorage load is to assess the extent of measured loss of load and compare it with the provisions embraced by the original design. This exercise is undertaken for groups of tendons of similar length and profile, rather than for individual tendons. Two tendons in each group are nominated for lift-off load measurement for each surveillance exercise, with, generally, a different pair of tendons being chosen on each occasion.

Results obtained from 'lift-off' measurements for the top cap tendon anchorages are shown in Figure 4. Results for other groups of tendons follow a similar pattern.

### Tendon Corrosion Examination

To confirm that the tendon corrosion protection system remains effective, two strands are removed from eight tendons on the occasion of each statutory surveillance exercise.

This operation demands that the tendons are first destressed. This is achieved by fitting the jacks onto the projecting strands at an anchorage with the jack reaction being carried on a detensioning foot which bears onto the female cone, in the same way as the stressing foot shown in Figure 2. The jack load is increased until the male cone is pulled from within the female cone. The male cone is then temporarily restrained in this position whilst the jack load is allowed to reduce and load in the tendon is released. The jack and anchorages are then removed from the unstressed tendon.

Before removing a strand from the tendon a new length of strand is butt welded to the end of the strand to be withdrawn. This strand is then winched out of the tendon and the new length of strand simultaneously drawn in to take its place. The withdrawn length of strand is then cut into manageable lengths and removed for examination.

The depth of any pits in the strand is measured by means of a micrometer probe, and in some cases, by sectioning. Further examinations, where considered necessary, are carried out using electron microscopy and chemical analysis of corrosion products. Metallurgical, chemical and physical property checks are carried out to determine the effect of different degrees of corrosion and pit depth on the strength, ductility, relaxation and stress corrosion cracking potential of the material.

Pitting corrosion has been found on some tendons but examinations have revealed no instances of corrosion which could be considered serious enough, when judged against the acceptance limits, to warrant replacement of tendons.

e)   Foundation Settlement

A survey of vessel foundation levels is carried out by an experienced surveyor using a precise optical level and invar staff. Closing errors are generally less than 0.1mm in these surveys. The allowable limits on PCPV foundation settlement and tilt are large compared to the movements that have been measured to date. The main concern from uneven settlement of the PCPV is associated with the structural strength of the reactor core and its supports together with ensuring the uninhibited passage of the reactor control rods under the influence of gravity.

f)   Vibrating Wire Strain Gauges

The original purpose of the vibrating wire strain gauges was to demonstrate that the vessel was behaving elastically and in accordance with design predictions during its proof pressure test. However strain gauges have proved of considerable value in understanding the operational behaviour of vessels and are now monitored regularly throughout the operational lifetime of each vessel. Readings are taken at intervals of not less than 1 month and at the start and end of each reactor shut-down and start-up cycle.

Continuous strain histories are maintained and typical plots comparing measurements with prediction are included in the relevant reports.

g)   Vessel Concrete Temperatures

Surveys of vessel concrete temperatures are made on a regular basis as part of the PCPV operating procedures. Examinations of these records have shown that temperatures have generally remained well within the expected levels.

The exceptions are in small areas where known departures from the general level of insulation or cooling exist, usually local to steam pipework penetrations. These areas were subject to additional assessment during commissioning of the vessels and special limits are specified in the operating rules together with a requirement for frequent monitoring to detect signs of significant deterioration in conditions at these points.

h) Main Reactor Coolant

Losses of carbon dioxide gas from the primary coolant circuit are regularly monitored to ensure that these do not exceed statutory limits.

In addition, surveys of carbon dioxide around penetrations and tendon anchorages are undertaken to detect any rise above the background level. In this way it is possible to detect whether there is any major leakage through the main vessel liner into the concrete structure.

i) Vessel Cooling Water System

Routine visual examination of the lower tendon anchorages and penetrations is carried out and any evidence of water in these areas is reported. A number of leaks from buried pipework have been found during inspections. In these areas the leaking pipe has been identified by systematic isolation of the PVCW system. The faulty pipe is then sealed, either by dosing with a proprietary sealer. Particular attention is paid to the effect of the leaking water on the condition of the prestressing system and additional examinations are made to check the condition of affected tendons.


CONCLUSIONS

Concrete surface examinations have shown that surface cracking on PCPVs is confined to drying shrinkage and thermal strain effects and no instances of structurally significant cracking have occurred.

The data obtained from periodic tendon anchorage lift-off load measurement of the prestressing system have shown that the design predictions of prestressing losses due to steel relaxation and concrete creep have been conservative.

A low rate of tendon corrosion, particularly pitting attack, has occurred. Chloride and contamination allied to moisture have been identified as the prime causes of pitting corrosion. Statistical analysis of the severity of corrosion, allied to mechanical and chemical testing, has been used to set acceptance standards which define the point at which replacement of corroded tendons may have to be considered. To date, replacement of tendons has not been necessary on any PCPV operated by CEGB. Indeed, there is little reason to believe that replacement of unbonded systems is likely to be required over the normal 20 to 30 year design lifetime of PCPVs provided present standards are maintained. It would appear, however, that corrosion cannot be entirely eliminated and inspection programmes should include an adequate degree of sampling to confirm freedom from stress corrosion cracking, and analytical support to determine pit depth/strength relationships.

Foundation settlement surveys have revealed that the allowable values of reactor vessel settlement and tilt are considerably larger than those likely to occur in practice and certainly within the accuracy of measurement of conventional optical surveying techniques.

Strain gauge readings have shown that predictions of operational deformations of the PCPVs have been borne out during actual service.

Concrete temperature measurements have shown that some short term temperature excursions due to temporary partial loss of vessel cooling water have occurred but the changes in vessel temperature arising from these faults have always been insignificant. Overall, the experience to date of the thermal performance of the PCPVs has been good and well within the limits required to maintain serviceability.

The long term effects of PCPV cooling water leaks on the pressure vessel structure have been assessed. To date no detrimental effect has been found and a stringent monitoring and leak sealing regime is being enforced. Small leaks affecting a limited number of tendons are of minor significance when compared to the large overall safety margins available from the total prestressing system.

On the infrequent occasions when operational incidents which might affect the vessels have occurred, these have been shown to have an insignificant effect on either safety or serviceability.

The overall conclusion gained from nearly 20 years of inspections is that the performance of PCPVs has been extremely good. These structures more than satisfy the stringent standards the public have a right to expect in the design and construction of nuclear installations.


ACKNOWLEDGEMENTS

This paper is published with the permission of the Central Electricity Generating Board and Taywood Engineering Limited.

REFERENCES

1.  "Proc. Conf. on PCPV", Institution of Civil Engineers, London, 1967.

2.  "Proc. Conf. on Design, Construction and Operation of PCRV and Containments", Univ. York, Institution of Mechanical Engineers, London, 1975.

3.  FIP Report 'An International Survey of in-service inspection experience with PCPVs and Containments for Nuclear Reactors' (1982), FIP Ref. 15.906, FIP, Wexham Springs, Slough.

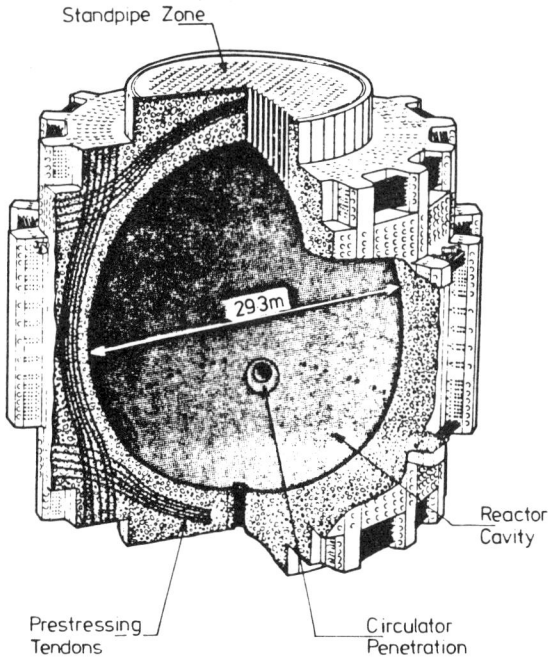

Figure 1  Wylfa Prestressed Concrete Pressure Vessel

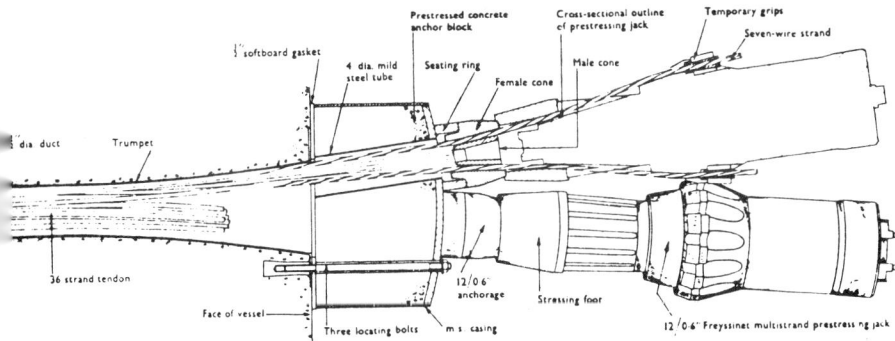

Figure 2  Wylfa PCPV Prestressing System

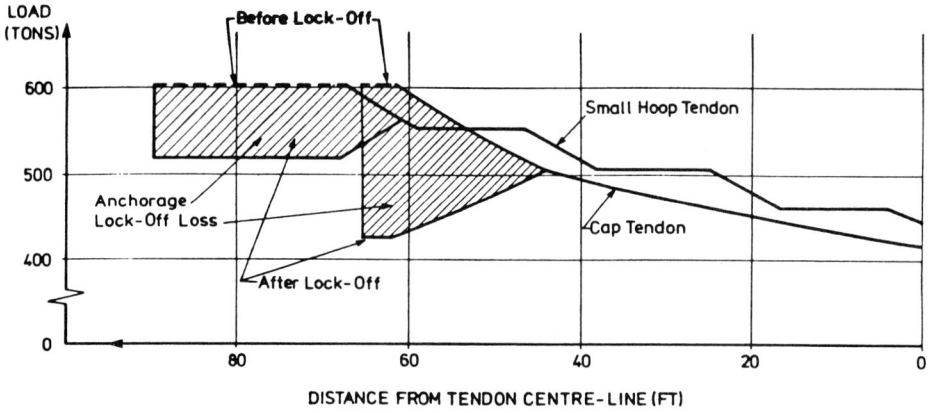

Figure 3   Typical Predicted Tendon Load Profiles

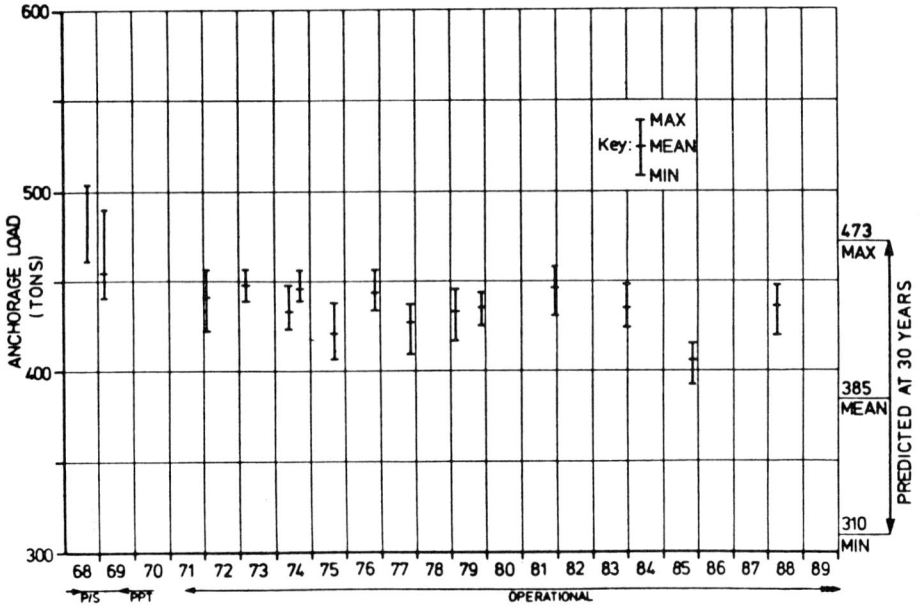

Figure 4   'Lift-off' Loads for Top Cap Tendons

P Dawson, Taywood Engineering, U.K.

Within the context of this conference, what we have reported to you is
a success story.   We have none of the dramatic illustrations of
damage, deterioration and failure which are given in other papers.
The Wylfa pressure vessels have behaved entirely in accordance with
the expectations for them at the time they were designed and they are
as fit for the purpose now as they were when they were put into
service about 20 years ago.

It may be worth reflecting for a moment on why this should be.   I will
suggest three reasons.

Firstly, they were designed by a highly skilled team of engineers,
ably supported by analysts and researchers who were fully aware of the
vital function of the vessels and the uniqueness of their design
concept, and determined to 'get it right'.

Secondly, the design period benefitted from a close and trusting
relationship between the designer and the client, the CEGB.  They were
as two rock climbers, roped together, each dependent for success on
the other's efforts and each willing and able to provide support,
encouragement and assistance to the other.  This is perhaps a valuable
lesson which the 1960's can teach the 1980's.

Thirdly, we were denied the present opportunities to use sophisticated
computer analysis.   Instead we were obliged to think carefully about
the suitability and relevance of all aspects of our design, and, in
many fundamental respects to 'keep it simple'.   I do not mean to decry
the substantial opportunities which computers have created for
engineers, but I am concerned that, too often, designers are seduced
by the analytical power and accuracy of the computer and become wedded
to designs containing basic flaws which would be revealed by a more
simple approach.   The modern designer risks being distanced by the
intervening computer from a 'feel' for structural materials, forces
and principles which are fundamental to good engineering.

A separate issue which I believe merits the attention of this
conference is the start date for the life of a structure.
Construction of Wylfa vessels commenced in 1964.   They were
prestressed about 3 years later and proof tested about a year after
that.  However, as far as the client was concerned, their 'life' only
started when the reactors which they contained were commissioned in
1969 and 1970.   At this time, the chemical ageing processes
influencing the properties of the structural concrete had been
continuing for five to six years.   Furthermore, for the particular
case of a prestressed concrete pressure vessel, the level of stress
imposed on the concrete due to 'prestress only' was generally higher
in the pre-commissioning period than subsequently when the vessels
were pressurised.   Such situations need to be taken fully into account
in establishing the time span of life for a structure.   The client may
think he is acquiring a virgin structure on the date he takes it over

from the contractor, but, in reality, the structure's clock started running on the day the first concrete was poured. For some structures this intervening period is measured in years, with corresponding consequences to its effective life.

Finally, a brief word on the engineer's concern with predicting the future. In this respect we are almost unique compared with other professions. Lawyers largely occupy themselves with society's past demeanours. Accountants sometimes seem to have difficulty in understanding the present. As Shaw said, if all the economists in the world were placed end to end they wouldn't reach a conclusion. But the very essence of engineering is to build safe and serviceable structures for the future. Society expects it of us, and often demands it of us. We are increasingly accumulating data and skills to respond to this challenge. I would caution, however, against our becoming too bold or too confident.

We need to be particularly prudent when working with new materials or new structural forms and must always take full account of the likely consequences of our getting it wrong. There are too many uncertainties in this world for us ever to dare to forecast the future with absolute certainty.


**T M Chrisp, Mott, MacDonald Group, UK**


i) The authors have indicated that Wylfa power station was designed in the late 1950's and early 1960's largely without the use of computers. Do the authors think that with the benefit of new computational techniques the design could have been bettered?

ii) The authors have explained how a number of prestressing tendons have, to date, been removed from the pressure vessel and inspected for corrosion. Exactly how many have been inspected and what percentage of the total number of tendons does this sample represent? What is the overall percentage of tendons where corrosion was recorded?

Author's reply

The Wylfa PCPVs were designed to respond to the functional requirements of the reactor system they contained, taking account of the then current states of concrete and prestressing technologies. Even had modern computational techniques been available it is doubtful whether the design would have been significantly different. The benefits which may have arisen from such availability are likely to have been a higher level of confidence in the estimation of stresses, movements, etc. and the opportunity to explore a wider range of operational circumstances. An analysis of the Wylfa vessels in the late 1970's, using finite element modelling substantially vindicated the stress regimes which were predicted at the time of the original design.

During the 1971 statutory inspection corrosion pits were discovered on the hoop pre-stressing tendons. As a result of this discovery, an extensive research programme was established to discover the causes and future effects on the integrity of the prestressing system. The results of this research work has been reported extensively elsewhere.

The tendons at Wylfa are in five main groups, samples from each of these groups was examined in the attempt to establish the extent of the corrosion by statistical methods. The worst effected areas being the hoop tendons which are exposed to the reactor hall environment between the ribs. Research work established that an 80/1000 pit did not reduce the strength of the strand below its G.U.T.S. this works resulted in a 40/1000 pit being established as the maximum pit, size for future assessments.

The inservice surveillance program for the prestressing strand was therefore designed so that a statistical assessment may be made which will give the mean increase in pit depth since the last inspection together with an estimate of there being a deeper pit within the tendon system, but outside the examination sample. The selection has generally been:-

Two strands from a tendon in each of the four tendon group.
Two strands from six of the hoop tendon group.

In addition to this the environment of the reactor hall was controlled to a relative humidity of below 33%.

The inservice inspection programmes have clearly demonstrated that corrosion if no longer a problem but inspection has continued on each reactor in alternate years in order to monitor the corrosion discovered in 1971.

# 20 Structural assessment of thick reinforced concrete pile caps using 1/5 scale models

**Gajanan M Sabnis,** Professor of Civil Engineering, Howard University, Washington DC, USA
**Raymond E Dagher,** Bridge Engineer, Century Engineers, Baltimore, Maryland, USA

Pile cap forms an important element between the column and the piles, through which the load is transmitted to soil. The problem is one of deep-slab (two deep beams). The work presented here is based on the tests of thirty - 1/5 scale models along with a select 1/2.5 scale large models. The results are compared with other researchers' work and recommendations are made in the design of caps. It is felt that the present investigation covers a wide range of variables and sheds light on strength of pile cap as a function of reinforcement quantity.

## INTRODUCTION

Pile cap is an important element in the structural foundation, which transmits the load from column to piles. Although an important element, it has drawn a little attention, due to the fact that it falls in between the structure above ground and the foundation below ground. Design provisions shown in ACI 318-83[1] may be used for thin pile caps, but not for thick pile caps unless suitable modifications are made. In the ACI 445, Shear and Torsion, attention was focussed to look into the behavior of deep members, such as deep beams, especially with continuity and pile caps. This paper is a culmination of research done at Howard over the last several years and papers published during that period [2,3,4]. In this presentation, over-all problem of structural strength assessment is tackled through various topics; these include, the present practice of pile-cap design, and experimental testing program using structural models designed to cover a wide range of parameters, including the percentage of steel.

## PRESENT PRACTICE OF PILE-CAP DESIGN

In order to review and understand the current practice, an investigation was conducted through personal interviews of local design engineers who have had experience with pile cap design. They showed a great interest in present research as they expressed concern about the real structural behavior of thick pile cap not being understood well due to the lack of substantial experimental data. In general these engineers expressed their satisfaction in the practice in that they used CRSI [5] handbook in conjunction with the ACI code; however, the main concern with ACI code was that "ACI does not give you the whole image about pile cap design and more experimental data is needed to strengthen the theory to design pile caps". They also were convinced that their designs of pile caps were over-designed or gave conservative results. Thus, it provided additional impetus to work on the research project on the topic.

## BRIEF REVIEW OF PREVIOUS RESEARCH

The most current research from Howard and University of Puerto Rico-Mayaguez has been presented in Refs. 2,3,4. The recommendation from the present research is based on them. The research started with basic tests to understand the failure mode and pattern, using gypsum concrete specimens. They indicated that the high strength shown by pile caps was mainly due to double shear cone failure. The failure pattern was verified through a Finite Element Model to indicate that the crack pattern can be reproduced. The general conclusion from this previous work was that there was very little strength advantage was achieved through reinforcement since the forces were transmitted as concrete struts of variable widths depending on their dimensions. The European work has been by Clark [6], Yan [7], Banerjee [8], Hobbs and Stein [9], Blevot and Fremy [10], and Whittle and Beattie [11]. They essentially developed a Truss theory and used a Beam-approach to determine the forces in the steel and stresses in concrete which were predominantly due to direct force action as in deep beam (two directions in the present case). A reduction of 10 to 25% was found using these methods. CRSI [5] has a comprehensive chapter on pile caps and has been used widely in the United States.

## VARIOUS CODES OF PRACTICE

ACI Code (318) does not have any provisions for pile caps. Eq. (11-27) of the Code reflect a conservative approach for additional strength in a deep beam action, which is followed E. (11-29) an additional increase is shown by arch action. Australian Code (CA2) allows design either by using beam action with critical section at a distance d/2 or by Truss analogy. British Code (CP 110) mentions only moment design of pile caps. no distinction is made between footing and pile cap, with a critical section at distance d from the face of column. French Code recommends the axial action of forces from column to piles, which is similar to a truss approach, but closer to a Strut approach as in Ref. 2. This code is used in several other countries in Europe.

## DETAILS OF TEST PROGRAM

The main purpose of this program was to investigate the behavior with varying amount of reinforcement, while other parameters, such as, concrete strength, steel depth were considered as secondary from the work reported in earlier papers. Furthermore, due to the difficulty of making large specimen and handling them with the limited available resources, it was decided to use scaled models throughout this test program at Howard. Work by Sabnis, Harris, White and Mirza [12] has demonstrated the successful use of scaled models in ultimate strength of concrete structures. A scale of 1/5 was used to suit the facility and the available reinforcement.

A total of 30 specimens were modeled from a 4-pile configuration given as a practical example in CRSI, with concrete dimensions as 13" x 13" x 6" and steel placed at a depth between 4" and 4.75" from the top surface. The column and piles were simulated with 3" steel cylinders. Details of specimen are shown in Figure 1. The concrete strength was measured using 2" x 4" cylinders and ranged from 4,080 to 5,400 psi. The reinforcement was deformed steel wires and their yield strength was 75 and 95 ksi. The steel ratio varied from 0.00171 (minimum steel amount) to 0.011. The reinforcement was arranged in a grid fashion (Figure 2) and layout was unchanged to reduce this variable. The bars were anchored to prevent bond failure and were instrumented to check if the steel would yield at failure.

RESULTS AND DISCUSSION

The results of these tests are classified and discussed in the areas of: crack formation, pattern and loads at cracking and failure; strains in reinforcement; shear cone failure and correlation with calculated loads. Cracks were formed at the tensile strength of concrete, but progressed further with additional load, based upon the reinforcement provided in these specimens. They formed at the bottom and spread to four sides with the load. The strain measured from the straingages indicated that yielding did take place at failure, although the failure may be classified as typical shear failure, as was observed from the removal of shear cone following the removal of specimen from testing. It should be noted that no bond or anchorage failure was observed.

Test results are presented in Table 1. In this table, test (failure) loads along with those calculated from the available methods/practice and the proposed method. Since all methods predict the loads on the low side, a lower bound was proposed in the present method. ACI and CRSI Methods use the deep beam design; however, ACI assumes the constant shear strength at $8bd\sqrt{f'_c}$ while, CRSI uses a parabolic variation of strength upto $32bd\sqrt{f'_c}$ as a/d reduces below a value of 2. On the other hand, truss analogy method assumes the pile cap as a space truss, in which the compression members are made of concrete, while steel takes care of tension. Previous research [4] indicates that this approach is better and recommended that a width of compression members are to be calculated as a 'strut' and the calculations based on their work are closer to experimental results. In the present method, this approach is further modified to take into account the additional test data with larger variation of steel amount. Figures 3 and 4 show the relationship between shear strength of pile cap and steel ratio as well as the concrete strength. They clearly show that steel %ge does not play much role until concrete is cracked, based on the minimum steel ratio of $200/f_y$. The proposed relation between the shear strength and $f'_c$ in psi. units may be represented as:

$$R = 100 - (f'_c)^{0.5}/100 \qquad \text{for } f'_c \leq 5,000$$

$$= 0.3 \qquad \text{for } f'_c > 5,000$$

CONCLUDING REMARKS

It is believed that the extensive test data presented herein is very valuable to assessing the strength of pile cap. Unlike the present ACI approach, there is a considerably more reserve shear strength that can be used to define the lower bound of strength and also the reinforcement that is needed for the design.

REFERENCES

1. ACI(1983), 'Building Code Requirements for Reinforced Concrete', published by American Concrete Institute, Detroit, Michigan, Pp.136.

2. Sabnis, G.M. and Gogate, A.B., "Investigation of Thick Pile Caps", Journal of ACI, Jan-Feb. 1980, Pp. 18-24.

3. Sabnis, G.M., "Experimental and Theoretical Analysis of Pile Caps", presented at the ACI Annual Meeting, Los Angeles, March 1983.

4. Jimenez-Perez, R., Sabnis, G.M. and Gogate, A.B., "Experimental Behavior of Thick Pile Caps", paper presented at the Structural Assessment Seminar at London, April 1986.

5. -"Pile Caps for Individual Columns" Chapter 13, CRSI Handbook published by Concrete Reinforcing Steel Institute, Chicago, IL, 1980.

6. Clark, J.L., "Behavior and Design of Pile Caps with Four Piles", Technical Report No. 42.489, Cement and Concrete Association, November 1973.

7. Yan, H.T., "The Design of Pile Caps", Civil Engineering and Public Review, May-June 1954.

8. Banerjee, A.C., "Design of Pile Caps", Central Building Research Institute, Roorkee (India), March 1973.

9. Hobbs, N.B. and Stein, P., "An Investigation into the Stress Distribution in Pile Caps with Some Notes on Design", Proceedings of the Institution of Civil Engineers, July 1957, Pp. 599-628.

10. Blevot, J. and Fremy, R., "Sur Pieus", Annales, Institute Technique du Batiment et des Travaux Publics (Paris), February 1967, Pp. 224-273.

11. Whittle, R.T. and Beattle, D., "Standard Pile Caps I and II", Concrete, January 1972 (Pp. 34-36) and February 1972 (Pp. 29-30).

12. Sabnis, G. M., Harris, H.G., White, R.N. and Mirza, M.S., "Structural Modeling and Experimental Techniques", Prentice Hall Inc, Englewood Cliffs, N.J. 1983, 600 pages.

## Table 1 Comparison of Experimental and Calculated Loads by Various Methods shown as Ratios w.r.t. Experimental Load

| Pile Cap No. | Experi. Failure {Kips} | ACI | CRSI | Sabnis & Gogate | Truss Analogy | Yan | Blevot | Proposed Method |
|------|------|------|------|------|------|------|------|------|
| RD1 | 62.0 | 1.68 | 1.55 | 1.78 | 2.48 | 3.12 | 2.54 | 1.21 |
| RD2 | 62.0 | 1.62 | 1.50 | 1.68 | 2.39 | 3.06 | 2.48 | 1.21 |
| RD3 | 69.0 | 1.77 | 1.64 | 1.82 | 2.61 | 3.37 | 2.73 | 1.35 |
| RD4 | 65.5 | 1.63 | 1.50 | 1.64 | 2.29 | 2.55 | 2.07 | 1.28 |
| RD5 | 52.0 | 1.17 | 1.08 | 1.12 | 1.67 | 1.92 | 1.56 | 1.02 |
| RD6 | 58.7 | 1.74 | 1.61 | 1.96 | 2.54 | 3.29 | 2.67 | 1.15 |
| RD7 | 64.7 | 1.62 | 1.75 | 1.86 | 2.55 | 3.42 | 2.78 | 1.26 |
| RD8 | 60.0 | 1.74 | 1.89 | 2.12 | 2.12 | 1.68 | 1.37 | 1.16 |
| RD9 | 61.0 | 1.62 | 1.75 | 1.86 | 1.98 | 1.61 | 1.31 | 1.18 |
| RD10 | 70.0 | 1.92 | 2.09 | 2.27 | 2.35 | 1.89 | 1.54 | 1.18 |
| RD11 | 74.0 | 1.80 | 1.95 | 1.96 | 2.21 | 1.85 | 1.51 | 1.35 |
| RD12 | 75.0 | 1.95 | 2.12 | 2.22 | 2.39 | 1.96 | 1.59 | 1.45 |
| RD13 | 73.0 | 2.03 | 1.87 | 2.10 | 2.29 | 1.89 | 1.53 | 1.41 |
| RD14 | 72.0 | 1.80 | 1.66 | 1.97 | 2.03 | 1.63 | 1.33 | 1.39 |
| RD15 | 67.5 | 1.72 | 1.59 | 1.78 | 1.52 | 1.11 | 0.90 | 1.33 |
| RD16 | 65.0 | 1.56 | 1.44 | 1.55 | 1.40 | 1.03 | 0.83 | 1.28 |
| RD17 | 69.7 | 1.85 | 1.71 | 1.97 | 1.64 | 1.17 | 0.95 | 1.37 |
| RD18 | 99.9 | 3.03 | 2.80 | 2.33 | 2.46 | 1.74 | 1.42 | 1.93 |
| RD19 | 72.0 | 2.10 | 1.94 | 2.36 | 2.12 | 1.21 | 0.99 | 1.42 |
| RD20 | 69.7 | 1.79 | 1.65 | 1.86 | 1.84 | 1.09 | 0.88 | 1.37 |
| RD21 | 85.0 | 2.13 | 1.97 | 2.08 | 1.77 | 1.31 | 1.07 | 1.64 |
| RD22 | 87.0 | 1.84 | 1.70 | 1.83 | 1.52 | 0.75 | 0.61 | 1.68 |
| RD23 | 70.5 | 2.26 | 2.09 | 2.70 | 1.94 | 1.07 | 0.87 | 1.39 |
| RD24 | 74.5 | 2.21 | 2.04 | 2.52 | 1.92 | 1.08 | 0.87 | 1.47 |
| RD25 | 99.0 | 1.84 | 1.82 | 1.76 | 1.82 | 0.71 | 0.58 | 1.91 |
| RD26 | 99.9 | 1.91 | 1.88 | 1.88 | 1.88 | 0.73 | 0.59 | 1.93 |
| RD27 | 99.0 | 2.72 | 2.66 | 2.81 | 2.66 | 1.00 | 0.82 | 1.91 |
| RD28 | 71.7 | 2.32 | 3.41 | 2.78 | 3.41 | 4.24 | 3.44 | 1.40 |
| RD29 | 66.0 | 2.00 | 2.51 | 2.27 | 2.51 | 2.10 | 1.71 | 1.29 |
| RD30 | 70.5 | 1.96 | 2.05 | 2.14 | 2.05 | 1.20 | 0.98 | 1.39 |
| Average | | 1.88 | 1.97 | 2.06 | 2.14 | 1.86 | 1.49 | 1.36 |

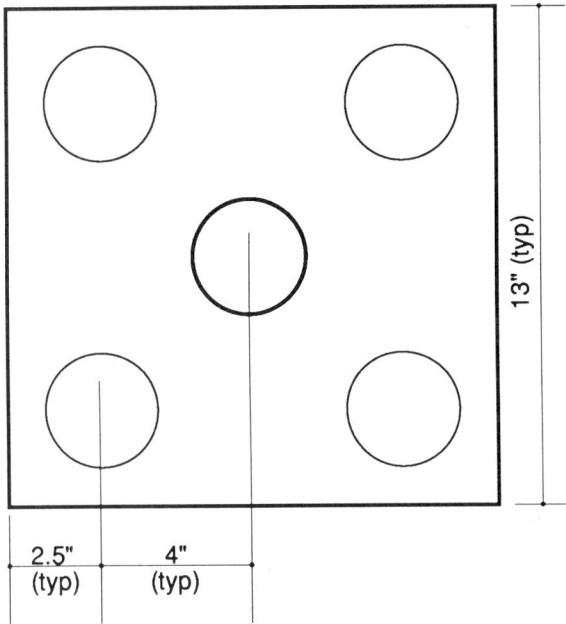

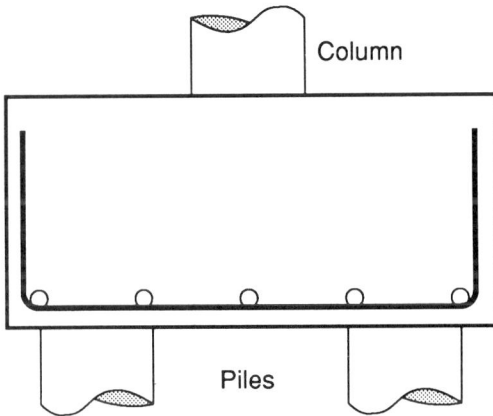

**Figure 1  Details of Pile Cap**

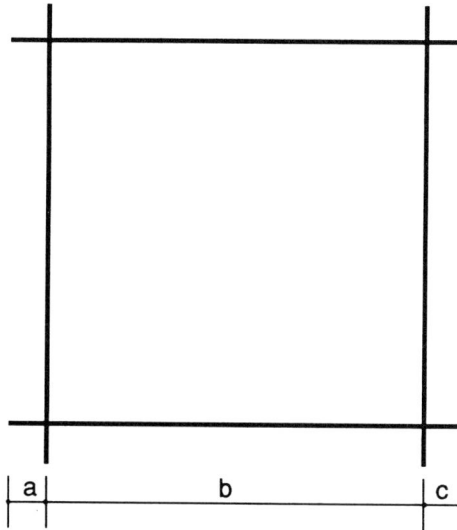

| a = c | 1.5" | 1" | 1" |
|---|---|---|---|
| b | 9" | 10" | 10" |
| No. of Bars | 3 | 5 | 6 |
| Spacing | 3" | 2.5" | 2" |

**Figure 2  Layout of Reinforcement**

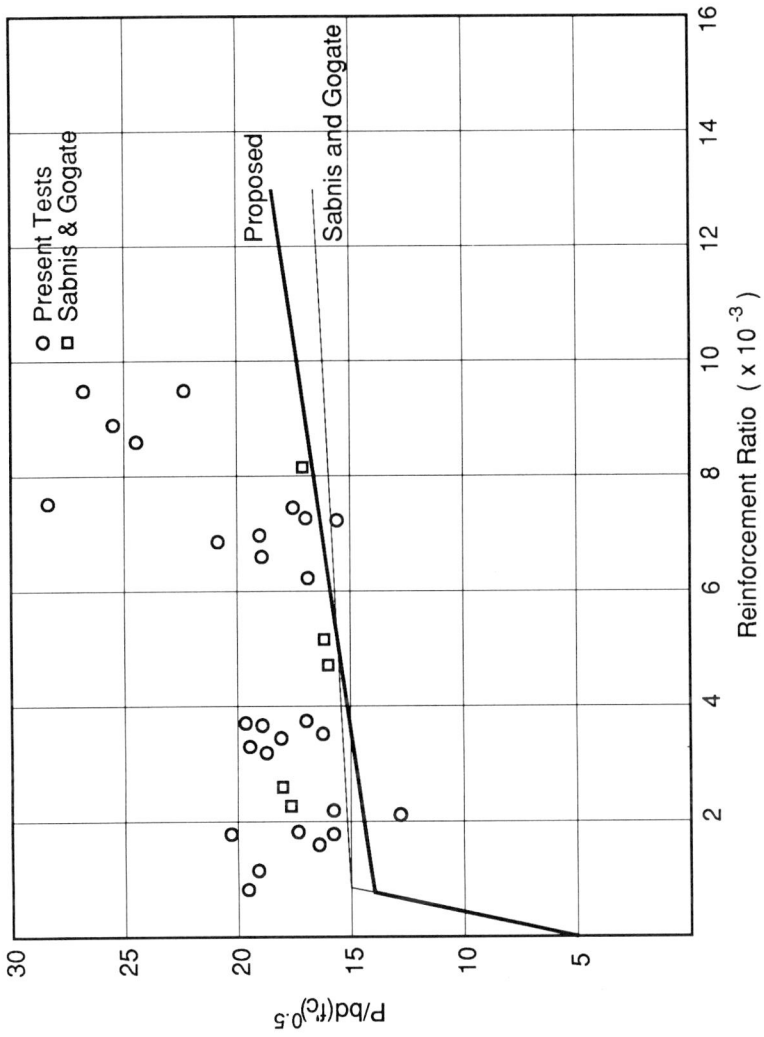

Figure 3 Ultimate Shear Strength vs. Flexural Reinforcement Ratio

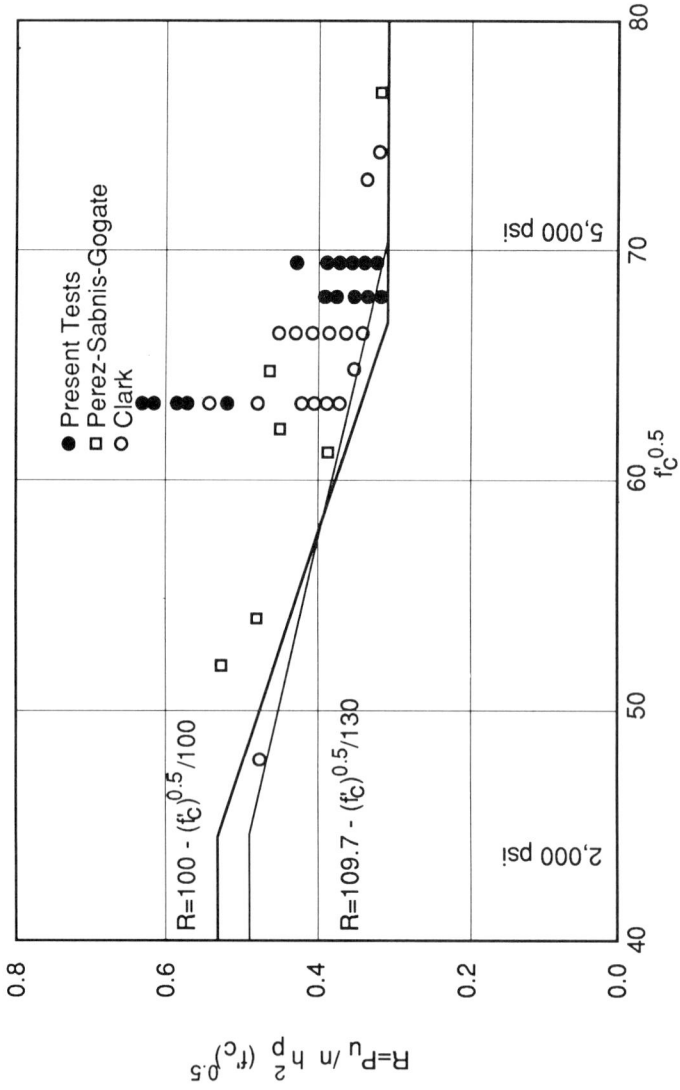

**Figure 4** Non-dimensional Strength Parameter "R" vs. Compressive Strength $f'_C$

# 21 The structural performance of damaged concrete elements

**D N Smith and L A Wood,** Department of Civil and Structural Engineering South Bank Polytechnic, London, UK

This paper reports the preliminary findings of a research project concerned with the structural performance of damaged concrete elements. Results are presented indicating the effect on strength of exposure of reinforcement in beams and longitudinal cracks in columns.

## INTRODUCTION

In recent years, the enduring quality of concrete has been brought into question by a wide variety of deterioration resulting in the main from chemical attack. The problems encountered have been accentuated by inadequate design or detailing, changes in materials, and the application of aggressive chemicals such as de-icing salts. A common feature of most of the differing causes of deterioration is that there is a reduction on the alkalinity of the concrete which allows oxidation of the reinforcing steel to take place. Unless arrested in the early stages, this oxidation process leads to cracking of the concrete and possible spalling of the cover to the reinforcement.

Much effort has been made recently to investigate the causes and effects on material behaviour of these chemical attacks and also on the means of reducing or eliminating such decay. Prevention is undoubtedly the best cure for the problem. In the meantime, however, many structures built in the last two or three decades will continue to deteriorate, and little work has been done to investigate the effects this damage has on their structural performance.

This paper aims to report the first phase results of a project which is investigating these structural effects.

## TEST PROPOSALS

In order to ascertain the effects of any chemical attack on the structural performance of an element, the ideal procedure would be to subject specimens to long term storage in an aggressive environment. This approach suffers from the double disadvantage of requiring a considerable period of time and of not being entirely reliable or consistent in the type or positions of the defects that result. In order to overcome these disadvantages, it was considered appropriate to actually form defects in specimens so as to be as similar as possible to those that would occur in practice. Clearly such an approach could not completely simulate the effects of chemical attack on concrete specimens. It would therefore be desirable at a future stage to compare the behaviour of real structural members with those in which defects had been mechanically introduced.

The project, the preliminary results of which are reported in this paper, is concerned with two types of defective structural elements, namely
(i)   reinforced concrete beams in which loss of cover and exposure of the reinforcement is simulated by forming recesses; and
(ii) axially loaded unreinforced concrete columns in which longitudinal cracks are pre-formed at the casting stage.


## BEAM TESTS

The overall aim of the tests on defective beams is to investigate the reduction in strength caused by loss of cover and exposure of the reinforcement for beams with varying extents of defect, and varying amounts of shear and longitudinal steel. In addition a long term aim is to investigate the proportion of strength that can be recovered by repairing such defects.

From previous work carried out to investigate the effects of defects on flexural behaviour (Ref 1), the loss of strength appears to be relatively small even prior to any repair being carried out. This is not unduly surprising since the variation of tensile stress in a bonded bar will not be great except in areas of high shear. For this reason, it was considered that the presence of defects would be more significant in regions where shear was the paramount action. The strength at such positions would be adversely affected both by the loss of area of concrete and a reduction in the effectiveness of dowel action. A series of beams have, therefore, been tested to investigate the effects on the strength of loss of cover and exposure of the steel.

## Beam Details

To date thirty-six simply supported beams have been tested in twelve sets of three. Each set consisted of one undamaged beam and two similar beams in which the cover was omitted and the reinforcement exposed over a variable length near one end. It was considered appropriate to expose the steel fully as this is the recommended procedure prior to repair (Ref 2). Fig.1 indicates the beam details. All beams tested were singly reinforced for bending and unreinforced in shear.

In the series of tests undertaken so far only the position of the single point load and the damaged length have been varied. The four load positions measured from the support varied from 300mm to 600mm in 100mm increments. The various extents of damage were taken as 200mm, 300mm, and 400mm.

## Test Details and Results

Each beam was simply supported over a span of 1700mm, and was loaded by means of a hydraulic jack with a single point load, the position of which was varied from beam to beam as noted previously. The central deflection was measured using a linear displacement transducer and a load-displacement graph was plotted for each beam.

The residual strength of each damaged beam was compared with the strength of the undamaged beam tested in the same set. Table 1 indicates the ratio of these strengths for the various cases.

Table 1   Beam Residual Strength Factors

| Defect Length (mm) | Load Position from Support (mm) | Damaged Strength / Undamaged Strength |
|---|---|---|
| 200 | 300 | 0.91 |
|  | 400 | 0.96 |
|  | 500 | 0.90 |
|  | 600 | 0.76 |
| 300 | 300 | 0.96 |
|  | 400 | 0.92 |
|  | 500 | 0.77 |
|  | 600 | 0.72 |
| 400 | 300 | 0.93 |
|  | 400 | 0.93 |
|  | 500 | 0.73 |
|  | 600 | 0.64 |

Typical major failure crack patterns for damaged and undamaged beams are indicated in Fig.2. Most beams appeared to fail through shear although a few of these loaded with the largest shear span showed indications of flexural failure.

The preliminary indications are, that as the extent of damage is enlarged so the loss of strength increases. The significance of load position on the loss of strength, however, appears to be somewhat greater. The loss of strength when the load is positioned nearest to the support is of the order of 5% whereas this figure rises to 35% when the load is positioned 600mm from the support. Fig.3 indicates the bounds for the residual strength of the damaged beams for differing load positions.


UNREINFORCED COLUMNS

On occassion longitudinal cracks have been observed in structural columns (Ref 3,4) probably due to the presence of alkali-silica reaction within the concrete. The development of such cracks may lead to reduced column stiffness and a much reduced load capacity as the propensity to a buckling failure mode is increased.

In order to investigate such effects a series of tests on 1.8m long by 75mm square section unreinforced columns has been undertaken. The absence of reinforcement, whilst perhaps unrealistic in most instances, has allowed the action of the concrete alone to be studied. The columns have been manufactured with varying lengths preformed central longitudinal cracks as shown in Fig.4. The cracks are approximately 2mm wide with smooth surfaces in order to negate the effects of any aggregate interlock.

Results of Axial Tests

The columns have been subjected to axial loading with pinned end conditions. Load-displacement curves have been recorded for all of the tests. In Fig.4 the envelope of ultimate axial load, normalised with respect to cube strength, is given with variation in the ratio of the length of the longitudinal crack to the length of the column. It is readily apparent that the crack must exceed 60% of the column length before appreciable, but rapid , reduction in load capacity occurred. However, further work is now in hand into the determination of the minimum shear connection necessary to ensure composite action of the column and hence the avoidance of a buckling failure.

CONCLUSIONS

All of the results to date are preliminary findings which have been obtained in order to enable the ensuing more rigorous and conprehensive research to be programmed to the best advantage.

Beam Tests

The strength of single point loaded beams in which cover is lost and the reinforcement exposed is reduced by a varying amount depending on the extent of the damage and the load position. This loss of strength which varies from 5% to 35% appears to be primarily affected by the load position rather than the extent of damage.

Further variables to be considered include reinforcement in the compression zone together with the inclusion of shear steel. It is likely that the inclusion of both these items will reduce the resulting loss of strength. It is also intended that the project will include an investigation into the strength of damaged beams after repair.

Column Tests

The preliminary results would suggest that the presence of a longitudinal crack alone does not lead to any major reduction in strength until the crack length exceeds 60% of the height of the column.

Further work is now in progress on the determination of the shear strength of micro-cracked concrete together with the minimum shear connection required in a cracked column in order to ensure composite action. The results of these investigations will lead to economic and efficient methods of repair together with a corresponding greater degree of confidence in the engineering decisions related to the strength of deteriorated structural concrete elements.

ACKNOWLEDGEMENTS

The work described forms part of a programme of NAB funded research within the department of Civil and Structural Engineering at South Bank Polytechnic into the rehabilitation of urban construction. The support of the Severn Trent Water Authority for the combined study with the Concrete Research Group at Queen Mary College into the column behaviour is gratefully acknowledged.

REFERENCES

(1)    I.Minkarah and B.C.Ringo 'Behaviour and repair of
       deteriorated reinforced concrete beams'. Transportation
       Research Record (1981)

(2)    P. Pullar-Strecker    Corrosion Damaged Concrete CIRIA
       (1987)

(3)    D.W.Hobbs    Alkali-Silica Reaction in Concrete Thomas
       Telford (1988)

(4)    Private Communication

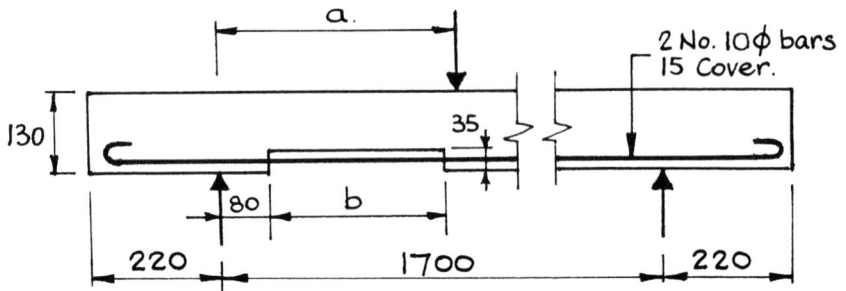

a = 300, 400, 500 or 600.
b = 200, 300, or 400.
Beam width is 75.

Figure 1. Beam Details

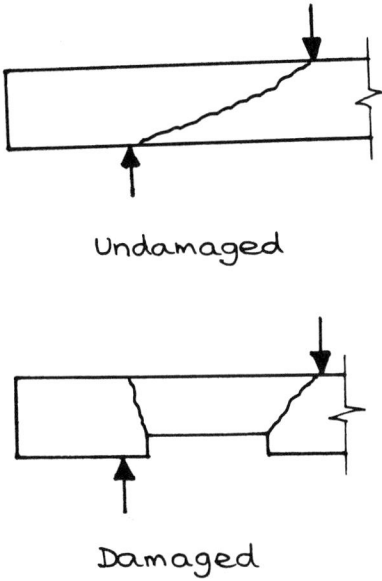

Figure 2. Failure Crack
Patterns.

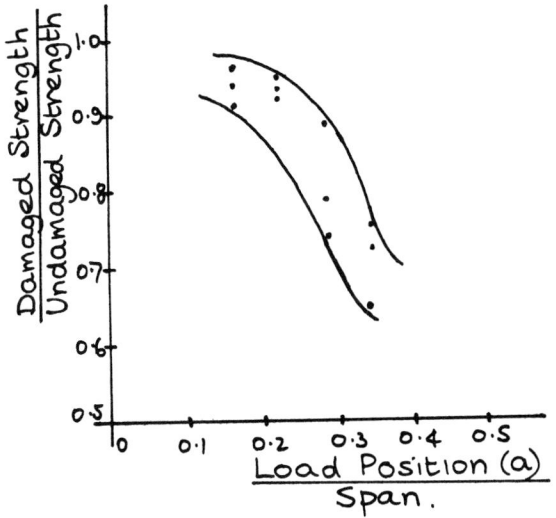

Figure 3. Variation of Residual
Strength Factor.

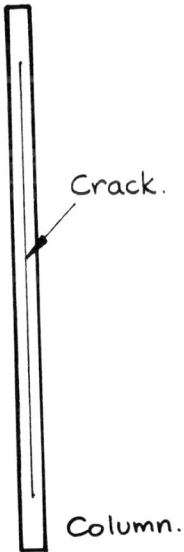

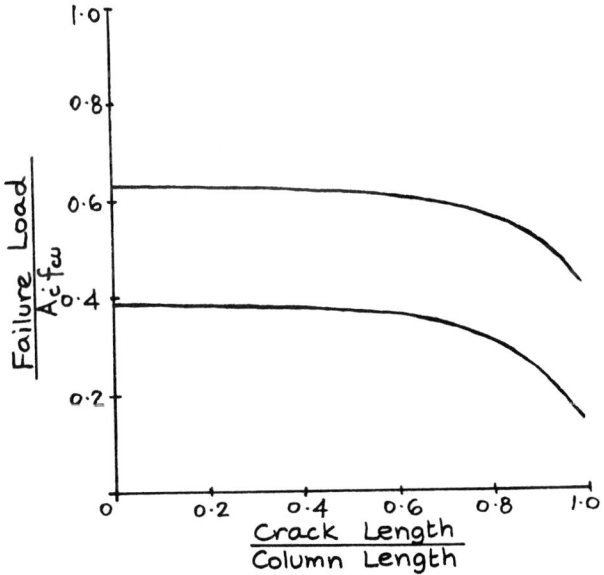

Figure 4. Details and Results for Columns.

187

S F Ray, Bingham Cotterell, UK


It was reported that some tests had been carried out on repaired beams
and the load at which debonding started shown on the load/deflection
curves.   Would the author describe the type of repair and how it was
made.

A useful development might be to test the bond and compare the
strength of different types of repairs using a simple pull-off test.
Please would the author comment on this proposal.


W D Biggs, University of Reading, Buro Happold, UK


In the presentation (though not in the published document) graphs were
shown of the stress-strain behaviour of a 'repaired' beam - one which
the i.e. exposed reinforcement has been re-covered with concrete.   My
notes show that the graph of Figure 1 was presented - the dashed curve
represents the 'repaired' beam and the curve first of all follows that
of the 'undamaged' beam and then abruptly, moves down to the 'damaged'
curve.

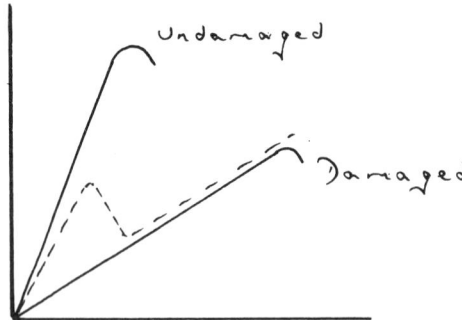

Such a curve appears on page 1 of any text book on linear elastic
fracture mechanics.   The area under the curve represents the strain
energy release rate associated with the propagation of a crack at
constant strain.   Have the authors considered this type of approach?

The authors further state that the phenomenon is a function of bonding
at the repair/original interface.   May be - but why should it
necessarily be so?   The repair concrete has, by definition, had less
curing time than the basic concrete so that a crack could start any
where in the repair concrete and propagate until it is stopped at
entry into the stiffer base concrete.   There is a mismatch of strain
energy between the two - so why describe it simply to bond strength?

R C de Vekey, Building Research Establishment, U.K.

What type of repair material was used - epoxy mortar, polymer-modified cement mortar or other?

Did, in the author's opinion, the repair crack before or after the serviceability limit for the test piece?

Would the repair function better if a feather edge was generated, thus putting the repair base junction mainly into shear?

## Author's reply

The authors thank the contributors for the interesting points they have raised and respond as follows.

To date, only a very limited number of repaired beams have been tested in order to determine the likely range of strength recovery that will result from any repair. Beams using the following types and methods of repair have been tested:

(i) A sand:cement mortar repair made to damage on the underside of the beam. The existing concrete faces were first coated with a cement slurry.

(ii) A polymer-modified mortar repair made with the beam inverted so that the damage was on the top surface. An acrylic emulsion bonding agent was used.

These two alternatives were felt likely to be lower and upper bounds to the degree of strength recovery.

A typical load/deflection graph for the repaired beams in comparison to the damaged and undamaged ones is indicated in Fig. 1. The sudden decrease in load was observed to correspond with the debonding of the repair material from the original concrete and occurred at approximately 30% and 70% respectively of the undamaged beam strengths for the two repairs referred to above. It is thus possible for the repair to debound within the serviceability range if the repair is poor. However, further work is necessary before load levels can be accurately quantified.

It is also possible for cracks to occur in the repair material itself. Such cracks, however, would tend to follow the lines of maximum principal tensile stress, as for any undamaged beam. If the bond of the repair was sufficiently strong, then the beam would behave similarly to an undamaged one. The significance of the strength of the bond at the repair/concrete interface is that it can introduce a weakness in a horizontal plane which may result in the separation of the repair from the beam. The beam will then act in a similar way to a damaged one.

The bond strength of the repair is, therefore, of considerable significance. It would seem worthwhile to test the adequacy of repair materials by performing simple pull-off tests. This would only give a comparative indication of the strength, however, as the most significant factor is likely to be the workmanship, particularly for repairs made to beam soffits.

Insufficient information is available to indicate whether or not a feather edge would improve the performance of the repair. However, since the bond of the repair is so important, it is felt that the treatment at the edge of the repair will not significantly affect its behaviour.

# 22 The Mary Rose dry dock dam

**D W Begg and J E Butler,** Department of Civil Engineering, Portsmouth Polytechnic, UK

With concrete construction techniques monitoring
the in-situ performance of full scale structures
still remains the only way of determining their
ability to meet their design requirements today
or at any time in the future.  This paper
describes some methods chosen to examine the
early life of this particular structure.  The
ability to monitor settlement, deflection,
strain and temperature during the construction
period and on subsequent loading led to the
application of a range of instrumentation where
often only one system would be used.

INTRODUCTION

As part of the continuing improvements to the Naval Heritage Area in
Portsmouth the Secretary of State for the Environment, acting through the
Property Services Agency proposed to provide a permanent reinforced
concrete dam to replace the existing steel caisson which closes off No. 3
Dock currently used for the Display of the Mary Rose.  To enable the
caisson to be removed a temporary cofferdam had to be constructed to the
seaward side of the caisson.  The caisson was then dismantled and
removed followed by the construction of the permanent dam.

DAM DESIGN

The dam is trapezoidal in shape·being 16.5m wide at its base and 21.5m
wide at the top.  The overall height is 9.6m although the maximum water
level is presently maintained at 7.5m (Fig. 1) by the caisson sealing
basin No. 1 from Portsmouth Harbour (Fig. 2).  The reinforcement required
to resist this hydrostatic force was found by a yield line approach to be
40mm dia bars at 250mm centres.  In order to gain improvements in strength
and long term durability replacement of 60% of the cement content of the
concrete with ground granular blast furnace slag was proposed.  This
should also allow a greater cementitious content with a reduced·water
content and also reduce the adverse effects of heat of hydration.  The
dam was assumed to be the same mass as the caisson it was replacing and
so the effect on this large masonry structure was neglected.  Settlement
monitoring would determine whether this was a realistic assumption.

Dam Performance Prediction

In order to check the in service behaviour, find the magnitudes and
positions of the maximum strains and displacements elastic analyses were
carried out using a simplified finite element model of the dam.

F-E Analysis

The structure was modelled using Eight noded thick facet shell elements
(Fig. 3). The load applied was an equivalent hydrostatic pressure over
the full height of the dam i.e. not allowing for any freeboard. The
supports were assumed to be simple along the two sides and the bottom.
Note the whole structure was modelled without utilising symmetry for the
benefit of presentation. The results of this analysis suggest that the
concrete stresses due to this loading will be less than 2N/mm² with a
maximum bending displacement of less than 3mm (Fig. 4 and 5).

STRAIN MONITORING

In order that the predicted performance could be compared with the actual
it was decided to incorporate strain gauges within the structure of the
reinforced concrete dam. Surface mounted Linear Variable Differential
Transformers (LVDT's) which were originally suggested as an inexpensive
method of short term monitoring were rules out due to their
susceptability to damage by the contractors or vandals. The type of
transducer chosen for this structure had to be both sensitive enough to
monitor the small strains expected whilst being sufficiently stable to
allow long term effects to be detected. It was felt that acoustic
strain gauges should be installed within the structure of the dam. Two
types of gauge were chosen; one to monitor reinforcement strains and the
other to monitor concrete strains. Both systems were to be used to
monitor the early life of the structure up to the initial loading of it
when the cofferdam is flooded and also for monitoring the long term
effects over many years.

Concrete Strains

The type of gauge used to measure the strains in the concrete is a
standard Acoustic Embedment gauge (Fig. 6). Its specification is as
follows:

        Wire Length          =   130mm
        Wire Diameter        =   0.0254mm
        Starting Frequency =   1000 Hz (approx.)
        Gauge Factor for Strain = $2.6 \times 10^{-3}$
        Temperature Measurement via change in resistance of
        pick-up coils

Reinforcement Strains

The type of gauge used to measure the strains in the reinforcement is a
Weldable Acoustic gauge (Fig. 7) with purpose built energisation and
pick-up coils which can be used to monitor dynamic strains[1]. This
dynamic ability would be desirable if the structure is subjected to wave
forces or a tidal range. Its specification is as follows:

        Wire Length          =   67mm
        Starting Frequency   =   1637 Hz
        Gauge Factor for Strain  =   $0.757 \times 10^{-3}$

Temperature measurement

In order to monitor the effect of cement replacement on the temperature
history and instantaneous thermal gradient across the dam section of the
dam during curing themcouples were used.  Also the ability to monitor
temperature at the embedment gauge sites was also possible.  The
temperature data is essential in understanding the thermal effects on
the measured strains.

DISPLACEMENT MEASUREMENT

For years surveying instruments have been used for engineering
measurement applications the recent appearance of the electronic
theodolite has fundamentally changed this.  The electro-optical reading
system of these instruments makes them ideally suited to direct computer
communication for real time data processing.  The results can be
available immediately for evaluation and display and can also be stored.
This elimination of the human operator reading and transcribing the
measured angles enormously reduces the time taken to log and process
the readings but more importantly makes mistakes and errors virtually
impossible.  These techniques have already found favour in aircraft
inspection antenna assembly[2] and the accuracy found[2,3] more than satisfied
that required for structural deformation measurement.
The instruments used for this work were KERN E2 electronic one-second
theodolites.  These instruments can be used to measure directions to an
accuracy of approximately 0.5".  At a distance of 10m this corresponds
to a length of 0.02mm.  This high resolution together with the ability
to monitor the structure without contact made this the ideal solution
for correlation of displacements with the analysis.
The two theodolites are set up at any suitable location to represent the
end points of a baseline.  To determine the length and height difference
of this baseline simultaneous sightings of a reference scale of
accurately known length set up anywhere in the object domain are made.
After this baseline calibration the points to be measured on the object,
in this case a Dam, can be sighted with both theodolites.  The results
of these sightings are transmitted to the computer thus allowing it to
continually calculate X, Y and Z coordinates.  As the Z coordinates can
be calculated by each theodolite observation of wrong sightings can be
immediately detected.
Settlement monitoring was accomplished using precise levelling techniques.

CONCLUSIONS

The approach described has several benefits besides offering an
opportunity to compare analytical and design predictions with those that
actually occur.  The temperature and dynamic strain measurements made
immediately following the casting of the dam concrete will provide a
more accurate record of material behaviour during early life than has
been possible previously.  The long term performance will also be able to
be examined for many years to come by using the embedded strain gauges
with the back up of displacement measurements should there be any
unanticipated change in performance.  It is also an opportunity to assess
the site performance of acoustic gauges which are capable of detecting
changes in strain occurring at a rate of 20 Herz.

ACKNOWLEDGEMENTS

This work was sponsored by the Property Services Agency under the direction of Mr. N. Awan and Mr. R. Bramwell. The embedment gauges were supplied by Strainstall Ltd., Cowes, IOW., all other technical support coming from the Department of Civil Engineering.

REFERENCES

1.  D.W. Begg & J.E. Butler, 'Into the 1990's with the Acoustic Gauge', Application of Advanced Strain Measurement Techniques.  Int. Conf. BSSM/SEM, August 1987.

2.  D.R. Johnson & G.F. Wasinger, 'An approach to Large Scale Non-Contact Coordinate Measurement', Hewlett Packard Journal, September 1980.

3.  H. Leitz, 'Three-Dimensional Coordinate Measurement By Intersection with Electronic Precision Theodolites', NELEX 82, September 1982.

ILLUSTRATIONS

Figure 1  Location Plan
Figure 2  Dam Elevation and Section
Figure 3  F-E Idealisation
Figure 4  Maximum Surface Stresses
Figure 5  Displaced Shape
Figure 6  Embedment Strain Gauge
Figure 7  Weldable Strain Gauge
Figure 8  Intersection Measuring Set up

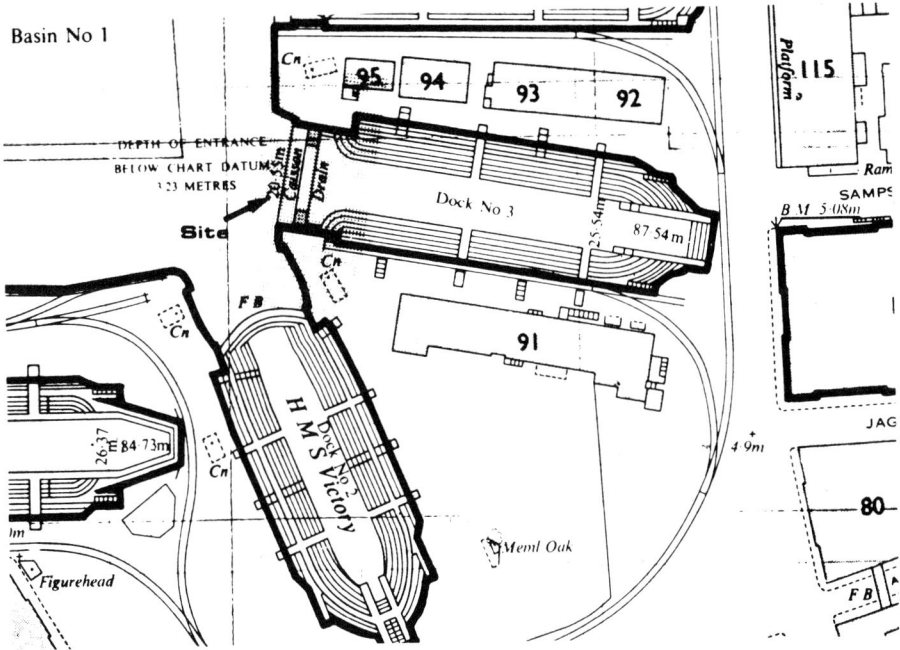

Figure 1   Location Plan

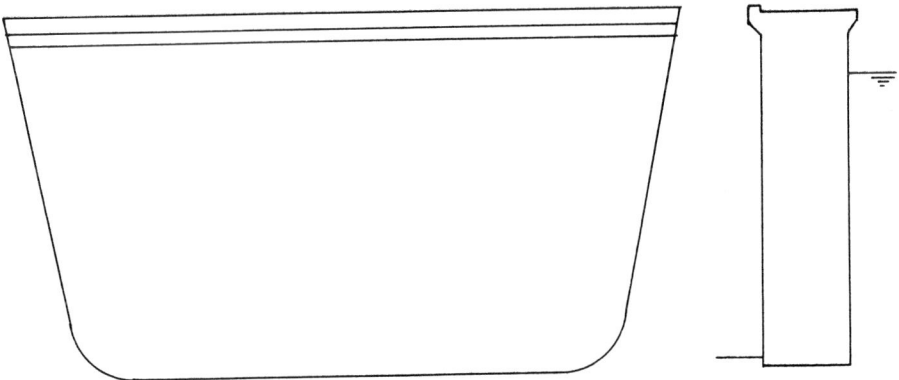

Figure 2   Dam Elevation and Section

195

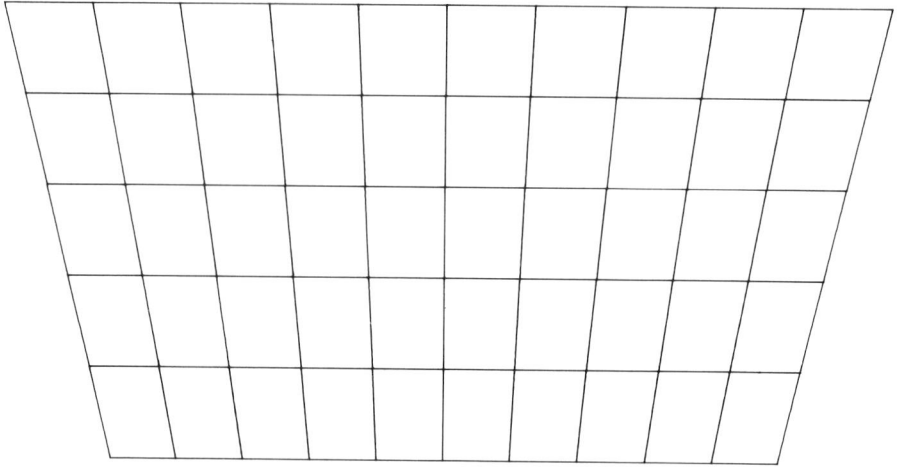

Figure 3  F-E Idealisation

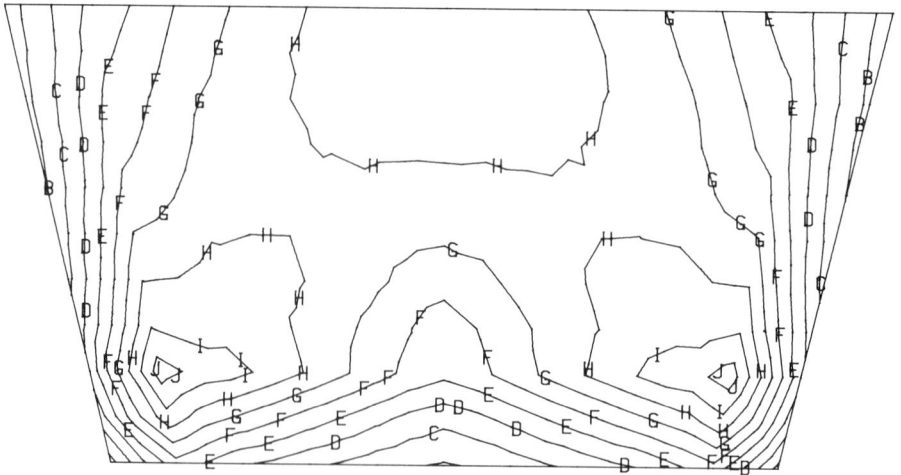

Figure 4  Maximum Surface Stresses

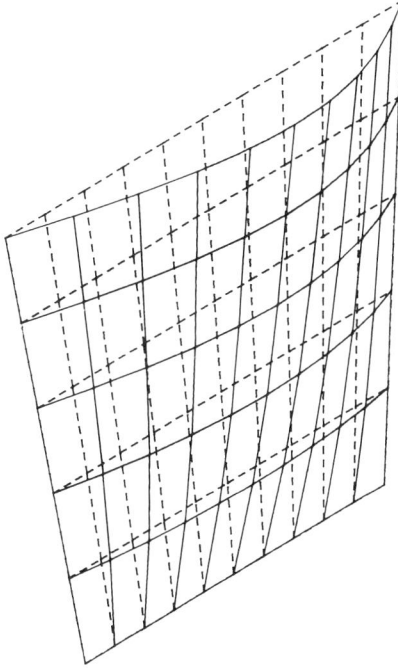

Figure 5   Displaced Shape

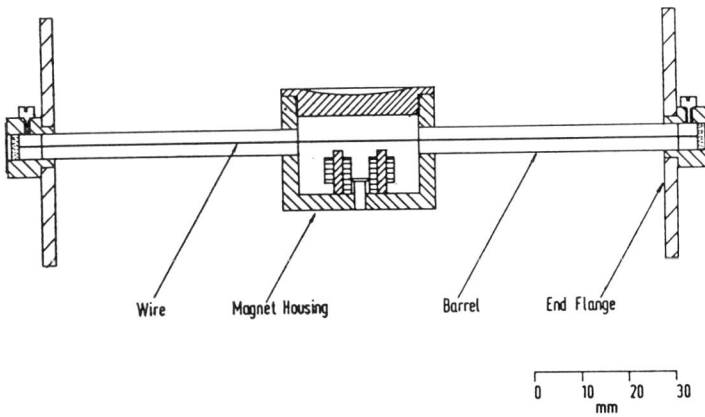

Wire          Magnet Housing          Barrel          End Flange

```
┌──────┬────┬────┬────┐
0      10   20   30
       mm
```

Figure 6   Embedment Strain Gauge

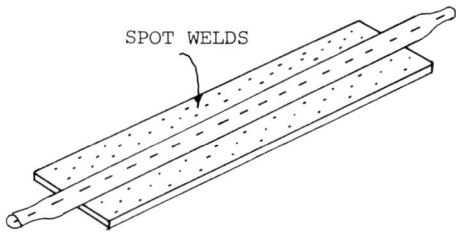

SPOT WELDS

Figure 7   Weldable Strain Gauge

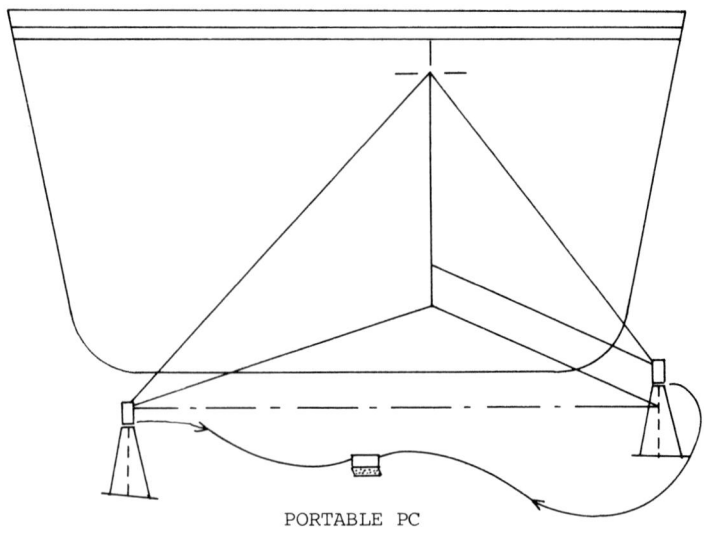

PORTABLE PC

Figure 8   Intersection Measuring Set up

# 23 Life expectancy of hull structures of boats

**B M Ayyub,** The University of Maryland, College Park, Maryland, USA
**G J White,** US Naval Academy, Annapolis, Maryland, USA
**E S Purcell,** US Coast Guard, R & D Center, Groton, Connecticut, USA

A methodology for the structural life assessment of a
ship's structure is suggested.  The methodology is based
on probabilistic analysis using reliability concepts and
the statistics of extremes.  In this approach, the estimation of structural life expectancy is based on selected failure modes.  For the purpose of illustration
one failure mode is considered in this study.  This is
plate plastic deformation.  Structural life, based on
this failure mode, for an example vessel is determined.

INTRODUCTION

The estimation of structural life is not a simple task.  Many factors
affect the life of a ship.  These factors include design parameters, design safety factors, design methods, ship type, structural details, materials, construction methods and quality.  Loads on the hull, including
ship weight, water pressure, waves, and engine and propeller vibrations
are important.  The maintenance practices and levels, and inspection methods also play a vital role in determining how long the structure of a ship
will survive.  But the final decision as to when a vessel is no longer fit
for continued service is a economic one.  It is not the intention of this
paper to cover the economic decisions which go into determining a ship's
fitness for further service.  Rather, a probabilistic-based approach for
evaluating the likelihood of reaching a desired service life is provided.

In order to demonstrate how this methodology might be used, an example
ship type is investigated.  The ship type used is a high performance semi-
displacement vessel.  This type was chosen for investigation because of
the recently published work [1] which provides the kind of information
needed to perform the analysis.

DEFINITION OF END OF STRUCTURAL LIFE

In order to demonstrate the approach being described here, a potential
failure mode was chosen.  The analysis that lead to choosing the ductile
yielding of individual plate panels as the potential failure mode is discussed in reference [2].  This potential failure mode represents a serviceability related mode of failure.  That is, the failures will not lead
directly to loss of the vessel, but may limit performance or endurance in

some manner. Relating this limited serviceability to a definition for end of useful structural life is one of the most difficult problems faced in performing an estimation of structural life.

As discussed earlier, the end of structural life is, usually, based on some economic factors. Once the cost of maintaining the structural system exceeds a specified budgeted amount, a decision must be made by the owners. Either the additional spending above budget is authorized, or the replacement of the structure is necessary. The scope of this study does not include economic analysis of the structural system of the vessel, or its maintenance schedule and cost. Therefore, in order to demonstrate a methodology for estimating the structural service life for the example vessels, a set of arbitrary constraints were selected to represent the owner's position on vessel replacement.

The end of structural life of the vessel is defined, for the plate deformation failure mode, as whenever more than five plate panels in a specified critical area need to be replaced at the end of any inspection and maintenance period (hereafter called inspection period). Plate panels are to be replaced when the ratio of plastic deformation to plate thickness is greater than or equal to 2.0.

This definition is based on the assumption that having to replace 6 or more plates in a critical region during any inspection period would cause the owner to use more resources than currently allocated for repair and steel replacement in the vessel's lifetime maintenance budget. The allowable deformation of the plate panel, $w_p/t = 2.0$ was chosen for the sake of demonstration. This value could have easily been chosen as more than 2.0, but realistically it would likely be less than 3.0. Defining a critical region introduces another level of complexity. Now there is a specific number of panels to consider, and end of life is in terms of an event where more than 5 out of the total panels in the region need to be replaced.

METHODOLOGY OF STRUCTURAL LIFE ASSESSMENT

Structural Reliability Assessment

The performance function that expresses the relationship between the strength and load effect of a structural member according to a specified failure mode is given by

$$Z = g(X_1, X_2, \ldots, X_n) \tag{1}$$

in which the $X_i$'s are the basic random variables, with g(.) being the functional relationship between the basic random variables and failure (or survival). The performance function can be defined such that the limit state, or failure surface, is given by $Z = 0$. The failure event is defined as the space where $Z < 0$, and the survival event is defined as the space where $Z > 0$. Thus, the probability of failure can be evaluated by the following integral:

$$P_f = \int\int \ldots, \int f_X(X_1, X_2, \ldots, X_n) \, dx_1 \, dx_2 \ldots dx_n \tag{2}$$

where $f_X$ is the joint density function of $X_1$, $X_2$, ..., $X_n$, and the integration is performed over the region where $Z < 0$. Because each of the basic random variables has a unique distribution and they interact, the integral of equation (2) cannot be easily evaluated. A probabilistic modeling approach of Monte Carlo computer simulation with Variance Reduction Techniques (VRT) can be used to estimate the probability of failure [3].

The strength (or resistance) R of a structural component and the load effect L are generally function of time. Therefore, the probability of failure is also function of time. The time effect can be incorporated in the reliability assessment by considering the time dependence of one or both of the strength and load effects [2].

Based on the load effect L(t) and strength R(t) which are time dependant, the probability of failure can be computed according to a specified failure mode. The resulting probability of failure $P_f(t)$ is function of time. Mathematically, the probability of failure can vary from zero to one. Realistically, the probability of failure varies from an initial (design) probability of failure based on design values to a final probability of failure at the end of useful structural life. The resulting mathematical variation of the probability of failure with time can be viewed as the cumulative distribution function of the structural life SL of a component according to a specified failure mode. Actually, the curve satisfies all the conditions of a cumulative probability distribution function. This relationship can be expressed as follows:

$$F_{SL}(t) = \text{Prob} \ (SL < t) = P_f(t) \tag{3}$$

where $F_{SL}(t)$ is the cumulative distribution function of structural life.


STRUCTURAL LIFE ASSESSMENT PROCEDURE

Limit State Equation

The limit state equation for the local plate deformation failure mode can be expressed as

$$g(x) = \text{Resistance - Still Water Load - Dynamic Load} \tag{4}$$

Each of the terms in the above equation are expressed in units of pressure. The still water load is the hydrostatic pressure at the depth of interest. The dynamic pressure is the extreme dynamic pressure based on the results from full scale experiments conducted on one of the vessel's of this class [1]. The resistance term is an empirical expression developed by Hughes [4] based on elastoplastic methods and is given as:

$$\text{Resistance} = \frac{F_Y^2}{E} \ \{Q_Y + T(R_w)[\Delta Q_0 + \Delta Q_1 R_w]\} \tag{5}$$

where
- $F_Y$    is the yield stress of the material
- $E$    is the modules of elasticity of the material
- $Q_y$    is the initial yield load calculated from elastic theory
- $DQ_0$    accounts for curved transition portion of load deflection curve (see Figure 1)
- $DQ_1$    accounts for straight portion of load deflection curve
- $R_w$    is the ratio of the deflection $w_p$ at a given loading to the

deflection at the completion of the edge hinge formation $w_{p0}$

$$T(R_w) = [1 - (1-R_w)^3]^{1/3} \quad \text{for } R_w \leq 1$$
$$= 1 \text{ for } R_w > 1.$$

This expression was calibrated with available experimental data [8]. It tends to provide a lower bound on stresses required for a specified deflection, but is extremely accurate for $w_p/t$ ratios of less than 4.

Extreme Load Effects

In this study, only loads and load effects in head-seas are considered. No other heading is considered because the stress records in reference [1] indicate that they result in much smaller stresses than the head-seas condition. For eight combinations of ship speed and sea state in the head-seas condition, tests and stress measurements at the locations of interest were performed in reference [1]. The case number and the speed/ sea-state combination are given in Table 1. The combination of high sea-state and high speeds was not considered because historical records indicate that this type of vessel is almost never operated under those conditions. The percentages that are shown in Table 1 for each combination represent the assumed percentage usage of the ship in the corresponding speed/sea-state combination. The total of the percentages in the table is about 20%, which is the expected percent usage of the ship in head seas.

The dynamic pressure characteristics were determined using the theory of statistics of extremes. In order to use the theory of statistics of extreme, the parent distribution for the measured stress needs to be defined. The parent distribution of the stress is defined as the probability distribution of the random variable which is defined as the maximum stress in each stress record. Based on this definition, the statistical characteristics of the parent distributions of stress for the eight cases were determined. A transformation from measured stress to uniform lateral pressure was performed using the approach shown in reference [1]. Essentially, that approach uses a semi-empirical formulation to determine a uniformly distributed pressure which causes the measured stresses in the plating. Then, the mean value of the maximum pressure can be determined using the theory of extremes. The results are summarized in Table 2. It is reasonable to assume that the maximum dynamic pressure has the same coefficient of variation (COV) as the maximum measured stress. The mean value and COV of the extreme pressure were, then, determined for a ship usage period of 15 years at a rate of 3000 hours per year and according t the percent use presented in Table 1. The selection of the usage period of 15 years and the 3000 hours of operation per year were for the purpose of illustration. The results are also summarized in Table 2. It is evident from the Table 2 that case 8 is the most critical sea state/ship speed combination. Therefore, for this case the statistics of the maximum and extreme pressures were determined using the usage periods of 0.2, 0.5 1, 2, 5, 10, 15, 50 and 100 years. The results are shown in Table 3.

The stress due to the still-water pressure component should not be considered in the statistics of extremes analysis. Because the stresses due to still-water pressure were not measured, the mean value of the still-water pressure was determined based on draft analysis and was found to be 2.667 psi. The coefficient of variation and distribution type of still-water pressure are assumed to be 0.20 and normal, respectively (after [3]). The total pressure applied to the plate is the still-water pressure plus the extreme dynamic pressure.

The statistical characteristics of the strength of the material used in the ship, and the dimensions of the plate of interest in this study are based on those given in reference [1] and are $\sigma_{Yield}$ = 47.8 ksi, Young's Modulus of 29,774 ksi, with a plate 11.75 in. wide, 23.5 in. long, and 0.161 in. thick. The COV's estimated for these strength parameters are 0.13, 0.038, 0.05, 0.05, and 0.01, respectively.

## ASSESSMENT OF STRUCTURAL LIFE ACCORDING TO PLATE DEFORMATION

The probabilities of plastic deformation ($P_{fp}$) of a plate according to the limit state of equation (4) can determined using Monte Carlo simulation with variance reduction techniques. Conditional Expectation with Antithetic Variates variance reduction techniques were used in the analysis. A computer program was developed for this purpose. The average simulated probabilities of failure ($P_{fp}$), and the coefficients of variation of the estimate of the probability of failure COV($P_{fp}$) for different usage periods of the ship are shown in Table 3.

The end of structural life due to plate plastic deformation was defined as plastic deformations more than twice the thickness of the plate in at least 6 plates within the critical region. The critical region was defined as that area of the vessel's bottom and side plating which experiences the most pressure as result of regular operation as well as extreme event such as bottom slamming. The selection was done based on a finite element analysis provided in reference [1]. This critical area of the vessel has a total of 28 plates. These plates are assumed to experience the same loading and have approximately the same strength characteristics, therefore, have approximately the same probability of failure. The vessel is assumed to have the following inspection and maintenance strategy: The ship is inspected at the end of the first year because it is the end of the warranty period, then the ship is inspected every I years. It is assumed that the probability of failure of a plate within the warranty period is the same as the probability of failure of a plate within a usage period that is equal to the warranty period. It is also assumed that any damage found during the inspection is corrected to no damage condition. In order to demonstrate the effect of inspection interval, values of I of one and two years are considered.

Given the probability of failure of a plate ($P_{fp}$) within the inspection period I as defined previously, the probability of failure of 6 plates out of 28 plates ($P_{f6/28,I}$) can be determined using the probability mass function of the binomial distribution. The binomial distribution is based on a Bernoulli sequence of trials, i.e., failure of plates, which are assumed to be statistically independent. Actually, the events of plate failure are statistically correlated with relatively small correlation coefficients. The probability of failure of the plates is function of the correlation level. Therefore, it can be estimated in the form of upper and lower limits which correspond to coefficients of correlation ($\rho$) of one and zero, respectively. Since the events of plate failure are statistically correlated with relatively small correlation coefficients, the probability of failure of the plates in plastic deformation is closer to the lower limit. The calculations for both limits were performed; however, the results of the lower limit are used in this study. The results of $P_{fp}$ and $P_{f6/28,I}$ for inspection intervals of one and two years, and for the lower and upper limits are summarized in Table 4.

The probability of failure of the ship due to plate deformation within i structural life (SL) depends on the inspection strategy of the ship. Fo the assumed inspection strategy, the probability of failure within life can be determined. Using the probabilities of plates failure from Table 4, the probabilities of failure of the ship in structural life due to plate plastic deformation $P_{fSL}$ were determined for I = 1 and 2 years, and SL = 1, 3, 5, 11,15, 21, 31, 51 and 101 years. The results for $\rho$ = 0 are shown in Figure 2.

The lower limit on the probability of failure in structural life assessment is a more accurate estimate than the upper limit. It is evident fre Figure 2 that by reducing the inspection interval, expected structural life can be increased. This is due to the fact that at the end of each inspection interval, any reported deformation damage is to be fixed befo: sending the ship for the next usage period. However, it should be noted that these results are highly dependant on the underlying assumptions. These assumptions include, for example, the number of hours of operation per year, the percent usage in each sea state/speed combination, loading conditions, definition of end of structural life, strength characteristics, etc.

CONCLUSIONS

The structural service life of the example vessel analyzed in this paper was determined for the plate deformation failure mode. The approach presented indicates that there is a 72% chance that the vessel will not sustain enough damage in 15 years of operation to constitute reaching the "end of structural life" as defined. This result is based on the definitions of "end of structural life" provided in this paper. The methodolog and example provided in this paper are meant to be a catalyst for furthe₁ discussion and investigation into the subject. While not meant to be taken literally, they do show one manner in which probabilistic methods can be use to address the subject of life assessment. The results of the approach are highly dependent on the assumptions with which the model works.

REFERENCES

[1] Purcell, E.S., S.J. Allen, and R.J. Walker, 'Structural analysis of the U.S. Coast Guard Island Class patrol boat,' Trans., SNAME, Vol. 96, 1988.

[2] Ayyub, B.M. and G.J. White, 'Life expectancy assesment and durabilit of the Island Class boat hull structure,' U.S. Coast Guard R&D Report, September, 1988.

[3] White, G.J., and B.M. Ayyub, 'Reliability methods for ship structures,' Naval Engineers Journal, ASNE, Vol. 97, No. 4, May, 1985, pp. 86-96.

[4] Hughes, O.F., 'Design of laterally loaded plating - uniform pressure loads,' Journal of Ship Research, SNAME, Vol. 25, No. 2, June, 1981, pp. 77-89.

Table 1. Combinations of Ship Speed and Sea State

| Sea State | Low (12 kts) | Medium (24 kts) | High (29 kts) |
|---|---|---|---|
| Low (1 & 2) | Case 1 (12 kts, 3 ft) 4.0 % | Case 2 (24 kts, 3 ft) 1.7 % | Case 3 (29 kts, 3 ft) 1.0 % |
| Medium (3 & 4) | Case 4 (12 kts, 8 ft) 4.7 % | Case 5 (24 kts, 8 ft) 1.3 % | Case 6 (29 kts, 7 ft) 0.7 % |
| High ( 5 ) | Case 7 (12 kts, 10 ft) 5.3 % | Case 8 (24 kts, 10 ft) 1.0 % | Not Considered |

Table 2. Statistical Characteristics of Pressure

| Case No. | "Max." Mean Value | "Max." COV | No. of Intervals | "Extreme" Mean Value | "Extreme" COV |
|---|---|---|---|---|---|
| 1. | 1.75 psi | 0.0993 | 216000 | 2.55 psi | 0.0177 |
| 2. | 1.89 psi | 0.0993 | 91800 | 2.71 psi | 0.0186 |
| 3. | 1.99 psi | 0.0993 | 54000 | 2.83 psi | 0.0192 |
| 4. | 6.17 psi | 0.0993 | 253800 | 8.99 psi | 0.0175 |
| 5. | 6.76 psi | 0.0993 | 70200 | 9.66 psi | 0.0189 |
| 6. | 3.07 psi | 0.0993 | 37800 | 4.35 psi | 0.0196 |
| 7. | 7.63 psi | 0.0993 | 286200 | 11.13 psi | 0.0174 |
| 8. | 13.37 psi | 1.0121 | 162000 | 74.30 psi | 0.0477 |

Table 3. Reliability Assessment for a Single Plate Panel

| Usage Period | No. of Pressure Intervals | "Extreme Pressure" Estimated Mean | COV | Probability of Failure $P_{fp}$ | $COV(P_{fp})$ |
|---|---|---|---|---|---|
| 0.2 years | 2160 | 60.49 psi | 0.0732 | 0.03004 | 0.0490 |
| 0.5 years | 5400 | 63.70 psi | 0.0657 | 0.05092 | 0.0401 |
| 1 year | 10800 | 66.02 psi | 0.0610 | 0.06765 | 0.0351 |
| 2 years | 21600 | 68.24 psi | 0.0569 | 0.09403 | 0.0294 |
| 5 years | 54000 | 71.07 psi | 0.0523 | 0.13950 | 0.0284 |
| 10 years | 108000 | 73.13 psi | 0.0493 | 0.17200 | 0.0244 |
| 15 years | 162000 | 74.30 psi | 0.0477 | 0.20310 | 0.0215 |
| 50 years | 540000 | 77.67 psi | 0.0435 | 0.27760 | 0.0155 |
| 100 years | 1080000 | 79.54 psi | 0.0414 | 0.32900 | 0.0121 |

Table 4. Probability of Deforming 6 Out of 28 Plates

| Inspection Interval (years) | Upper Limit, $\rho = 1$ $P_{fp}$ | $P_{f6/28,I}$ | Lower Limit, $\rho = 0$ $P_{fp}$ | $P_{f6/28,I}$ |
|---|---|---|---|---|
| I = 1 | 0.06765 | 0.06765 | 0.06765 | 0.009895 |
| I = 2 | 0.09403 | 0.09403 | 0.09403 | 0.042719 |

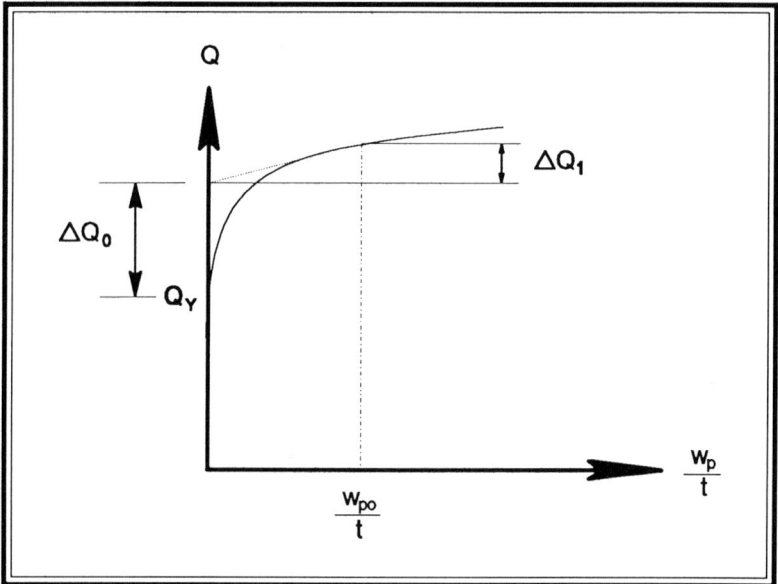

Figure 1.   Load vs. Permanant Set Plates of Finite Aspect Ratio [4]

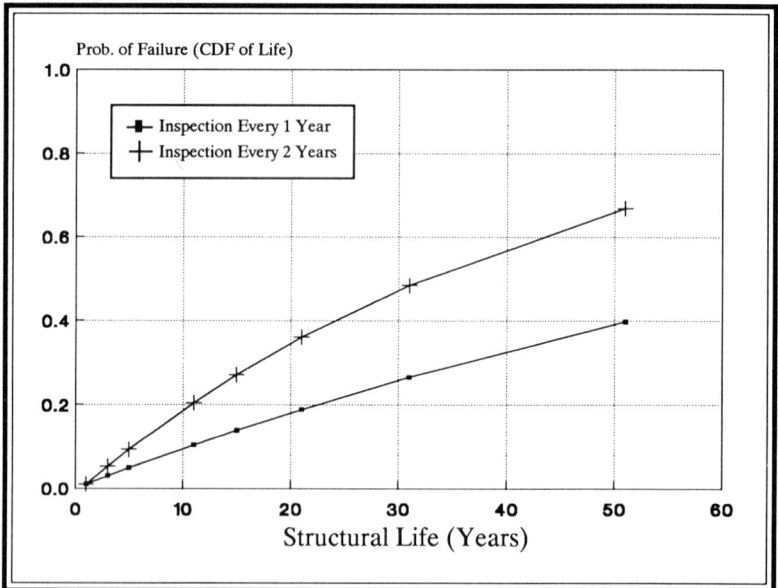

Figure 2.   Lower Limits on Structural Life Based on Plate Deformation

# 24 The life of steel structures in the marine environment

**Dr Colin J Billington,** Director, The Steel Construction Institute, UK
**Dr Ray G Guy,** Associate Director, Billington Osborne Moss Engineering Ltd, UK

A major generic design study funded by the Department of Energy and British Steel is in progress to prepare steel intensive solutions for tidal power generation barrage schemes. An early part of this study concerned the assessment of the long term performance of a variety of steel types and components in the marine environment to meet the specified design life of the barrage structures of 120 years. The corrosion performance of steel in this environment is not generally well understood and as a consequence misinformed opinion has on occasion influenced material selection to the disadvantage of steel. However there is a great deal of data available as a result of major testing programmes carried out in the UK, USA and France to provide real time corrosion performance of a variety of materials. The tests cover unprotected, coated, clad and cathodically protected steels in atmospheric, splash zone, tidal zone and fully submerged marine environments.

This paper describes the role of these tests in establishing the durability performance of steel and the choice of cost effective protection systems. The relationship between test results and long term monitoring of complete structures is described by particular reference to the performance of steel components in the La Rance tidal barrage in France.

## INTRODUCTION

All structures suffer from deterioration during their life. Where the rate of deterioration is such that the structure would become unserviceable during its operational life, maintenance is required. Many structures are designed with in built maintenance requirements; on the other hand many are built in the expectation that maintenance will not be required during the operational life. Some materials, notably steel, degrade at known and calculable rates in certain environments and therefore require protection or maintenance or both. Other materials, notably reinforced and prestressed concrete, present much more difficult analytical and chemical properties, which are less amenable to calculation and transference of experience from one situation to another. Such materials have frequently produced surprises of major and even catastrophic proportions, (eg HAC, chlorides and sulphates in the Middle East, concrete cancer).

Where deterioration is understood and predictable its costs (ie costs of prevention and/or maintenance), are assessed at the conceptual stage and allowed for in calculating the commercial return from the project. Where deterioration is less well understood and more random in occurrence, there is a tendency not to allow properly for it in assessing project viability. This approach is compounded by the separation, in many organisations, of budgetary control of capital expenditure from maintenance (operational) expenditure.

The steel versus concrete debate has been brought sharply into focus particularly in respect of maintenance and durability by consideration of major capital schemes with a long operational life which provide public services but which the present government regime expects should be privately financed (eg major bridges, power generation, water supply, leisure amenities).

One such application is for tidal power generation barrages where design lives of 120 years have been specified and where the Steel Construction Institute has been undertaking generic design studies funded by the Department of Energy and British Steel. These studies have been undertaken because it was found that steel was not being considered seriously or properly for use as the principal construction material by several of those involved in potential barrage schemes. On investigation it appeared that, in several instances, this was due to entrenched attitudes typified by a comment at the ICE Conference in 1986 to the effect that "steel corrodes in the marine environment and therefore I will not construct a steel barrage." This comment was made without assessments of corrosion rates, cost of protection and overall cost comparisons on an equal basis.

The SCI decided that it was most important to put the record straight, particularly in respect of durability and therefore have collected test data and operational experience of steel structures in the marine environment and carried out detailed studies of steel components of the La Rance power generation barrage in France, which has seen in excess of 20 years operation. This paper describes the SCI generic studies, the marine atmospheric corrosion test site and the monitoring sites for splash zone and sub-sea behaviour. It also describes the corrosion protection systems used at La Rance and the excellent operational performance and finally describes the systems proposed and costed in the SCI generic studies.

UK CORROSION TEST AND MONITORING SITES

In marine applications there are several separately identifiable zones or levels where corrosion of unprotected steel takes place at different rates. These are from top downwards:

Atmospheric zone:      zone not wetted by wave or tidal action
Splash zone:           wetted zone above mean high water level
Inter-tidal zone:      between mean high and mean low water levels
Low water zone:        between low neap and low spring tides
Fully immersed zone:   zone never exposed to the atmosphere
Below sea-bed zone:    not exposed directly to the sea.

Corrosion data collection consists both of taking measurements on specially deployed steel samples and on actual structures. In the UK the majority of atmospheric data has come from steel samples whereas the partially or fully submerged measurements have generally been taken on steel structures, the majority of which have been various forms of steel piling.

There are two marine test sites maintained by British Steel - at Rye, Sussex which is an atmospheric site situated some 15m from the high water mark, and at Skinningrove, Teeside which provides facilities for total immersion, inter-tidal and splash-zone testing. The location of the structures which are monitored in-situ is shown in Figure 1. There are thirteen sites in all, including locations close to the potential Severn and Mersey barrages.

The Rye test site is fully monitored for corrosivity, atmospheric pollution, rainfall, airborne chloride and sulphur dioxide and hours of sunshine. Corrosion rates, the durability of coatings and the performance of protection systems, are measured and can be compared with those at other inland exposure sites.

On the in-situ structures detailed corrosion studies were carried out from the atmospheric zone down to sea-bed level. In a few cases extracted piling has enabled underground corrosion rates to be measured. Residual steel thickness measurements were made using ultrasonic thickness meters. In some cases it was possible to cut out samples to confirm the ultrasonic thickness measurements. Additionally, holes drilled to provide samples for chemical analysis also gave check measurements. Samples of sea water were taken for chemical analysis and water temperature and oxygen content were also measured. Chemical analyses were also made of corrosion products. 95% maximum probable corrosion rates and average corrosion rates measured over three years from 1980 - 1982 are:

|  | 95% maximum probable rate mm/year | Average rate mm/year |
|---|---|---|
| Splash zone | 0.18 | 0.09 |
| Intertidal zone | 0.11 | 0.04 |
| Lower water zone | 0.18 | 0.09 |
| Immersion zone | 0.14 | 0.05 |
| Below sea-bed | 0.05 | 0.02 |

These rates are for unprotected structural steelwork.

## EXPERIENCE AT LA RANCE AND OTHER MARINE SITES

The use of steel in a marine environment is not unique. Although largely a concrete construction, La Rance tidal barrage in France has many steel components. As this barrage has been operating for more than 20 years, it provides the best example of the performance of steel in such conditions. Similarly, steel has been used in the W.T. Love generating station in the USA. The power plant structure was fabricated in a shipyard and installed to replace a weir, part of the Greenup Dam on the Ohio River. Steel has also been used extensively in structures in The Netherlands as part of the main North Sea defence system. Finally, steel offshore gas and oil rig structures have been installed worldwide over many years and over the last 20 years more than 300 have been installed in the North Sea.

These structures have all been designed to required durability limits with great success and illustrate the ability of steel to perform in such environments.

### Experience at La Rance

The barrage has major electromechanical components in steels, eg bulb turbine generators, liners, sluice gates, lock gates etc. These components are in excellent condition after more than 20 years service. This has been achieved essentially by the use of suitable painting systems and impressed current cathodic protection and these methods are expected to provide satisfactory corrosion protection on an indefinite basis.

Paint trials of many different systems and from many manufacturers were carried out for nine years prior to the selection of the system. Trials of a similar period were carried out to determine the choice of alloy material for the turbine blades. The cathodic protection system for the bulb generators was designed by calculation and was checked by a scale model in the laboratory to assess the correctness of the design.

### Sluice Gates

The sluice gates, made from mild steel, were originally designed to be used without cathodic protection, relying solely on the paint coating for corrosion resistance. Within one year paint breakdown of the coating was observed and a cathodic protection system was retrofitted. The cathodic protection is of an impressed current type using a platinised titanium stand-off anode, each anode providing 40A. Permanent, fixed, reference electrodes are fitted and potential measurements are made every fifteen days. Plans are being made to fit an automatic potential monitoring and control system next year.

## Bulb Turbine Generators

Two alloys were used for the turbine runners; twelve are in aluminium bronze and twelve are in 17/4 stainless steel. Twelve flush-mounted impressed anodes of platinised titanium set in polyethylene are sited in the stainless steel around the runners. Twelve more anodes are placed on the outside near the moveable blades. A further twelve anodes are sited on the bulb itself. Each group of thirty six anodes is supplied from a transformer/rectifier with 40 - 50 A and the design current density on the steel is 170 mA/m$^2$. Three fixed reference electrodes are placed midway between the anodes and the design potentials are between -800 and -1200 mV (Ag/AgCl). Adjustment of the current output to each set is made manually to ensure the correct working potential. The anodes are replaced at the rate of 8 - 10 each year, 1% failure per annum - an excellent record.

The bulb turbine unit shows no evidence whatsoever of any corrosion of the mild steel even though most of the paint system has now completely gone.

It is clear that the combination of mild steel and cathodic protection in this environment has been most suitable and has provided and will continue to provide corrosion protection indefinitely while the cathodic protection is maintained.

## Lock Gates

The lock gates are made from mild steel (A37) and are painted with tar epoxy. The choice of tar epoxy is not surprising since this system is commonly used at the present time for underwater corrosion protection because of its good protective properties, its compatibility with cathodic protection and its reasonable mechanical properties. The lock gates are still in good condition.

## Cathodic protection using sacrificial anodes

Zinc anodes are extensively used for various accessories such as, turbine-bay stop locks, chains, lock gates, floating cables, isolated structures, piling, etc., and are inspected, together with the structure potential, and changed where necessary as part of the regular maintenance programme.

## Experience from North Sea Offshore Structures

Considerable experience is available from operators of North Sea platforms, some of which have been operating for in excess of 20 years. Three basic types of subsea protection systems are used; sacrificial anodes, impressed current and a hybrid of the two. From operators' experience, several conclusions can be made.

Sacrificial anode systems (normally aluminium anodes) provide consistently good and trouble free corrosion protection for offshore structures. Care at the design stage is essential to ensure satisfactory distribution of anodes in complex regions. A design requirement is the high initial current density for polarisation, suggesting longer, thinner anodes.

The hybrid system can perform very well and with good design and monitoring systems can require no modifications or retrofitting. However, an impressed current system requires a higher level of maintenance than sacrificial systems.

All offshore operators monitor their protection systems, showing a desire to improve future designs as well as to ensure continuing even protection to existing structures. This feedback of protection system performance has resulted in the creation of a design guide to be published shortly. Although one or two early structures required retrofitting of anodes, recent designs have shown entirely acceptable protection of exposed steel surfaces for offshore structures.

## GENERIC STUDIES ON TIDAL BARRAGES

The SCI studies commenced in May 1987, with an original project duration of eighteen months, subsequently extended to twenty one months. The studies have been performed at a time when various site specific work had been carried out at potential barrage sites, mostly concentrating on the use of large concrete caissons to form the barrage.

Most studies had been carried out at the Severn and Mersey sites, although now further studies are in hand for such sites as Cardiff, Humber and Conway. The only specific study carried out into the potential for steel constructions was for the Severn barrage (1). This work was inconclusive because, as it did not allow direct comparison with the concrete designs, its impact was lost. Designers of barrages had no available data for steel designs on which to base their feasibility studies, hence the need to conduct a generic study into the most effective use of steel in tidal barrage schemes became apparent.

### Methodology

The general principle adopted in the study was to design and cost a selected central case for each component, of which a steel turbine caisson is just one example. Variations to design and hence costs were prepared for parameter variations, for example, the turbine draft tube length. The costs considered the fabrication, installation, in-service use and service life of each component. The final full-life costs were presented in graphical format.

The nature and appropriateness of steel designs are fundamentally related to the foundations, seals and installation aspects, all of which have been developed considerably in recent years in the offshore industry. The work in this study took a detailed look at such aspects and, where appropriate, fed back the results into the caisson designs. Where suitable, piled foundations were evaluated, sometimes changing the form of the caisson significantly from traditional concrete gravity caissons. The full work content is presented in Figure 2.

In order to develop steel designs and resulting costs for typical barrage elements, it was considered necessary to consider three potential barrage sites. Padstow, Mersey and Severn were selected as being three potential sites for which design data are available and which are representative of small, medium and large barrage schemes respectively. In practice, the particular sites are relatively unimportant in themselves as the results can be used to provide best estimate costs for structures at any site.

The work performed for the design of caissons as a direct substitution of the concrete caissons indicated that the construction cost was dependent on the structural volume. Hence, alternative configurations were considered which reduced the structure volume. These alternatives were fully costed and presented in a format to enable direct comparison with the results for the look-alike caissons.

To enable rapid consideration of many alternative forms of construction, the design process was automated and incorporated into a computer program BAGPUS (Barrage Generic Program for the Use of Steel). The program takes each design variable in turn, including material and fabrication rates, and produces a technically acceptable design. By investigating a number of variations, among them different types of construction techniques such as a steel-concrete-steel composite, an "optimum solution" may be found.

## Material Durability

Steel has been used successfully for many years in a corrosive environment for ships, offshore oil and gas platforms, jetties, hollow walls, lock gates, etc. It becomes therefore, not a question of if steel can be used in the construction of a large tidal power barrage, but what type of protection system should be employed.

## Corrosion

The basic protection system selected for the caissons and other structures comprises:-

1. Paint to all exposed steel surfaces in air
2. Stainless steel cladding in the splash zone
3. Impressed current cathodic protection to draught tube and exposed surfaces below water level
4. Sacrificial aluminium anodes cathodic protection to base, joints between adjacent caissons, ballast compartments and gates
5. Temporary paint system to all bare steel areas.

The design life is assumed to be 120 years but any chosen system requires some maintenance during this time. The impressed current system allows for monitoring during the barrage life, but this is relatively minimal and the gates will require changeout of the anodes after about 25 years. The drain on the impressed current system allows for the surface areas of gates, bulb turbines, etc.

## Erosion

The erosion of elements in a sediment carrying turbulent flow through a draught tube is a primary consideration for a tidal barrage. Whatever the construction material, erosion must be considered in the design. For a concrete structure there will be many steel or other metal components, for example the runner blades, the bulb and sections of the venturi.

An investigation was made into the experiences of various organisations with structures in erosive flow media. Several interesting observations can be made:

1. Whatever the construction material, the single most detrimental factor to degradation by erosion is boundary discontinuities in high velocity regions.

2. The rate of erosion can only be assessed on a site specific basis because it depends on the type and amount of material carried (in addition to the velocity) which can vary from site to site.

3. The rate of erosion is an order of magnitude different between cathodically protected steel and freely corroding steel.

4. It is uncertain how much sediment will be carried through a tidal barrage in the years after construction, making estimates of erosion uncertain.

212

## Study Results

Costs for caissons were developed for many alternative configurations during the course of the study. The breakdown of costs into constituent parts is shown for typical caisson types in Figure 3. It can be seen that cathodic protection (CP) comprises approximately 10% of the structure cost, although this does vary slightly from scheme to scheme.

## CONCLUSIONS

The work demonstrated that steel may be designed for use in a marine environment with adequate design life for strength and durability. As well as being able to compete with other forms of construction for look-alike designs, it is shown that when the attributes of steel, such as its high strength to weight ratio, are applied to the designs, more economical configurations are possible. The nature of a steel construction, having good floating characteristics thereby reducing installation constraints, permits more efficient designs which may not be possible with the lower strength to weight structural materials such as concrete.

The results of the study warrant serious consideration by those involved in tidal power barrage schemes. Only if this is done can a full commercial evaluation be made for developers or sponsors.

## REFERENCE

(1) Steel caissons for the Severn Barrage, Report to the Department of Energy, Energy Technology Support Unit, Harwell, by Yard Limited, YM4223 1984.

213

Figure 1.   LOCATION OF UK TEST SITES FOR
IN-SITU CORROSION INVESTIGATIONS

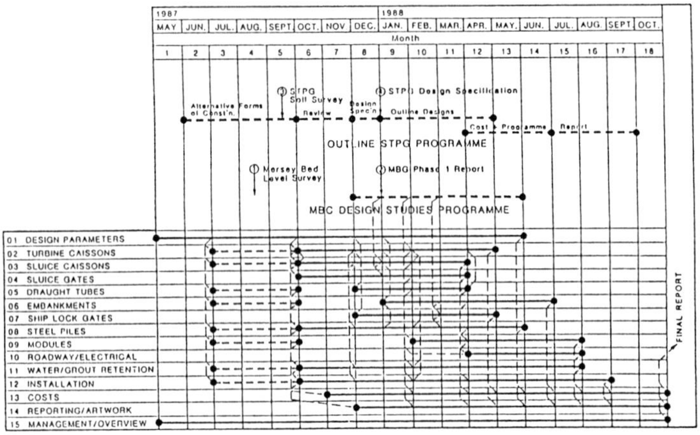

Figure 2.   STUDY PROGRAMME

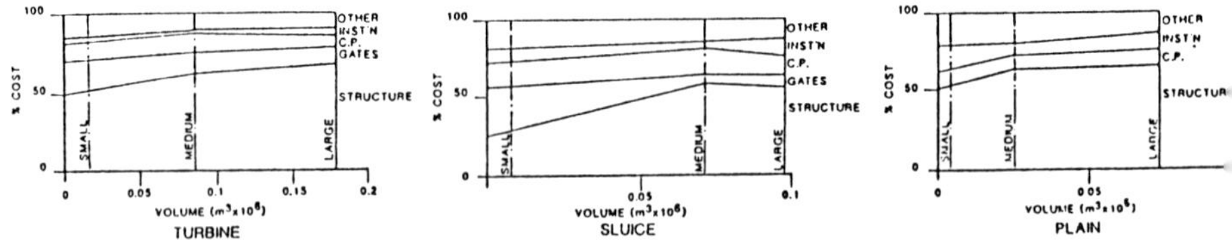

Figure 3.   BREAKDOWN OF COSTS FOR CAISSONS

R C de Vekey, Building Research Establishment, U.K.


If I might be allowed a touch of whimsey - what is the power consumption of the La-Rance cathodic protection system (plus the power to make sacrificial anodes) as a proportion of the power output of the system?

Author's reply

The cathodic protection (CP) system installed consumes 20Kw of power and this can be compared with the barrage rated output of 240MW. The CP uses 0.35% (1/30th of 1%) of the total annual power production of the barrage; this is probably the best comparison to use as the barrage does not generate continuously.

The power to make sacrificial anodes cannot be rated in the same way as they are made independently of the barrage site and would really be assessed on capital cost. When costs of anodes are compared with generated power costs, they again are shown to be relatively insignificant. I hope these comments are of assistance to Mr de Vekey.

# 25 Time dependent strength properties of materials

**L Sentler,** Lund Institute of Technology, Lund, Sweden

Most structural materials are mainly evaluated in terms of their short term characteristics even though the long term behaviour, which reflects an actual use, might be different. Both the short term and the long term characteristics are addressed based on the actual failure behaviour of materials in terms of strain limitations. Because these limitations in a material depend on the existence of flaws of random size and location, a stochastic viscoelastic strain characterization is utilized.

## STRENGTH CHARACTERIZATION OF MATERIALS

The properties of materials and especially the strength properties can be dealt with in two principally different ways. The traditional way within the engineering profession is to use material models which are the logical consequence of a calculation theory. This is the case for the theory of elasticity and the theory of plasticity. Such material models have many advantages because of their simplicity which makes the calculation of stresses and strains a relatively simple matter in most applications. But the behaviour of real materials is often much more complex than that given by these simple material models. This is the case when time dependent damage accumulation which affects the service life has to be considered (1). In such applications it is desirable to use more advanced material characterizations, preferably based on actual material behaviour. In doing so it is also natural to make use of statistical concepts because of the nature of the problem.

## THE DEFORMATION AND FAILURE BEHAVIOUR

### The material structure
The constituents within a material and the bonding between them determine the basic material characteristics. Of significant importance is the bonding which can be of two types, strong primary bonds which are the result of attraction forces between atoms, or week secondary bonds, which keep molecules together. Dependent on the type of bonding certain basic material characteristics can be explained. For engineering purposes, however, it is often more appropriate to make a slightly different subdivision in terms of metallic bonds and other bonding mechanisms.

The bonding present for metals is of primary type and it involves all the valence electrons from all the atoms in the aggregate being shared equally between all atoms. A peace of metal therefore can be seen as a collection of positively charged ions bonded together by freely moving cloud of negative valence electrons. The special with this type of bonding is that it permits planes of ions to be moved along certain planes without causing any disruption in the bonding mechanism. This behaviour will make the material tougher and it is often referred to as a strain hardening.

In other materials than those with the metallic bonding mechanism the constituents are kept in their places in a much more rigid way. In these materials it is not possible with a slip in the material and if a bond is broken there will be a disruption in the bonding mechanisms. This will create small micro cracks in the material which will weaken it and this behaviour is sometimes referred to as a strain softening.

The property of a material will initially depend on its constituents and the bonding in between them. But no material is perfect. A material will contain flaws, both inherent with random size and location, and additional imperfections introduced in the production of some structural unit. These flaws will in particular influence the bonding within a material and cause a failure to occur at a much lower stress leve. than for a theoretically perfect solid. But the nature of these flaws is such that larger flaws will be present for increasing volumes being affected and existing flaws tend to grow in time when a sufficiently high stress is present. In addition there will also be a certain randomness in the material characteristics. Since the flaws in a material determine the strength characteristics it seems rational to make use of this knowledge in the characterization of the strength of a material.

The strain response
A body subject to stress will deform. This is reflected in a reduced or an increased distance between the constituents in the body. Dependent on the type of stress or the orientation of the constituents in the material relative to the applied stress this will cause two types of deformation or strain, namely volumetric strain or distortion strain. The effect of these two types of strain is different, but since they depend on the same type of bonding forces within the material, they are coupled.

A volumetric strain occurs when the constituents in a material are separated from each other in the direction of a stress. This is the initial response when an external normal stress is applied. But because the orientation of the constituents in a material never is in perfect symmetry there will also be a distortion strain. The form of response in terms of volumetric strain and distortion strain which can be expected depends on the character and the magnitude of the stress system which is acting on the material. In this respect there is a significant difference between the short term behaviour and the long term behaviour. A volumetric response seems only to be stable at low stress levels. For higher stress levels there will be a smooth transition in time from volumetric strain to distortion strain for an external normal stress. In the case of an external shear stress there will be a distortion strain response directly.

The short term strength
Most information about the strength and the failure behaviour of materials are obtained in tests where the stress is increasing continuously until a failure occurs. It is a well known fact that the failure stress in such experiments is inversely proportional to the size of the specimen. This reflects an increasing possibility of finding a major flaw for increasing volumes under stress and it is most often explained with the Weibull strength theory (2). There will also be a time dependence of similar type (3), which is illustrated in Fig. 1.

The long term strength
In reality no structural member is loaded to failure deliberately. Instead it is subject to a stress which may be more or less constant or it has a certain variability in time. The response of a member subject to such loading conditions might be very different from that obtained in a short term loading test (4).

A member subject to a constant, sustained stress will, after the instantaneous strain at loading, respond with creep. At low stress levels only with primary creep, but for an initial stress above a certain threshold stress, primary creep will be followed by secondary creep and eventually by tertiary creep under certain conditions. The domains where different types of creep can be expected are visualized in Fig. 1.

Primary creep is a delayed volumetric strain which is recoverable at unloading. Secondary creep, however, which is a form of distortion strain, is not recoverable after unloading. Secondary creep therefore is a damage accumulation in the material, which, if the loading time is sufficiently long, will lead to a rupture. This reflects the growing importance of flaws in time. A similar effect will also arise for variable loading conditions. A damage accumulation will reduce the available distortion strain capacity and decrease the rupture stress. It is thus important to consider this behaviour in an appropriate manner so that future predictions of the structural behaviour can be made.

## THE FAILURE CHARACTERIZATION

The minimum strain principle
In a characterization of the strength of a material the strain behaviour is important and especially the strain at failure (5). The available strain capacity depends initially on the bonding within a material. For a body subject to a one dimensional state of stress the strain behaviour will not only depend on the stress level, $\sigma$, but also on the amount of stress introduced in a body, V, and the duration of this stress, D, because of the existence of flaws in the material. In general terms this can be expressed as

$$\epsilon = g(\sigma;V,D) \tag{1}$$

where g(.) is a strain function. The strain in a material can take place in two different ways, as volumetric strain or as distortion strain. The strain function in eq 1 therefore has to be expressed in terms of these two components as

$$\epsilon = g_1(\sigma_v;V,D) + g_2(\sigma_d;V,D) \tag{2}$$

where $g_1(.)$ and $g_2(.)$ are volumetric strain and distortion strain functions respectively. In this formulation the applied stress in terms of a volumetric stress $\sigma_v$ or a distortion stress $\sigma_d$ are the dependent variables, and the volume V under stress and the duration D of this stress are independent variables.

A failure may occur as a volumetric strain failure, which reflects a continuously increasing stress, or as a distortion strain failure, which reflects creep to failure. In the first case such a failure will be directly dependent on existing flaws and thus size dependent while it in the second case will depend on the growth of existing flaws and thus time dependent. The ultimate strain at failure, however, is not determined by the average material behaviour as it is expressed in eq 2. Instead, due to the existence of flaws in the material, the ultimate strain will reflect the weakest part of the material. This behaviour is considered in the most rational way within the statistical theory of extreme values. By the theorem of multiplication of probabilities the statistical distribution of the minimum value, or no failure, of the two strain components can be expressed as

$$F(\sigma) = 1 - \left[1 - F_1(\sigma_v;D)\right]^{nV} \left[1 - F_2(\sigma_d;V)\right]^{mD} \tag{3}$$

where $F_1(.)$ and $F_2(.)$ are the cumulative statistical distribution functions of volumetric strain for a volume V and the distortion strain during a duration D. The arguments in this distribution will be $g_1(.)$ and $g_2(.)$ in eq 2 which have to meet certain continuity requirements in space and in time. Based on a linear strain response for a small stress, these functions are assumed to have the form

$$g_1(.) = c_{01}\sigma_v(D/D_0)^{1/h} \qquad g_2(.) = c_{02}\sigma_d(V/V_0)^{1/k} \qquad (4)$$

where $D_0$ and $V_0$ are reference values of duration and volume respectively; h and k are time and size parameters respectively; $c_{01}$ and $c_{02}$ are constants. For nV and mD large, eq 3 can be written in terms of the asymptotical extreme value distribution of type 3, the Weibull distribution as

$$F(\sigma) = 1 - \exp(-c_{01}nV((D/D_0)^{1/h}\sigma_v)^k - c_{02}mD((V/V_0)^{1/k}\sigma_d)^h) \qquad (5)$$

where the same size and time parameters as above are used. An inspection of the two strain terms expressed in extreme value form reveals, if the time aspect is considered, that the volumetric strain reflects primary creep and the distortion strain secondary creep. The secondary creep behaviour is similar to that in the Norton power law. From experiments it is known that primary creep and secondary creep only take place over certain thresholds, here called $r_1$ and $r_2$, where $r_1 < r_2$ for normal structural materials. By taking this into account, and at the same time normalizing eq 5, the following result is obtained

$$F(\sigma) = 0 \qquad\qquad\qquad\qquad \sigma \le r_1 \qquad (6)$$

$$F(\sigma) = 1 - \exp\left(-\frac{V}{V_0}\left[\frac{D}{D_0}\right]^{k/h}\left[\frac{\sigma_v - r_1}{c_1}\right]^k\right) \qquad r_1 < \sigma \le r_2 \qquad (7)$$

$$F(\sigma) = 1 - \exp\left(-\frac{V}{V_0}\left[\frac{D}{D_0}\right]^{k/h}\left[\frac{\sigma_v - r_1}{c_1}\right]^k - \frac{D}{D_0}\left[\frac{V}{V_0}\right]^{h/k}\left[\frac{\sigma_d - r_2}{c_2}\right]^h\right)$$
$$r_2 < \sigma \qquad (8)$$

where $c_1$ and $c_2$ are new constants which normalize the expressions so that proper statistical distributions are obtained. The failure behaviour, which is given by eq 8, indicates two extreme possibilities. A failure may occur essentially because of volumetric strain, which will result in a volume dependence of the same type as that given in the Weibull theory. For a distortion strain failure there will be a time dependence of similar type. Each part in eq 8 can be visualized as curves in the strain − time domain which will have the appearance shown in Fig. 1.

Most materials have the ability to absorb both volumetric strain and distortion strain. The transition from one domain to another will therefore not necessarily lead to a complete failure. This is illustrated in Fig. 1 with curves reflecting a short term loading condition and a creep to failure condition. But the transition from one domain to another means that the failure behaviour will be different. The final failure will be brittle if all distortion strain is used up, which is the case in creep, or ductile if the volumetric strain is used up first and a distortion strain capacity remains. Thus, a brittle or a ductile failure behaviour is not a material characteristic but it depends on the loading conditions.

Damage accumulation
There is a significant difference between volumetric strain and distortion strain. This is most evident after creep if both primary and secondary creep have taken place. After an unloading the primary creep will be recovered in delayed form but not the secondary creep. The distortion strain which takes place during secondary creep is a permanent damage accumulation. Since a material has a certain distortion strain

damage accumulation capacity from the beginning secondary creep will reduce the rupture stress.

Under long term loading conditions a damage accumulation will occur if the stress level is above a certain threshold stress. By only considering the second term in eq 8, which reflects distortion strain, such a damage accumulation can be expressed as

$$\sum_{i=1}^{n} \frac{D_i}{D_0}\left[\frac{V_i}{V_0}\right]^{h/k}\left[\frac{\sigma_{di} - r_2}{c_2}\right]^h = -\ln(1 - p) \tag{9}$$

which is a damage accumulation procedure where i refers to a number of events during which a sustained stress has been effective and distortion strain has taken place, and p is a chosen failure probability.

The damage accumulation can be of two types dependent on the material and the loading conditions. Most metallic materials have the ability to deform by slipping. Other material do not have this ability in general and instead there will be a crack growth in the material. But the loading conditions are also very important since a viscous slip in a material only can take place under relatively steady state stress conditions. This is especially noticeable for variable loading conditions when fatigue takes place and the manifestation always is a crack growth. The damage accumulation procedure in eq 9 is similar to the Palmgren − Minor damage accumulation principle but based on a better material characterization. For variable loading conditions special considerations in the evaluation of $V_i$ and $D_i$ are required (5).

THE STRENGTH

In many applications it is desirable to have expressions of the strength instead of the strain. This can be obtained in terms of the mean value and the variance of $\sigma$ in eq 8. The mean value is obtained from

$$E[R] = \int_r^\infty R\, f(R)dR = r + \int_r^\infty (1 - F(R))dR \tag{10}$$

where $\sigma$ is substituted with R, the rupture stress, and r is the appropriate threshold stress. Because $\sigma$ can be expressed in terms of volumetric stress or distortion stress there will be two solutions. For uniform stress conditions in space and in time such solutions are obtained with the help of a suitable variable transformation. For a volumetric strain failure this results in

$$E[R_v] = r_1 + c_1\left[\frac{V_0}{V}\right]^{1/k}\left[\frac{D_0}{D}\right]^{1/h}\int_0^\infty \exp(-u^k - \mu u^h)du \tag{11}$$

where $R_v$ is the volumetric failure stress. The corresponding expression for a distortion strain failure is

$$E[R_d] = r_2 + c_2\left[\frac{V_0}{V}\right]^{1/k}\left[\frac{D_0}{D}\right]^{1/h}\int_0^\infty \exp(-u^h - 1/\mu u^k)du \tag{12}$$

where $R_d$ is the distortion failure stress. The size and the time dependence in eq 11 and eq 12 are similar and the main difference lies in the two integrals. The behaviour of these two integrals is determined from the indicator $\mu$ which will depend on the relation $\sigma_d/\sigma_v$. For $\mu = 0$, which reflects a situation with only volumetric strain, the integral in eq 11 can be expressed as $\Gamma(1 + 1/k)$ where $\Gamma$ is the gamma function. In the other extreme when $\mu = \infty$, which reflects pure distortion strain, the integral in eq 12 can be expressed as $\Gamma(1 + 1/h)$. For values of $\mu$ increasing from zero the integral

220

in eq 11 will decrease in value while the integral in eq 12 will increase in value. This reflects the gradual transition from volumetric strain to distortion strain.

The size and the time dependence in eq 11 and eq 12 are similar. By making another variable transformation it is possible to express both equations in a similar form as

$$R = R_0 \left( a + b \left[ \frac{V_0}{V} \right]^{1/k} \left[ \frac{D_0}{D} \right]^{1/h} \right) \tag{13}$$

where $R_0$ is a reference strength which corresponds to the volume $V_0$ under stress and the duration $D_0$ of this stress, and $a$ and $b$ are new constants whose sum equals one. The mean failure stress given by eq 13 is shown in Fig. 2 where it can be seen that for increasing volumes under stress and for increasing durations the failure stress will decrease. The corresponding variability of the failure stress expressed in terms of the coefficient of variation, the c.o.v., is

$$c.o.v. = \frac{b \left[ \frac{V_0}{V} \right]^{1/k} \left[ \frac{D_0}{D} \right]^{1/h}}{a + b \left[ \frac{V_0}{V} \right]^{1/k} \left[ \frac{D_0}{D} \right]^{1/h}} G(k,h,\mu) \tag{14}$$

where $G(k,h,\mu)$ involves gamma functions which may be expressed in closed form under certain circumstances. This is the case for $\mu = 0$ when $G(.) \simeq \pi/(k\sqrt{6})$ or for $\mu = \infty$ when $G(.) \simeq \pi/(h\sqrt{6})$.

## APPLICATIONS

### General
All materials are viscoelastic in nature because they contain flaws. There will thus be a size and a time dependence in their strength characteristics. But the size and the time dependence will differ significantly between materials as well as the available volumetric strain and distortion strain capacities. For a complete description of the strength characteristics it is necessary with more information than what normally is the case.

### Parameter estimates
The main parameters in this viscoelastic formulation of the strength of materials are the size parameter $k$ and the time parameter $h$. These parameters are most easily determined from tests where the failure is either a volumetric strain failure or a distortion strain failure. This can be done directly from eq 13 if different volumes under stress or different durations have been utilized respectively. But it is also possible to use eq 14 if information about the variability in the test results are available. The parameters $a$ and $b$ will depend on the type of failure. For a volumetric strain failure it is reasonable to assume that $a$ is equal to zero because primary creep seems to take place at any stress level. In a distortion strain failure the value of the parameter $a$ will take a value between 0 and 1 which is typical for a specific material. This value reflects the shear strength of a material and it appears in applications as the threshold below which no damage accumulation takes place, for instance in fatigue.

### Time dependence in short term tests of concrete
In experiments where different rates of loading have been used it is possible to demonstrate the time dependence. The result from such an experiment (3) is shown in Fig. 3 where specimens of size 100 x 100 x 300 mm were loaded to failure at two different strain rates. It can be seen directly that the mean rupture stress is decreasing when the duration is increasing. If it is assumed that the rupture is mainly

a volumetric strain failure, then the parameter a will be zero, and the variability dependent on the size effect. From the difference in the rupture stresses the time parameter h can be estimated to be around 25 from eq 13, and the variability indicates that the size parameter k is around 17 based on eq 14.

Time dependence in long term tests of concrete
Very few tests have been performed where concrete specimens have been subject to a sustained stress. In one of these tests (7), which is shown in Fig. 4, durations of up to 30 years have been included. The reference strength is the rupture strength in a standardized test with a duration of approximately 2 minutes. If it is assumed that the threshold stress above which distortion strain takes place is $0.25\ R_0$, which reflects the shear strength of concrete (8), then the time constant h can be estimated to around 30 from eq 13 for this particular concrete.

REFERENCES

1   Sentler, L. 'Service life predictions of concrete structures' Durability of Building Materials 5 (1987), pp 81 − 98.

2   Weibull, W. 'A statistical theory of the strength of materials' Proceedings of the Royal Swedish Engineering Research, No. 153, (1939).

3   Mihashi, H. and Wittmann, F.H. 'Stochastic approach to study the influence of the rate of loading on the strength of concrete' Heron, Vol. 25, No. 3, (1980).

4   Pomeroy, C.D.(Ed.) Creep of engineering materials, A Journal of Strain Analysis Monograph, Mech.Eng.Publ.Ltd., London, (1978).

5   Sentler, L. A strength theory for viscoelastic materials, Document D9:1987, Swedish Council for Building Research, Stockholm.

6   Neville, A.M. Properties of concrete 3rd ed. Pitman Publ Ltd., London, (1977).

7   Troxell, G.E., Davis, H.E. and Kelly, J.W. Composition and properties of concrete, McGraw−Hill (1968).

8   Graf, O., Albrecht, W. and Schäffler, H. Die Eigenschaften des Betons, Springer Verlag, (1960).

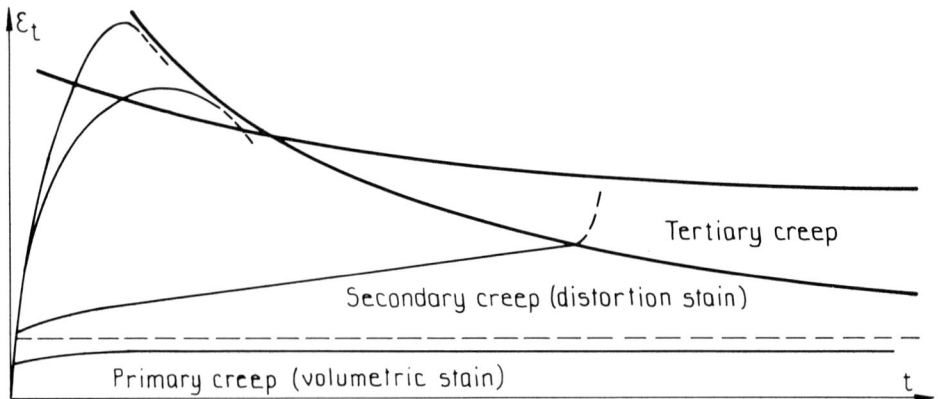

Figure 1   The domains where different types of stable strain and creep are expected. The heavy curves reflect volumetric and distortion strain failure bounds and the broken line the threshold below which no distortion strain damage accumulation takes place.

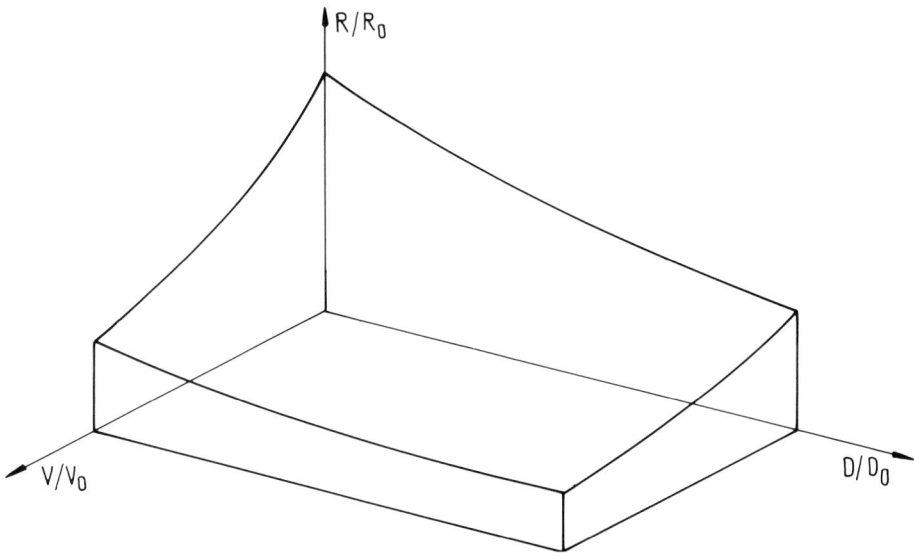

Figure 2    The rupture stress as a function of the volume under stress and the duration of the stress expressed in normalized form.

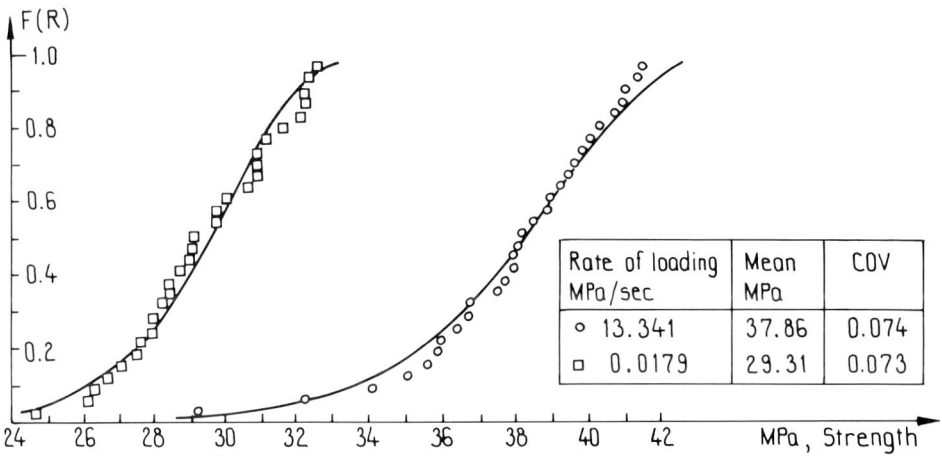

| Rate of loading MPa/sec | Mean MPa | COV |
|---|---|---|
| o  13.341 | 37.86 | 0.074 |
| □  0.0179 | 29.31 | 0.073 |

Figure 3    The influence of the rate of stress on the rupture stress in short term loading.  From (3).

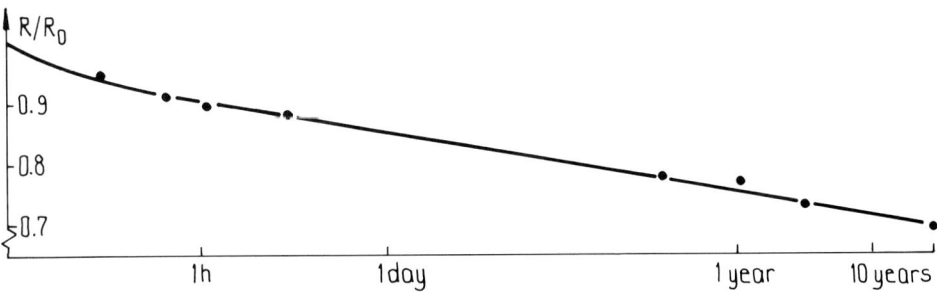

Figure 4    The reduction of the failure stress as a function of the duration of a sustained stress.  From (7).

J W Smith, University of Bristol, UK

Please would Dr Sentler explain why the general expression for strain behaviour includes volume as an independent parameter, if flaws are considered to be uniformly distributed.

In addition, in the fatigue behaviour of metals and polymers there is often observed to be distinct stages of (i) crack initiation and (ii) crack propagation. Some people believe that (i) is actually the same as (ii) but at a sub-visible scale. Does Dr Sentler's theory unify the behaviour over both stages?

## Author's reply

The strain induced in a body is directly dependent on the stress. But the strain response will also depend on an effective duration of the stress and an effective volume which is affected by the stress because of the influence of flaws. Mathematically, this will reflect an integral surface formulation where the stress is the dependent variable, and duration and volume are the indepedent variables which form a failure surface. This independence is more formal than real since both the effective values of duration and volume will depend on the flaw characteristics of a material. For a volumetric strain response the effective values are given by

$$V = \int \int \int \left[ \frac{\sigma_v(x,y,z;t)-r_1}{c_1} \right]^k dxdydz \qquad D = \int \left[ \frac{\sigma_v(t;x,y,z)-r_1}{c_1} \right]^h dt \quad (15)$$

where the integration is made over the appropriate intervals. Similar equations can be derived for the influence of the distortion strain. (These equations are not included in the original paper due to lack of space).

The crack growth which is observed in strain softening is similar to creep deformation in terms of primary and secondary creep which is observed in strain hardening, primarily for metals, under relatively steady state stress conditions. For this reason, it is appropriate to talk about a crack initiation phase, which reflects primary creep, and a crack propagation phase, which reflects secondary creep. The Paris-Erdogan law for crack propagation, the Norton creep law and the last part of eq. 8 are all of similar type and they reflect a distortion strain in the material.

**R C de Vekey, Building Research Establishment, U.K.**

If the philosophy outlined was applied to masonry would the statistical flaw distribution be affected by the joints?

Most flexural applications for masonry involve transient loading, e.g. wind, seismic. I presume that the time dependent element would not apply to this case?

Author's reply

For composite materials the stochastic strength theory can be applied to each of the materials involved. Because we consider a minimum value formulation the weakest material for a given volume under stress and the corresponding duration will be decisive. In the case of masonry the mortar in the joints will normally determine the strength for long term loading conditions. This is because the strength of mortar normally is lower than for bricks. But this may not be the case for very short loading durations, like for wind gusts, for which the strength of mortar will increase more than for bricks based on eq. 13 because of a more pronounced size and time dependence. With a known time behaviour of the stress the effective duration can be evaluated from eq. 15.

# 26 Chloride-induced corrosion of steel in concrete – investigations with a concrete corrosion cell

**P Schiessl and M Raupach,** Institute for Building Research, Technical University of Aachen, West Germany

For ascertaining the significant parameters influencing the chloride induced corrosion of steel in concrete a concrete corrosion testing cell has been developed. The macrocell currents which are in proper proportion to the rate of metal removed by corrosion can directly be measured between two steel electrodes in concrete. The results show that the processes caused by the chloride induced corrosion of steel in concrete are amenable to measurement in a way relevant to the conditions in actual practise.

## 1 INTRODUCTION

Although the reinforcement generally is protected against corrosion by the alcalinity of the concrete cover, some reinforced concrete structures, e. g. buildings in the oceans, bridges, parking decks or supporting walls, have been damaged due to the action of chlorides during the last few years.

The mechanisms of corrosion of steel in concrete are very complexe, especially in the case of chloride induced corrosion. To be able to estimate the durability of reinforced concrete structures, it is necessary to know the corrosion mechanisms and the main parameters influencing the critical chloride content and the corrosion rate.

During the last years a concrete-corrosion cell has been developed, that gives answer to relevant questions with respect to corrosion of steel in concrete, especially to questions in connection with chloride induced macrocell corrosion (1), (2).

## 2    ELECTROCHEMICAL PRINCIPLES
## 2.1  Protective effect of concrete

Reinforcing steel embedded in concrete is durably protected against corrosion. As known, this is due to the alkalinity of the pore solution in
concrete (pH $\geqslant$ 12,5). In this range of pH-values a so called passive film
of oxides which prevents the anodic dissolution of iron is formed on the
surface of the steel. If the pH-value falls below a value about 10 due to
the carbonation of the concrete or if the chloride content at the steel
surface exceeds a critical limit value, the corrosion protection gets
lost (depassivation). If sufficient oxygen and moisture are available
corrosion of the reinforcing steel may occur.

## 2.2    Corrosion of steel in concrete

Corrosion of steel in concrete is an electrochemical process which can be
devided in two sub-processes. Fig. 1 shows schematically the processes in
the case of depassivation due to chlorides. The actual dissolution of
iron takes place at the anode. Iron ions enter into solution as $Fe^{++}$ and
the free electrons pass through the steel to the cathode. There the free
electrons react at the concrete-steel interface with water and oxygen to
hydroxyl ions which migrate through the concrete back to the anode and
form rust products. Normally, this rust products can be found near the
anode. The corrosion process is comparable to a battery with a short circuit.

If anodes and cathodes are very small and placed closely side by side
(microcell elements) the progress of corrosion occurs evenly along the
steel surface (general corrosion). This can happen e.g. when the front of
the carbonation of concrete has reached large areas of the steel surface.
Chloride induced corrosion generally leads to small anodes and large
cathodes which can even be placed wide apart. These macrocells often have
high corrosion rates at the anodes (pitting corrosion).

## 3  DESCRIPTION OF THE CONCRETE CORROSION CELL

The features of the concrete corrosion cell are shown in Fig. 3. At
first, a layer of concrete containing chloride is placed in an air- and
watertight vessel. After the hardening of the concrete, a chloride-free
second layer of concrete is placed upon the first. Steel electrodes are
installed in these two concretes. The electrodes are conductively connected, enabling the macrocell current flowing between them to be measured.

The electrode installed lower down works as the anode, provided that the
chloride content is above the critical value. Besides the chloride content, the lack of oxygen due to the location of this electrode supports
the anodic action. The second electrode is installed near the upper surface of the chloride-free concrete and, being protected by the alkalinity
of the concrete, it only can work as the cathode.

The macrocell current is directly measurable in the external conductive connection between anode and cathode. The rate of corrosive removal of metal is in proper proportion to the electric charge carried:
One ampere-hour corresponds to 1.04 g of metal removal at the anode.
The relation between macrocell current and steel removal at the anode has been investigated in (1). It was found out, that there is a very good agreement between the calculated (intregration of macrocell currents) and the directly measured weight loss.

## 4  EXPERIMENTAL RESULTS
### 4.1  Basic informations

The macrocell currents have been measured with an instrumentation that allows to store the currents of 64 corrosion cells on a personal computer every minute. The electrical resistance of the instrumentation is 10 $\Omega$, so that the influence on the macrocell current can be negleted.

The investigations described in this article have been carried out with 400 cm$^2$ large cathodes and 20 cm$^2$ large anodes placed in a distance of 7,5 cm from one another. The influence of the geometrical design of the corrosion cell has been investigated in (1).

The concrete mix design is 300 kg/m$^3$ OPC 35 F, w/c = 0,60 and aggregate grade curve AB 16 if there is no special remark.

Generally the temperature was 20 ± 1 °C and the relative humidity 80 ± 5 %. The curing of the concrete corrosion cells has been carried out in a room with a relative humidity near 100 % for two days.

### 4.2  Interpretation and assessment of the measured macrocell currents

The usual behaviour of the macrocell currents is schematically shown in Fig. 3 and can be devided into four stages.

- Stage 1: The macrocell currents are caused by the potential differences between the two electrodes after concreting. The electron-flow is poss-ible, because the passive layer has not competely been developed in this period.

- Stage 2: A few hours after the casting of the concrete the chloride in-duced depassivation of the steel surface at the anode leads to a steep increase of the macrocell current. In this stage the first pits start to grow with a high rate of iron removal. The reasons for this behav-iour are very complexe and will not be considered here. A detailed de-scription of the present state of knowledge on these matters is given in (3).

- Stage 3: In this period the macrocell-currents decrease. On the one hand it is due to the progress of hydration which influences the chemi-cal composition of the pore solution, expecially the pH-value, and the porosity of the concrete. And on the other hand it is due to chloride diffusion, binding processes and the decreasing moisture content of the concrete which is a consequence of drying. The conditions stabilize in this stage.

- Stage 4: The third stage leads to a stage of slow and nearly constant decrease of the macrocell current. This stage cannot exactly be separated from the third stage on account of the same causes.

The investigations in (1) show that the increase of the electrolytic resistance in the course of time is not the only reason for the decrease of the macrocell current. Furthermore the increase of the polarization resistances of the anode and the cathode plays an important part.

In the following diagram 100 microampere (µA) correspond to a corrosion rate of 911 mg iron removal per year. This is equivalent to 58 µm of even removal per year, but the local rates of removal in the pits are much higher. It has to be considered that this data are only relevant for the geometry of the corrosion cell described above.

## 4.3  Effect of chloride content and of temperature

The macrocell currents of four corrosion cells with different amounts of chloride are shown in Fig. 4. About 75 days after having cast the concrete the air temperature was raised. This led to a significant increase of the macrocell currents. It can be concluded, that the corrosion behaviour of steel in concrete is considerably influenced by the temperature.

After the stabilization of the macrocell currents they increased almost proportionally to the content of added salt. The addition of 0,5 % of chlorides relative to the cement content didn't initiate corrosion of the reinforcement under the described conditions.

## 4.4  Effect of water cement ratio

Fig. 5 shows the influence of the water cement ratio on the weight loss which was calculated by integration of the curve of the macrocell currents versus storage time. As to be expected, the reduction of the water cement ratio led to smaller weight losses of the steel, because the permeability of the concrete decreased.

## 4.5  Effect of the water content of the concrete

A special test with a corrosion cell, which was stored in air up to an age of about two years, is represented in Fig. 6. The concrete surface was kept covered with water for 100 days, so that any acces of air to the concrete was prevented. Due to the decrease of the electrolytic resistance the macrocell current increased very rapidly after the application of water and remained very high during the first two weeks. Later the lack of oxygen at the cathode causes a decrease of the macrocell current until corrosion stops about two months after the water application. After having removed the water the corrosion cells were stored at 55 % r. h. The macrocell current increased again, because oxygen had reached the cathode. This test shows that the water content of the concrete has a decisive influence on the corrosion rate of steel in concrete.

## 5 CONCLUDING REMARKS AND FURTHER OUTLOOK

The tests reported here show that the recently developed concrete corrosion cell is eminently suitable for the investigation of fundamental parameters affecting the corrosion of steel in concrete.

With this technique it is possible to answer many open questions associated with corrosion and corrosion protection of reinforcement (1), (2). For example, the effectiveness of coatings on reinforcing bars as preventive corrosion protection has already been successfully investigated with this method.

Further research with partly modified design of the testing cell and ambient conditions is to be carried out. More particulary, the following influencing parameters will be investigated:

- Critical chloride content and corrosion rate due to chloride penetration from the exterior, as a function of the concrete mix composition and the ambient conditions.

- Effect of the quality of the concrete cover – thickness and permeability – on the steel (including the influences due to curing).

- Effect of the concrete mix composition, particularly the type of cement, additives and admixtures.

- Corrosion mechanisms, particularly cathodic action in cracked concrete. It would appear especially important to study the effect of continuous splitting cracks through which chloride-bearing water can seep (e.g., cracks due to restrain stresses in parking decks).

- Danger of macro corrosion cell formation in the vicinity of local refurbishment work.

## Literature

(1) Schiessl, P. ; Schwarzkopf, M.: Chloridinduzierte Korrosion von Stahl in Beton. In: Betonwerk und Fertigteiltechnik 52 (1986), Nr. 10, S. 626-635

(2) Schiessl, P. ; Raupach, M.: Chloridinduzierte Korrosion von Stahl in Beton. In: Beton-Informationen 28 (1988), Nr. 3/4, S. 33-45

(3) RILEM TC 60-CSC. Corrosion of Steel in Concrete. Draft State of the Art Report, Nov. 1988

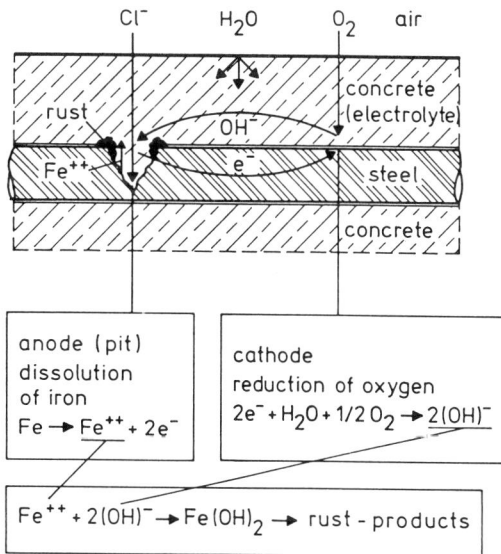

Figure 1: Schematic representation of chloride-induced macrocell-corrosion of steel in concrete

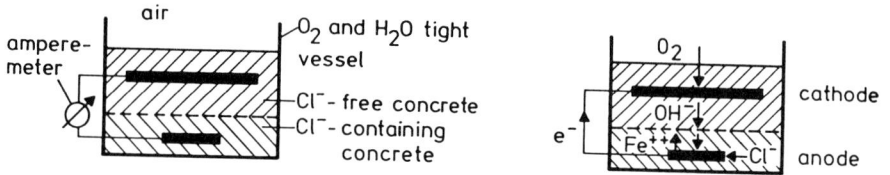

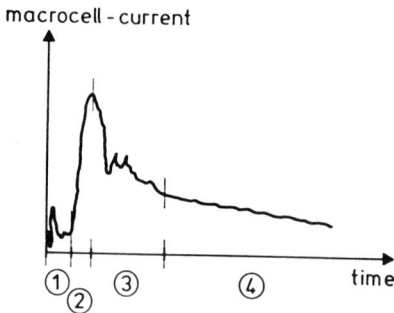

Figure 2,3: Concrete corrosion cell and behaviour of the measured macrocell currents (schematically)

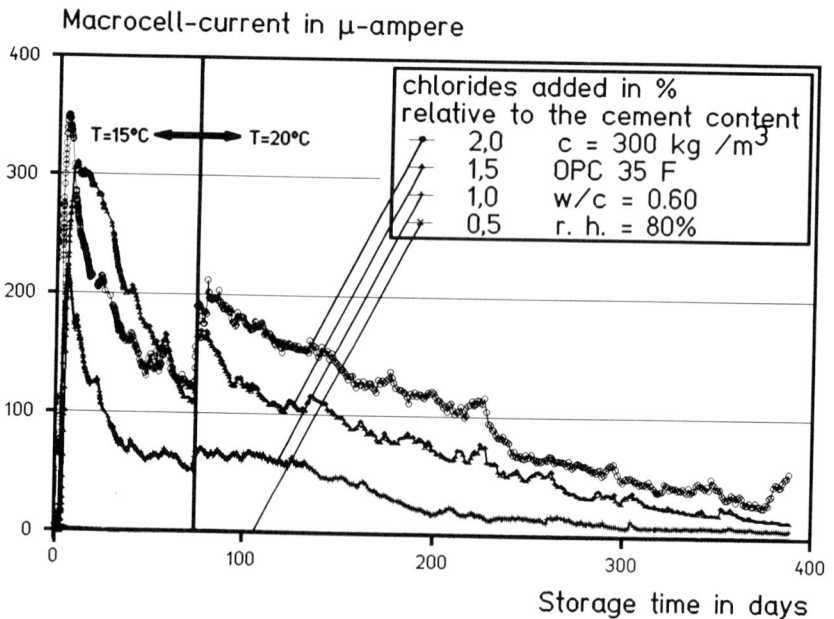

Figure 4: Effect of chloride content and of temperature on the macrocell current

Calculated weight-loss in mg

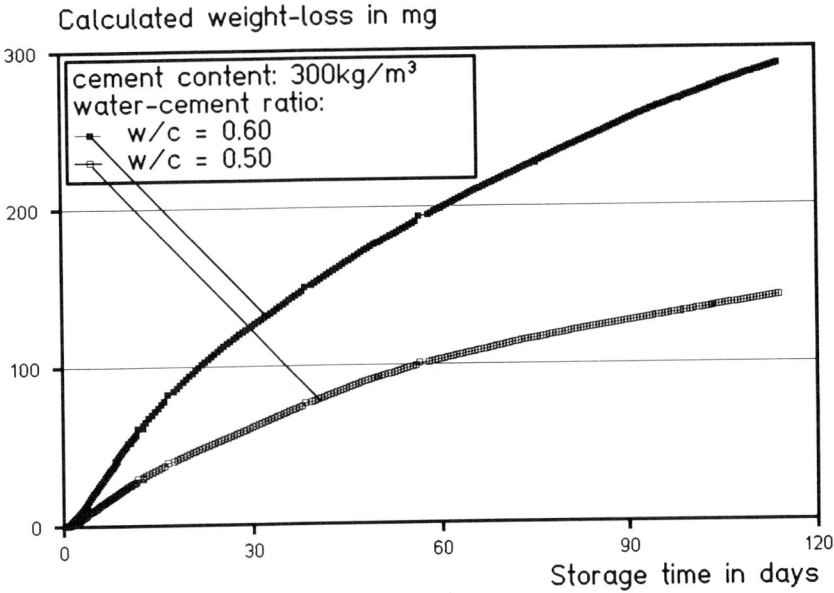

cement content: 300kg/m³
water-cement ratio:
  ■— w/c = 0.60
  □— w/c = 0.50

Storage time in days

Figure 5: Effect of w/c-ratio on the calculated weight-loss of the steel

Macrocell-current in μ-ampere

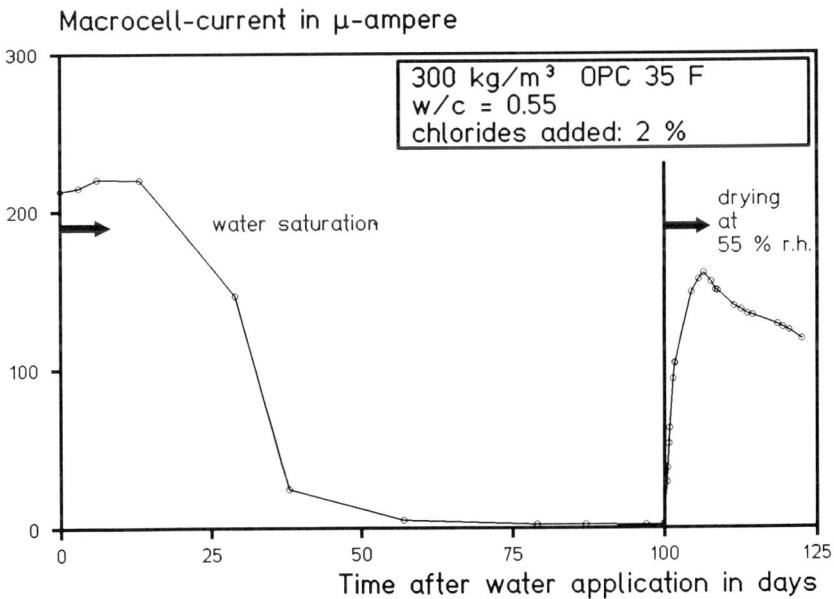

300 kg/m³  OPC 35 F
w/c = 0.55
chlorides added: 2 %

water saturation

drying
at
55 % r.h.

Time after water application in days

Figure 6: Effect of the water content of the concrete on the macrocell current

233

# 27 Concrete cover design

**D Bjegović, R Milčić, V Ukrainčik,** Civil Engineering Institute, Zagreb, Yugoslavia

In normal exposure conditions of structures, the most aggressive media causing reinforcement corrosion are carbon dioxide
and chloride ions. This is the consequence of inadequate
quality and insufficient depth of concrete cover. The penetration of aggressive media into the concrete depends on the pore
distribution system, and consists of three processes: absorption, diffusion and percolation. The introduction of calculation procedures based on well-known physical laws would advance
the engineering approach to the problems of concrete structure
durability. In this paper four typical engineering tasks are
defined, and two examples from practice are given.

## 1. INTRODUCTION

In normal exposure conditions of structures, the most frequent aggressive
media causing reinforcement corrosion are carbon dioxide and chloride ions
Very often this is the consequence of inadequate quality and insufficient
depth of the concrete cover. The current codes of practice express the
requirements for the quality of cover by the series of recommendations
regarding its depth, crack width, concrete, constituents, mix design,
concreting and curing. In reference (1) has been presented a draft proposal for a procedure for designing concrete structures in chemically aggressive environments. The corrosive media are treated as loads, and the
intensity of these loads is defined by means of partial coefficients.
On the other hand, the material parameters determining concrete durability are defined in terms of its permeability. They can be numerically
set as potential values when designing the structure, and then checked
during the construction and after the completion. The ingress of aggressive media into concrete depends on the pore system and could be
described by three processes: absorption, diffusion and percolation.
Each of the three processes can be defined by well-known laws. On the
basis of these relationships the design parameters can be calculated, the
expected aggressive media concentration per concrete depth can be estimated, and reinforcement cover can be prescribed (material parameters and
concrete cover depth). Pore structure influencing the material parameters relevant for the transport of aggressive media through the concrete
can be influenced by technological measures. (Fig.1).

## 2. DESIGN PROCEDURE - ENGINEERING TASKS

For engineering purpose, in calculating the durability and designing the structure regarding the corrosion protection of the reinforcement, four typical tasks (Fig.2) can be solved:

1. Diagnosis of the condition of the existing reinforced concrete structure. By chemical analyses of samples taken from the structure, the rate of flow of aggressive media ($Cl^-$, $CO_2$, etc.) per cover depth and around the reinforcement is obtained. This and the actual age of the structure (t) serve as a basis to calculate the diffusion coefficient (D), absorption (A), permeability coefficient (k), and to forecast the structure service life (t);

2. Calculation of concrete cover permeability parameters D, A, k for the expected structure service life (t) and the given dimensions of the cover (c);

3. Designing the concrete cover depth (c) for the expected structure service life and the given parameters D, A, k;

4. Parameters D, A, k and the cover depth (c) are optimized for a given service life (t).

## 3. TWO EXAMPLES

### 3.1. Example for solving the first task

It was necessary to evaluate the condition of the concrete bridge structure located 60 m above sea level, after t=10 years in service. The concrete cover depth is 3 cm. The prevailing process of penetration of chlorine ions was diffusion. The results of chemical analyses of concrete samples taken from the structure are listed in Table 1. The second Fick's law was supposed to control the process. Its analytic form is:

$$C_{x,t} = C_o \left( 1 - erf. \frac{x_c}{2\sqrt{D_{Cl} \cdot t}} \right) \tag{1}$$

$C_{x,t}$ - the chloride level at distance $x_c$, after time t, for an equilibrium chloride level $C_o$, at the surface
$D_{Cl}$ - the chloride diffusion coefficient
erf - the error function, calculated by rational approximation

The equation (1) is solved numerically with computer programme SUMT (3), and calculated $C_o$=2.204%, $D_{Cl}$=4.647.$10^{-9}$ cm²/s.

The values obtained for $C_o$ and $D_{Cl}$ are used in the programme DIFCL (computer programme developed at the Civil Engineering Institute Zagreb) This programme gives Cl ions distribution per cross section of a unit for the actual age of the structure (Fig.4). The distribution curve can provide the data to determine the depth of critical concentration of Cl ions ($C_{cr}$). In this case, for the actual age $t_o$=10 years, $C_{cr}$ is at the depth of 2.3 cm.

Forecast of Cl ions distribution per depth is done supposing that $D_{Cl}$ does not change with time. Following the Masuda's (2) theoretical

235

solution for increase of Cl ions in the boundery layer, for t=15 years, $C_O$=3.08%. The distribution of Cl ions is shown in Fig.4. It is visible that the concentration around the reinforcement is greater than critical.

TABLE 1. Concentration of chloride ions $C_i$ at different depths $x_i$, as measured

| $x_i$ (cm) | | $C_x$ (%) | $x_i$ (cm) | | $C_x$ (%) |
|---|---|---|---|---|---|
| 0 | - 1 | 1,43 | 1 - | 6 | 0,11 |
| 0 | - 2 | 1,595 | 1 - | 6 | 0,91 |
| 0 | - 1,7 | 1,98 | 2 - | 7 | 0,165 |
| 0 | - 2,5 | 0,55 | 2,5 - | 11 | 0,91 |
| 0 | - 5,5 | 0,385 | 0,5 - | 6 | 0,77 |
| 0 | - 0,5 | 3,41 | 2 - | 7 | 0,22 |
| 0 | - 2 | 0,825 | 0,5 - | 6 | 0,11 |
| 0 | - 0,5 | 0,88 | 2,5 - | 8 | 0,19 |
| 0 | - 2,3 | 1,21 | 2,5 - | 8 | 0,22 |
| 0 | - 2,5 | 1,43 | 3 - | 8,5 | 0,11 |
| 0 | - 3 | 0,44 | 3 - | 8,5 | 0,11 |
| 0 | - 3 | 0,77 | - | | - |

3.2. Example for solving the second task

Prefabricated walls of a containment for concentrated sea water shall be casted in a standard formwork, and the thickness of the cover is defined, c=3 cm. The production quality is high and $D_{Cl}$ could be supposed as $1 \times 10^{-9}$ cm$^2$/s. The service life should be 10 years. The difficult problem is in assuming the Cl ions concentration in the boundary layer. It should be noted that in this calculation $C_O$ is the apparent chloride content of boundary layer. However, $C_O$ depends on surrounding environment, micro conditions and concrete skin structure. In this example a high concentration (10%) was taken because this is the maximum detected Cl ions concentration in our experience.

In order to solve this task DIFCL computer program was used, yielding Cl ions distribution (Fig.5) for different $D_{Cl}$ and two levels of critical concentrations of Cl ions. The results are plotted in Fig.6. It is visible that supposed two critical concentrations do not give great differences in material parametar $D_{Cl}$.

CONCLUSION

According to current regulations, design of durability regarding the reinforcement protection to chloride action consists in prescribing technological measures relating to the depth of the protective cover, crack width, concrete mix design and curing. The introduction of calculation procedures based on well-known physical laws starting from environment conditions defined as loads, and the relevant material parameters would advance the engineering approach to the problem of concrete structures durability.

From the design point of view, four possible tasks can be defined (Fig.2) where calculation parameters are given on different levels of information (measured, required, allowed, calculated).

The examples solved show that the existing data are still insufficient both in assigning the loads and in the variations of material parameters. The most reliable sources are analyses of samples from existing structures complemented by laboratory research on concrete specimens.

## 4. REFERENCES

1. D.Bjegović, V.Ukralnčlk, Model for an engineering approach to the design of concrete structures in chemically aggressive environments, Materials an Structures, 1988, 21, 198-204

2. Y.Masuda, Penetratlon Mechanlsm of Chlorlde lon into Concrete, Chapman and Hall, Proceedlngs of the First International RILEM Congres, From Materials Sclence to Construction Materials Engineering, Volume Three, Durability of Constructlon Materials, 935-942

3. Fiacco,A.V. and Mc.Cormlck, G.P., Nonlinear Sequential Unconstrained Minimization Techniques John Wiley and Sons, Inc., New York 1968.

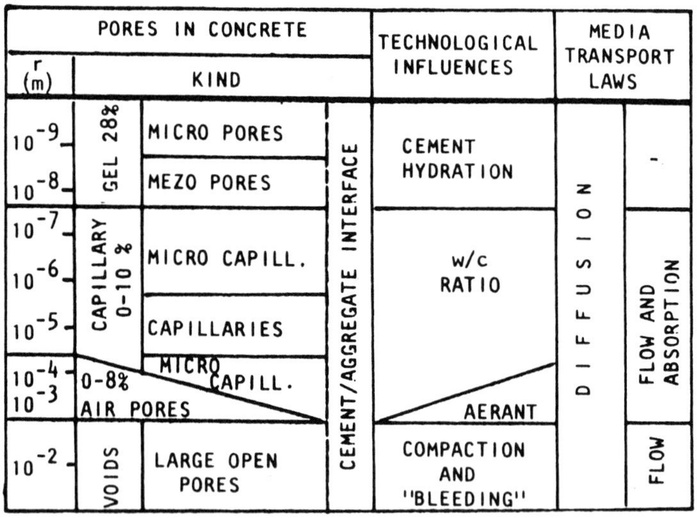

| PORES IN CONCRETE | | | | TECHNOLOGICAL INFLUENCES | MEDIA TRANSPORT LAWS | |
|---|---|---|---|---|---|---|
| r (m) | | KIND | | | | |
| $10^{-9}$ | GEL 28% | MICRO PORES | CEMENT/AGGREGATE INTERFACE | CEMENT HYDRATION | DIFFUSION | - |
| $10^{-8}$ | | MEZO PORES | | | | |
| $10^{-7}$ | CAPILLARY 0-10 % | MICRO CAPILL. | | w/c RATIO | | FLOW AND ABSORPTION |
| $10^{-6}$ | | | | | | |
| $10^{-5}$ | | CAPILLARIES | | | | |
| $10^{-4}$ $10^{-3}$ | 0-8% | MICRO CAPILL. AIR PORES | | | | |
| | | | | AERANT | | |
| $10^{-2}$ | VOIDS | LARGE OPEN PORES | | COMPACTION AND "BLEEDING" | | FLOW |

Fig. 1  Pore distribution and technological influences

| ENGIN. TASKS / VARIABLES | I | II | III | IV |
|---|---|---|---|---|
| c | KNOWN (x) | GIVEN | DESIGNED | DESIGNED |
| D, A, k | (xx) | DESIGNED | (x x) | |
| $C_o$; $C_x$ | (x) | GIVEN | GIVEN | GIVEN |
| t | | GIVEN | | |

x   as measured          o   as required
xx  as calculated        oo  as allowed

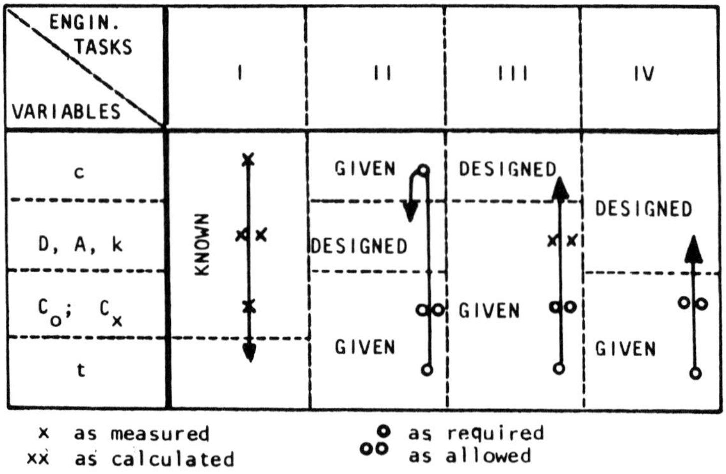

Fig.2  Scheme of four tasks

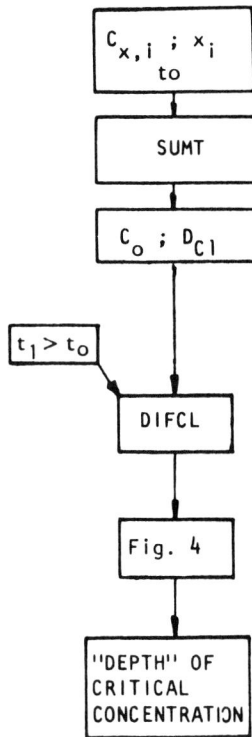

Fig.3  The flow diagram for the first task

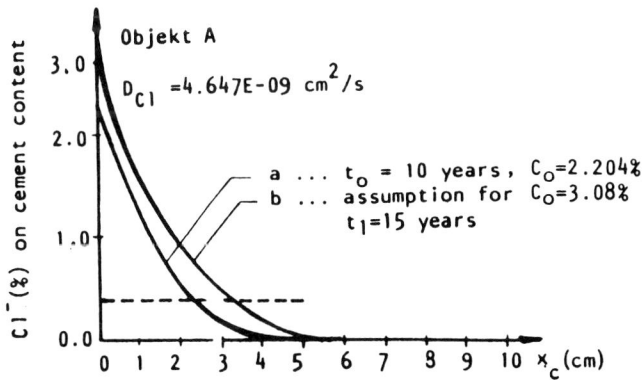

Fig.4  Distribution of $Cl^-$ in cross section

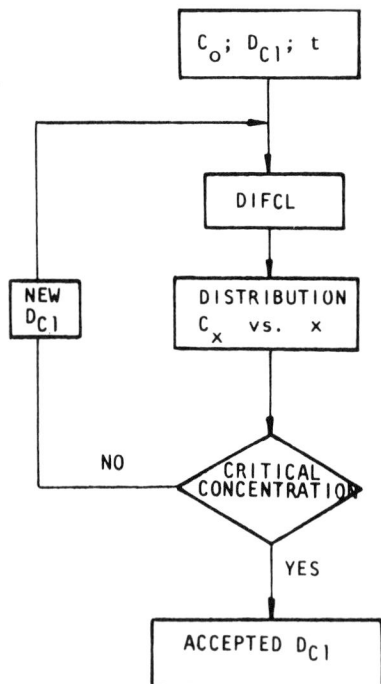

Fig.5  The flow diagram for the second task

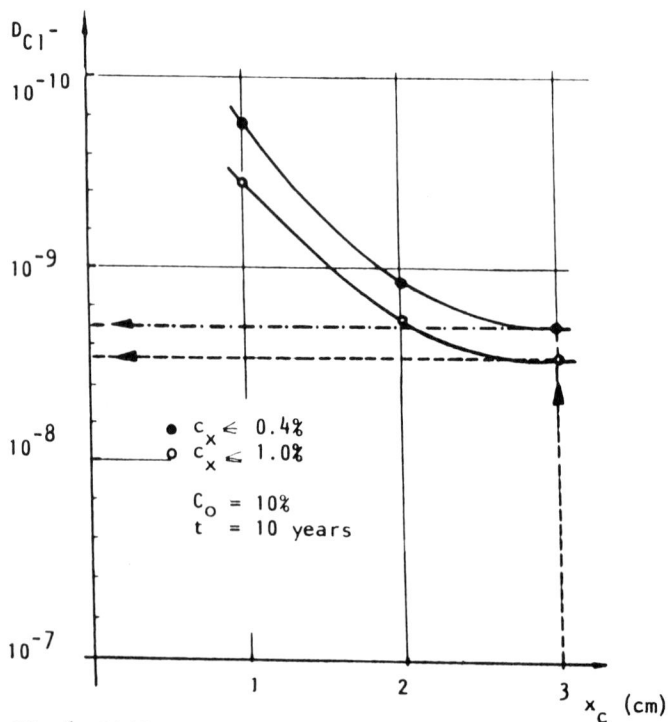

Fig.6  Diffusion coefficient curves

**W D Biggs, University of Reading, Buro Happold, UK**

These three should be considered together for reasons described later.
It was, with some hope, that paper 25 appeared to be concerned with
strain(which is at least measurable) rather than stress (with which
engineers seem to be obsessed). A failure mode based upon a limiting
strain criterion is, scientifically, more attractive. But no, at the
end of the day everything comes back to stress via the use of a
modulus whose precise nature is not described. Is it the
instantaneous modulus, the relaxed modulus or what? Furthermore I
cannot see the necessity for such mathematical complexity, the
viscoelastic linear solid incorporating a retardation time (which
appears to be subsumed in the authors exponent h) and the non
recoverable creep due to progressive crack formation by almost any
probabilistic cumulative damage law e.g. Miner, Webull etc.)
containing an exponent which takes account of the probability of a
larger crack (or cracks) being found in larger specimen (the authors
exponent k).

Paper 26 was interesting but seems to have missed one important point
- namely that the initial current appears at first sight to be a
linear function of the chloride content. If this were to be so a
cell, such as that described could be of some value in making a quick
assessment of the liability of a structure to rebar corrosion. Have
the authors considered this?

Paper 27 describes a precise, but complex, method of determining the
depth of diffusion but why involve the more complex second Flick's law
when the simpler first law, is good enough? If this is integrated and
the diffusion depth is set to $X=0$ at $t=0$ we find the simple answer
that the depth of penetration x sis proportional to t. Applying this
to the situation described for $t=15$ years I obtain a penetration depth
of $x= 3.87$mm.

This particular group of papers reinforced the sense of unreality
already created by some of the earlier papers. As a materials
scientist for some forty years now I suppose I must applaud a striving
for precision. As a specialist consultant I require quick workable
answers to determine what, if any, remedial action should be
recommended. Calculations or even measurements purporting to give a
result to two decimal places do not help - very often a calculation
giving a result to an order of magnitude is enough to make decision.
And when one considers the approximations inherent in every
engineering design one cannot but feel that if the time and effort
spent in carefully dotting the i's and crossing the t's in the
laboratory were spent in producing guidelines for practiced decision
making (often involving very large sums of money) life would be
easier.

Every scientist learns but does not, - unfortunately, often need the
works of William of Occam (d 1347), - 'Entia non sunt multiplicanda
praeter necessitatem' or, freely translated - do not choose a

complicated hypothesis if a simpler one will serve. Occams 'razor' is clearly in need of re-sharpening.

### Reply by M Raupach

The results of testing more than 100 concrete corrosion cells show that the maximum value of the corrosion rate appearing about one day after concreting cannot be used as a value indicating the following corrosion rates.

On the one hand, this is because the macrocell currents often staggered during the first days due to the quick changing of the properties of the concrete during hydration and the very complex mechanisms from depassivation to stabilisation of the pits.

On the other hand, to consider long term effects such as the pozzolanic reaction of e.g. fly ash with the cement, the testing period has to be sufficiently long. To test the influence of the chloride content of the concrete on the corrosion risk, a testing period of about one week seems to be sufficient.

Actually, we are running tests with modified cells to be installed in structures in order to continuously monitor the corrosion risk.

# 28 In-situ measurement of permeability of concrete cover by overpressure

**H W Reinhardt,** Darmstadt University of Technology, West Germany
**J P G Mijnsbergen,** Delft University of Technology, Netherlands

Various methods have been applied to concrete
for assessing the permeability of concrete
cover. A new (in—situ) method is presented,
which is less affected by moisture and which is
quick in use.

## 1. INTRODUCTION

After numerous years of exclusive assessing of concrete
quality by mechanical testing it has become clear that
physical properties of concrete are important as well and
that durability of structures is mostly a question of
lacking physical quality and geometrical precision. Whereas
mechanical testing has been standardized this is not yet
true for physical testing. Of course, there are objective
testing methods for the determination of the permeability of
concrete and its diffusion resistance against gases and
ions, however, these methods are confined to laboratory
circumstances. Furthermore, they intend to measure the
properties of the bulk concrete whereas durability is
predominantly governed by the properties of the surface
layer of concrete which is approximately equal to the
thickness of the concrete cover to the reinforcement.

In order to promote in—situ testing as a quality control and
to measure relevant physical properties, an attempt was made
in developing a new method of measuring the permeability of
concrete at the surface of a structure. Of course, there are
several methods already available — see for instance the
overview in (1) and (5) — but all methods have considerable
shortcomings as far as the in—situ applicability, the
sensitivity to moisture content, and the accuracy are
concerned.

243

In the following, the results of a comparative study are presented and discussed. Furthermore, a new testing device will be described and the results of tentative measurements will be given.


## 2. CONCRETE USED IN THE TESTS

Transport of fluids and gases takes place in the pores of the concrete, mainly in the capillary pores. The amount of capillary pores depends strongly on the water—cement ratio, the water content, and the degree of hydration. Since the pore system is also influenced by the type of cement, the cement also influences the transport rate. Despite these many variables, only water—cement ratio and curing conditions were varied in the present study.

Portland cement of average composition (class A according to Netherlands Standard) was used. The water—cement ratio was 0.4 (plus superplasticizer), 0.5 and 0.6. The cement content amounted 280 kg/m³. The maximum aggregate size was 16 mm while the amount of fine (≤ 0.25 mm) was 135 1/m³. The grading curve was adjusted to Fuller's parabola. After 28 days, the cube compressive strength ranged between 31 and 54 MPa.

Prismatic specimens of 150 x 150 x 600 mm³ were manufactured for the physical testing. To vary the moisture conditions, the samples were covered by plastic sheets during two days, and subsequently cured during 0, 4, and 26 days respectively in a fog room. After curing, the specimens were stored at 20°C and 50 % r.h. during three months. Additionally, three specimens were stored at 80 % r.h. Nine specimens were returned after six weeks to the fog room and three days to 50 % r.h. before testing. This has been done in order to check the sensitivity of the methods to the moisture conditions.


## 3. TESTING METHODS USED

### 3.1 Oxygen diffusion

Oxygen diffusion was measured in stationary state on concrete disks of 30 mm thickness and 100 mm diameter. The disks were dried at 65°C for one day and subsequently at 105°C until the mass loss was less than 0.1 g/day. Several disks with water—cement ratio of 0.5 were wetted and stored at 80 % r.h. for eight weeks. They were not dried before testing.

Fig. 1 shows the testing arrangement schematically. The specimen is clamped to the testing cell and sealed by an

0—ring. Oxygen enters at one side, nitrogen at the other. By use of a manometer it is checked whether there is an absolute pressure difference. The oxygen concentration at the nitrogen side is measured by an oxygen meter (Thermo Analyser, by Thermo—lab).

## 3.2 ISAT

The method developed by Levitt (2) is described in BS 1881, part 5. The principle is that water which is in contact with the surface of the concrete (structure) penetrates into it by capillary forces. The amount of water penetrated is measured by a capillary tube after 10 to 120 minutes of capillary suction.

There are some doubts about the practical applicability of the method to existing structures. First, 120 minutes seems to be a too long a period of time if many places on a struc-ture have to be examined, and second, how sensitive is the method to the moisture content of concrete. Since the skin of the structure is most susceptible to moisture fluctua-tions by rain and wind and, on the other hand, the skin is the governing part of the structure with respect to dura-bility, this aspect needs more consideration.

## 3.3 Figg's method

According to the improved Figg method (3,4), a hole of 10 mm diameter and 40 mm depth is drilled into concrete and sealed by silicone rubber. A needle is pierced through the rubber and the pressure in the hole is lowered to 45 kPa. After reaching this pressure the time is measured which is needed to raise the pressure again to 50 kPa. This time is a measure of the permeability of concrete in the surrounding of the drill—hole.

The same question as for the ISAT method raises· whether the moisture content has a large influence on the result. Furthermore, the time of testing in case of high strength concrete may be too long for economic practical use. It can also be asked whether the pressure increase is exlusively due to permeation of air through the skin of the structure — which is intended — or whether the air flows from the inner of the concrete to the drill—hole.

## 3.4. High pressure method

To overcome some inconveniences or practical limitations a new method has been developed which resembles at a first glance the Figg method. However, instead of low pressure, high pressure is used. This difference has at least three advantages:

- high pressure is not limited to 1 bar so that the most suitable time of measurement can be adjusted
- the air will escape exlusively or predominantly through the skin of the structure
- high pressure permeability is assumed to be less affected by moisture in concrete than at low pressure

The method works as follows (Fig. 1): a hole of 14 mm diameter and 40 mm depth is drilled and sealed by a rubber bung through which a hollow steel bolt leads to a high pressure reservoir containing nitrogen. The high pressure in the reservoir is reduced by a valve to 1010 kPa. As soon as this pressure is reached in the drill—hole the valve is closed. The time which elapses during the pressure decrease from 1000 to 950 kPa is measured with a digital manometer. This time is the technological measure for the permeability of the concrete of the skin of the structure.

In order to detect the sensitivity of the methods and to check the influence of various parameters a comparative study has been carried out.


4. RESULTS

The oxygen diffusion coefficient can be considered as a reference since this is the quantity which is determined in a physically sound manner. The diffusion coefficient is called $D_{eff}$ ($m^2/s$) since no distinction is made between the various mechanisms of transport which may occur in wet and dry concrete. As Fig 2a shows $D_{eff}$ increases almost linearly as a function of the water—cement ratio. The highest values are reached if there was no curing of the specimens after demoulding. With increasing number of days of curing, $D_{eff}$ decreases. The storing conditions after curing have been varied between 50 and 80 % r.h. during 6 months. The open symbols in Fig. 3 indicate that the storage conditions have a minor but still significant influence on the diffusion coefficient.

As far as the oxygen diffusion coefficient is concerned, it can be concluded that this method allows to dinstinguish between the three variables — water—cement ratio, curing, drying conditions — clearly.

The results of the ISAT testing are shown in Fig 2b. They show also a clear distinction between the three water—cement ratios and between the three durations of curing. At low water—cement ratio the differences are smaller than at high water—cement ratio. However, this seems less important since a low absorption rate means a good concrete quality with respect to durability.

A testing series has been carried out with concrete which was stored in water during six weeks until three days before testing. Then, the specimens were stored at 50 % r.h. until testing. This procedure should have resembled a concrete surface which may have been wetted during a period of rain and subsequently dried. The results of Fig. 3 show that the absorption rate has decreased by a factor of 3 to 5 and that the differences within the range of water—cement ratios and curing duration became very small. From such tests it may be concluded that a very impermeable concrete has been tested whereas the true reason for this phenomenon is seen in the high moisture content of the concrete.

The same type of concrete has been tested with the improved Figg method (Fig. 2c). It can be seen that the measuring time is rather long when a low water—cement ratio concrete is tested and that the differences between various curing durations become small. With higher a water—cement ratio the measuring times decrease considerably and the curing duration becomes a significant influence on the measuring time, i.e. the method becomes more sensitive.

During the test, concrete with various moisture contents has been tested and it appeared that a higher moisture content did not always cause longer measuring times which is rather unexpected. Repeated measurements taken at the same hole have shown that the measuring times became longer with the number of measurements. This phenomen should not occur if the air permeated through the outer layer of concrete. However, it seems that the same effect has been detected as found in (4), i.e. that the air flows preferably from the interior of the concrete to the drill—hole.

The new method using high pressure has been applied to three concretes with three water—cement ratios and three curing periods. Fig. 2d shows that the measuring time decreases with increasing water—cement ratio and with curing time as has been expected. It is also clear that the differences be—tween the three curves is significant for all water—cement ratios. Compared to the other methods it may be stated that the new method takes less time — for preparing the measure—ment and for the measurement itself — and distinguishes more clearly between the variables which affect concrete perme—ability. Furthermore, it has been seen during testing that repeated measurements lead to the same results within narrow scatterbands. Visual inspection during the tests has shown that the compressed air flows through the outer layer of the concrete to the surface (which can be seen when the test is performed under water).

## 5. COMPARISON OF TEST RESULTS AND QUALITY INDICATION

At the moment only qualitative indications and comparisons will be made with respect to concrete quality. Similar to what has been suggested in (1) for the ISAT method, all results will be ranged in low/average/high permeability. If all test results are taken, the following table is suggested (Table 1).

Table 1  Comparison of various methods and quality indication

| method | low | average | high |
|---|---|---|---|
| oxygen diffusion coefficient $(10^{-8} m^2/s)$ | 0—10 | 10—15 | > 15 |
| ISAT (10 min) $(ml/m^2 \cdot s)$ | <0,25 | 0,25—0.50 | > 0.50 |
| Figg (s) | > 250 | 50—250 | < 50 |
| high pressure (s) | > 25 | 5 — 25 | < 5 |

If this classification is used it can be seen from Fig. 2 that all methods lead to the same result. What differs from method to method is the preparation of the measurement, the measuring time, the effect of moisture content and the power of discrimination.


## 6. CONCLUSION

A new in—situ method for the determination of the permeability of the concrete cover has been developed and compared with other existing methods. From the results can be concluded that the new method which uses high pressure nitrogen is suitable for all concrete qualities and is less affected by moisture than other methods. Therefore, it is recommended to use this method in practice in order to get aquainted with all prominent features, especially in order to get experience with the handling and the reliable judgement of the results.


## 7. ACKNOWLEDGEMENT

The support received by Mr J.H. Croes, Mr J. Luijerink, and Dr. Cornelissen is gratefully appreciated.

REFERENCES

(1) The Concrete Society. <u>Permeability of concrete and its</u>
    <u>control.</u> Chameleon Press Ltd., London, Dec. 1985,
    130 pp.

(2) Levitt, M. <u>Non—destructive testing of concrete by</u>
    <u>initial surface adsorption method</u> Symp. on Non—
    destructive testing of concrete and timber. London
    1969, pp 23—26

(3) Figg, J.W. <u>Methods for measuring the air and water</u>
    <u>permeability of concrete</u> Mag. Concrete Res.
    25 (1973), no. 85, pp 213—219

(4) Cather, R. <u>Improvement to the Figg method for</u>
    <u>determining the air permeability of concrete.</u> Mag.
    Concrete Res. 36 (1984), no. 129, pp 214—215

(5) Dhir, R.K., Hewlett, P.C., Chan, Y.N. <u>Near—surface</u>
    <u>characteristics of concrete: assessment and develop—</u>
    <u>ment of in—situ test methods</u> Mag. Concrete Res. 39
    (1987), no. 141, pp 183—195

FIGURES

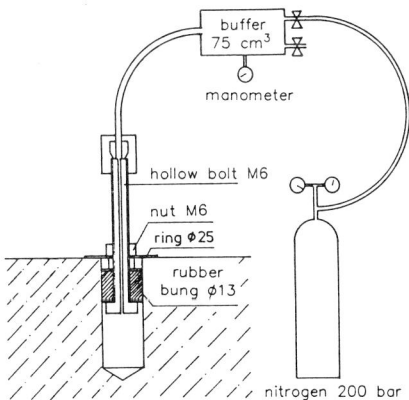

Figure 1   Schematic of overpressure method

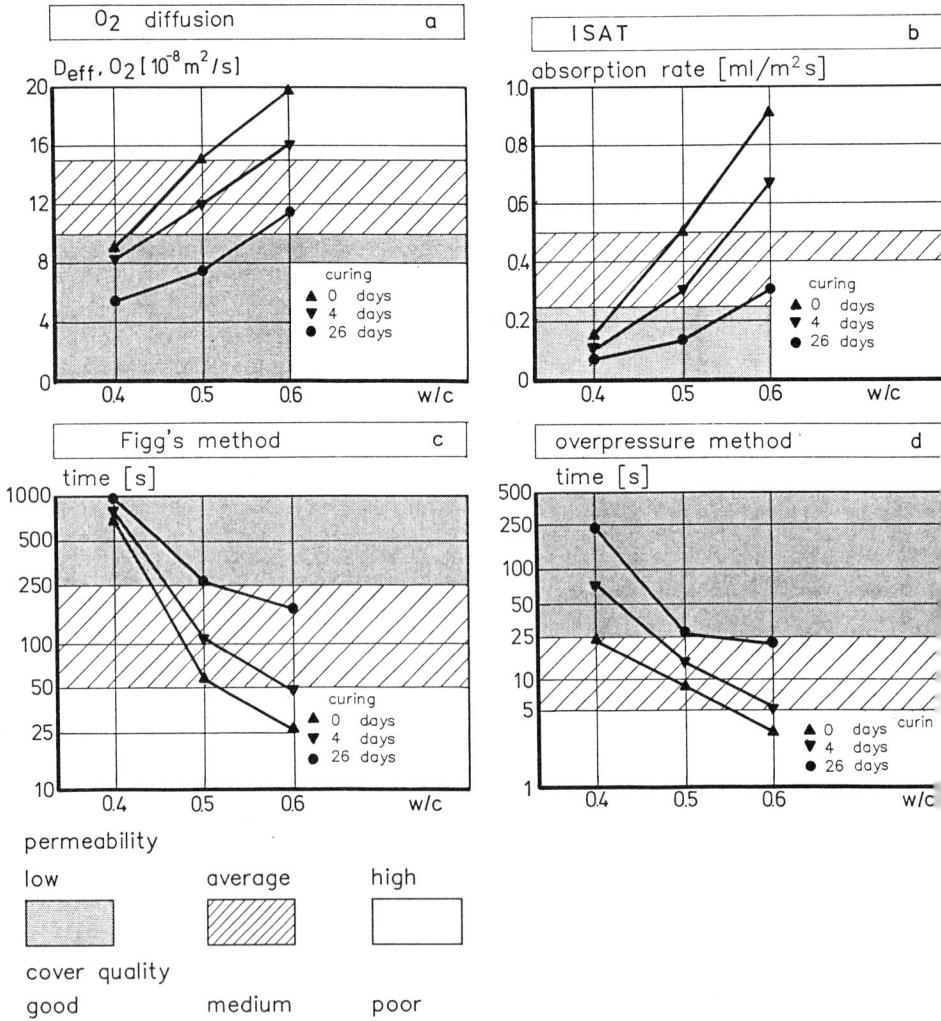

Figure 2    Comparison of results by the four methods used and
quality indication

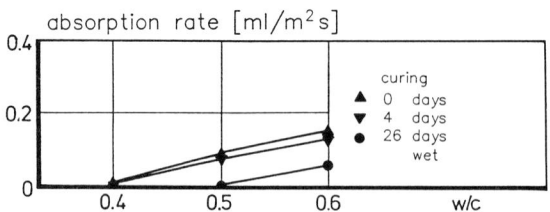

Figure 3    Absorption rate (ISAT) as function of water—
cement ratio and curing duration

R C de Vekey, Building Research Establishment, U.K.

Is the method described usable for highly porous gap-graded concrete as used for many lightweight aggregate concrete blocks in the U.K?

## Author's reply

The proposed method is mainly aimed at the investigation of the quality of the outmost layer of concrete which is usually the cover of the reinforcement. The moment being, we tested only structural concrete with dense aggregates. We assume that structural lightweight concrete with a dense matrix, i.e. no more than normal air voids due to entrapped air, can be investigated with the new device. Highly porous concrete (no-fines concrete) has such a big permeability that the time elapsed during the pressure decay is very short and the measurement will be unreliable. Therefore, we think that the method is not suited for application to the concrete blocks you mentioned.

N Clayton, Building Research Station, U.K.

I would like to put the following questions in relation to the authors' in-situ method for measuring concrete permeability using high pressure gas.

Firstly, does a significant amount of gas permeate through the rubber/concrete interface? In tests carried out some years ago using a very similar method, we found that this was a major problem. Is a special type of rubber and/or a particular method of drilling the hole necessary?

We also found that the method of fixing often caused cracking of the concrete and sometimes complete separation of the concrete would occur when gas pressure was applied. Are the authors sure that no weakening of the concrete is caused by their method?

## Author's reply

The seal between concrete and hollow tube is airtight as has been found in tests under water. The rubber material is a piece cut from a flexible tube which is commonly used in chemical laboratories. If the seal is not airtight, the pressure decay is so fast that this is immediately recognized. In that case, the nut of the hollow bolt has to be turned again pressing the rubber stronger against the wall hole. The holes have been made by a hand-held drilling machine without special instructions.

There wasn't any case of cracking or fracture of concrete due to gas pressure. Due to your comments, we will pay special attention to this question when we shall investigate more types and grades of concrete.

**J G M Wood, Mott MacDonald, UK**

We are all faced with the problem of determining the quality of concrete in relation to durability. The three main objectives are:

(a) determining trends and predicting the chloride ingress and corrosion rates in deteriorating structures,
(b) evaluating the potential performance of new concretes in structures subject to severe environments,
(c) providing a basis for quality control of concrete for severe environment.

Graph 1 gives typical results of chloride ingress in a Bulk Diffusion test we have been developing in Mott MacDonald as a cost effective robust test procedure. It shows the concentration of chloride ions in the surface layers of a concrete core vacuum saturated prior to immersion in a strong salt solution 5M Cl⁻ (177,250ppm) for a period of 42 days at 40°C. This can be carried out on any reasonable sized 75mm or 100mm diameter core and the depth of chloride penetration to the as-cast faces and the cust-faces on the core surface can be determined for periods of 28 days upwards. With high quality concretes the rate of pentration of chlorides is very low and conventional drilling methods of determining chloride profiles are not sufficiently precise. We have developed, with the assistance of David Lawrence of the C&CA, a technique of chloride profile determination by grinding off 1-3mm thick layers and collecting the dust with a vacuum filter. The analysed profile is shown as the histogram in the graph. We also use SEM techniques, as these give a better indication of the mode of chloride ingress, (e.g. is the chloride moving preferentially down surface microcracks, along the boundaries between aggregate particles and the paste or through porous aggregate). We have found that SEM gives some distortion of the magnitude of chlorides due to evaporation during surface preparation, although it provides a useful indication of the profile. The measured chloride ingress profiles are compared to calculated profiles using a range of assumed concrete properties. The chloride ingress equations are calculated using the program CHLORPEN, which we have developed specifically for this application and for predicting the behaviour in structures. This considers:

i) Permeability flow under a pressure head

ii) Ionic diffusion.

iii) The effective binding of chlorides by $C_3A$.

iv) The chloride exposure environment.

v) The concrete mix details.

vi) Time of exposure.

vii) Temperature gradients through sample.

252

viii)    Internal evaporation.


The equations can take account of changes in properties (e.g. permeability and diffusion and free water content) with time as hydration develops.

We feel that this Bulk Diffusion Test, which has been related to more detailed permeability, diffusion, wick action and paste diffusion tests, provides a good basis for assessing concrete properties, for extrapolation of deterioration trends in existing structures, for mix design trials and for routine quality control during construction.

A fuller discussion on the methodology is given in the forthcoming paper(1).

Professor Reinhart's test procedures are more appropriate to the normal building works where carbonation rather than chloride ingress is the risk.   I suspect his methods would be difficult to apply to very low water cement ratio mixes with pfa and slag, for which permeability is exceptionally low.   The approach we have adopted is consistent, but in some respects more fully developed, than that outlined by D Bjegovic.   I would welcome the authors comments on our approach.

1)  Wood J G M, Wilson J R and Leek D S.   Improved Testing for Ingress Resistance of Concretes and Relation of Results to Calculated Behaviour.   To be published in 3rd International Conference on Deterioration and Repair of Reinforced Concrete in the Arabian Gulf, B.S.E., October 1989.

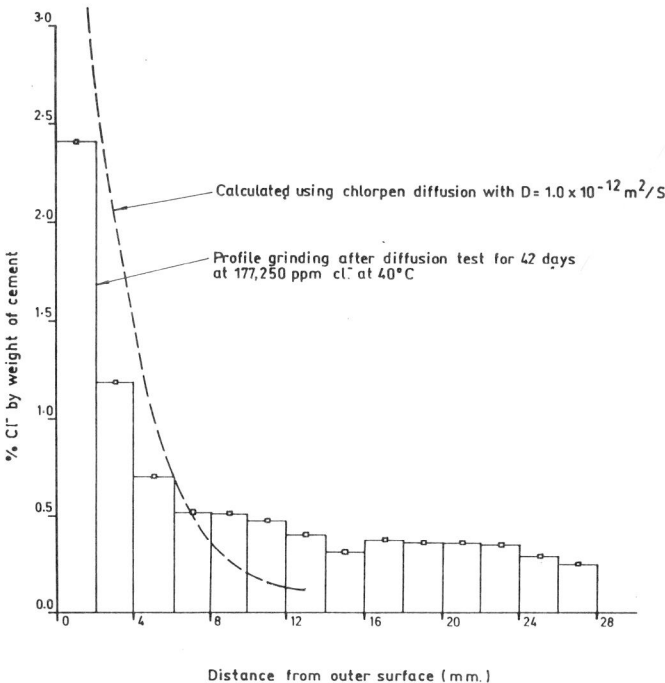

Calculated using chlorpen diffusion with $D = 1.0 \times 10^{-12}\,m^2/s$

Profile grinding after diffusion test for 42 days at 177,250 ppm cl⁻ at 40°C

% Cl⁻ by weight of cement

Distance from outer surface (mm.)

Author's reply

The bulk Diffusion Test as developed by Dr Wood combines several
positive aspects: it is robust, it simulates very severe exposure
conditions, and it is theory supported. In my opinion, it is a method
which is suitable to compare a wide range of concretes with respect to
their physical properties which are important for corrosion protection
of the reinforcing steel. Together with the program CHLORPEN, the
test may lead to a reliable prediction of the life of a concrete
structure. However, from the comment it is not clear how the steps
VII) and VIII) are related to the test since the specimens are kept at
constant temperature and in full saturation.

Opposite to Dr Wood's test, which is carried out in the laboratory,
our permeability test is conceived as an in-situ test. Emphasis is
laid on quick testing with the possibility of checking many places at
·a concrete surface within a reasonable time. The results of the test
give always and only an average measure of permeability of the depth
tested. In our opinion, this is a relevant quantity with respect to
rate of carbonization and oxygen supply in case of corroding
reinforcement. It may be correlated with chloride ingress if
appropriate assumptions can be made with respect to chloride binding
of cement and pore size distribution of the concrete. However, these
relations may be weak and require more research.

The moment being, our method has been applied to usual concrete of
buildings and bridges. We agree with Dr Wood that low water cement
mixes with pfa, slag and silica fume are difficult to distinguish.

In our opinion, the two methods, Dr Wood's Bulk Diffusion Test and our
permeability test, can support each other in the common effort to
improve durability of concrete structures.

# 29 Concrete shrinks, problems don't

**A T Tankut and U Ersoy,** Middle East Technical University, Ankara, Turkey

High tension can develop due to shrinkage if deformations are restrained. Three case studies are reported. In the one-storey three bay frame, relatively rigid side spans restrained shrinkage deformations of the wide central span and caused significant cracking. In the multi-storey frame, the axial tension due to shrinkage caused considerable reduction in the shear capacity and led to critical shear cracking. In the grain silo, the restraining effect of the rigid foundation caused appreciable vertical cracks in the silo walls.

INTRODUCTION

Shrinkage and temperature drop cause shortening in reinforced concrete members. In beams of some frames, high axial tension develops if such contractions are restrained by other members of the frame, such as columns or walls. Stresses induced by shrinkage and temperature drop may lead to severe cracking if members have not been reinforced and detailed accordingly. The restraint caused by such boundary members should be represented by springs where spring constants are estimated by considering the deflections of the restraining members.

Three case studies reported in this paper are related to structures located in Central Anatolia where shrinkage may be very critical if due care is not provided, since the climate is dry and significant temperature changes take place.

CRACKING IN ONE-STOREY FRAMES

Cracks were observed in the beams of a one-storey framed structure of a slaughter-house[1]. Cracks were reported to have developed during one year following the completion of construction and remained unchanged since then. The building consisted of one-storey, three-bay frames with inclined beams as shown in Figure 1. The maximum height of the frame at the crown was 11 m. Roof slabs were supported by cross-beams which spanned from one frame to the next. Snow and wind loads had been considered in the design besides dead load.

Cracks observed in the frame beams were at the centre span and they were about 1.0 mm. wide. These cracks extended full depth of the beams and even penetrated into the roof slabs. Cracks were generally observed near the points of inflection and were perpendicular or slightly inclined to the beam axis, Figure 1a.

There were similar cracks in the cross-beams; almost perpendicular to the beam axis, extending full depth and penetrating into the roof slab and even running across the slab in some cases from one beam to the next, Figure 1b. However, cross-beam cracks were even wider, reaching 2 mm. at some locations, and they were generally located near the midspan.

All the indications explained above led to the conclusion that these cracks had been caused by shrinkage and temperature drop. Calculations based on an equivalent temperature drop of 40K, revealed that considerable axial tension would be produced by shrinkage, since the large centre span was restrained by the two adjacent short spans supported by relatively stiff columns. Although considerable steel was present in the frame beams, the reinforcement was not distributed properly, most of it being concentrated at the bottom and top of the beam. The authors have the impression that the crack width could have been kept within allowable limits if more longitudinal bars were provided at intermediate levels.

In cross-beams, neither the amount nor the distribution of the reinforcement was adequate. Unusually low steel ratios had been used and no intermediate steel had been provided. The wider cracks were the result of inadequate reinforcement.


CRACKING IN MULTI-STOREY FRAMES

The multi-storey frames of an artificial fertilizer plant were studied to investigate the widespread cracking observed in the beams. The structural system is illustrated in Figure 2. Two different types of cracking were noticed; (i) shrinkage cracks and (ii) combined shear and shrinkage cracks.

Cracks of the former type were observed near the mid-span and were perpendicular to the beam axis. These cracks were not considered critical, since a reasonable amount of steel (top, bottom and longitudinal web) was available.

The latter category consisted of 1.5~2.0 mm. wide typical shear cracks located at 1.0~1.5 m. from the column faces. Formation of shear cracks consistently at the same location was due to the fact that a large portion of the bottom steel was cut off at the same section and high stress concentration initiated cracking. These inclined cracks could not be explained by shear alone, since the shear capacity of the section was much larger than the shear force transferred. However, an approximate shrinkage analysis revealed that shear stresses could easily lead to cracking when combined with axial tension caused by shrinkage corresponding to a very reasonable shrinkage strain value of $\varepsilon_{cs}=0.00037$.

Beams having combined shear and shrinkage cracks were strengthened by steel brackets serving as additional external stirrups to improve the shear capacity. These brackets consisted of two (at the top and bottom faces of the beam section) structural steel channel sections attached by two (at the sides) 16 mm diameter high strength steel bars. These brackets were equally spaced in the critical shear zones at either ends of the frame beams and were prestressed by tightening the nuts at the ends of the steel bars.

## CRACKING IN SILO WALLS

Vertical cracks observed in the walls of a grain silo[2] caused concern, and the authors were asked to investigate the problem. The building consisted of six bins arranged in a row. Each bin had a square cross-section, 7.30 m x 7.30 m and the height was about 25 m from the foundation level. Wall thickness was 300 mm and each corner was stiffened by a 600 mm x 600 mm column. Ring beams at the 6 m level were used to support the bunkers, Figure 3.

The cracks which had been noticed approximately one year after completion of construction, were extending from the foundation level to the level of the ring beams. The crack width was around 1.5 mm at the ground level and decreased gradually towards the top.

A site investigation and review of the design calculations led to the conclusion that the cracks were caused by imposed deformations attributed to shrinkage and temperature drop. The walls were not free to shorten due to fixity at the foundation level and restraining effect of the boundary members, namely columns and interior walls. The restraint was obviously maximum at the foundation level and rapidly decreased moving upwards. An approximate analysis assuming a shrinkage coefficient of $\varepsilon_{cs}$=0.00028 and a temperature drop of 15K indicated 34.5MPa and 0.25MPa tensile stresses at the foundation and ring beam levels respectively. These approximate values were sufficient to explain the cracking in the silo walls which had inadequate horizontal reinforcement.

## RESTRAINT ANALYSIS

To estimate the streses caused by shrinkage and to determine the required steel, restraint caused by boundary members should be taken into consideration in terms of spring constants estimated on the basis of flexural rigidities of the restraining members. The simple case illustrated in Figure 4 sets an example. Considering the deformations shown, the following compatibility equation can be written,

$$2\delta_c + \delta_b = \delta_f$$

where,  $\delta_f = \ell\varepsilon_{cs}$      free shortening due to shrinkage

$\varepsilon_{cs}$      shrinkage coefficient

257

$$\delta_b = \frac{T\ell}{AE} \qquad \text{beam elongation}$$

$$\delta_c = m\,\frac{Th^3}{EI} \qquad \text{column deflection}$$

$$T \qquad \text{tension caused by shrinkage}$$

$$k = \frac{T}{\delta_c} = \frac{EI}{mh^3} \qquad \text{spring constant}$$

An approximate elastic analysis based on the above compatibility equation and neglecting the effect of steel can produce sufficiently accurate results. Once the axial tension T is known, steel required to take care of shrinkage can easily be determined. This steel should be distributed across the depth of the beam for an efficient use. The unfavourable effects of the axial tension T on the shear strength should also be taken into consideration.

For wall type structures, a similar method of analysis can be used if the wall is divided into a number of horizontal strips and a series of springs is introduced as shown in Figure 5. The spring coefficients should be chosen considering the deflections of the restraining boundary members. Obviously, the spring coefficients will be inversely proportional with the deflections of the boundary members at strip levels.

## CONCLUSIONS

In the light of the case studies briefly presented in this paper, the following conclusions can be drawn.

a. Shrinkage problems can be very serious in dry climates where significant temperature changes take place.

b. Significant tension can develop due to shrinkage if the member is restrained by stiff boundary members.

c. Shrinkage cracks can easily exceed the acceptable limits if adequate steel is not provided or is not distributed properly.

d. Axial tension due to shrinkage may cause significant reduction in the shear capacity.

e. Axial tension can be estimated easily by an approximate elastic analysis and can be used to determine the required steel.

f. A well detailed reinforcement can lead to uniform crack distribution and acceptable crack width.

## REFERENCES

1. Ersoy, U. and Tankut, A.T., 'Report on cracking in Slaughter house in Malatya', report presented to E.B.K., November 1983 (in Turkish).

2. Ersoy, U. and Tankut, A.T., 'Report on silo cracking', report presented to Ege Bıracılık ve Malt Sanayii A.Ş., June 1983 (in Turkish).

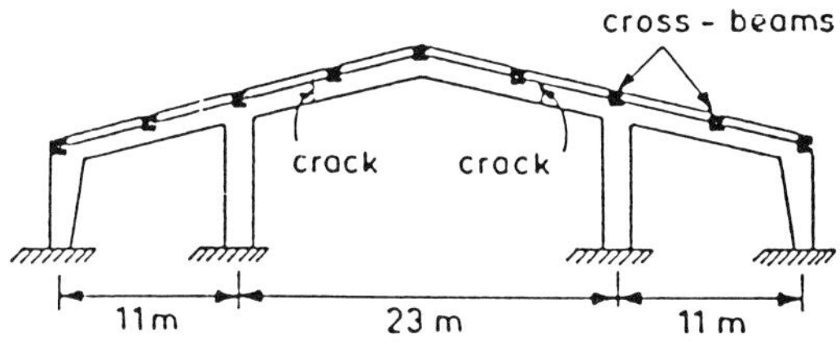

a. Typical Frame

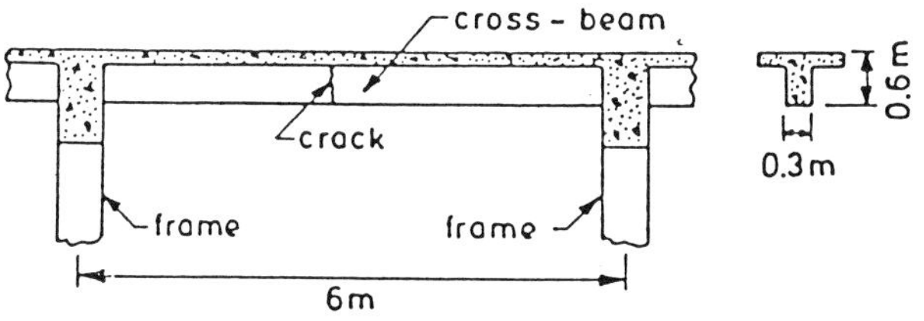

b. Typical Cross-Beam

Figure 1 One-storey frame

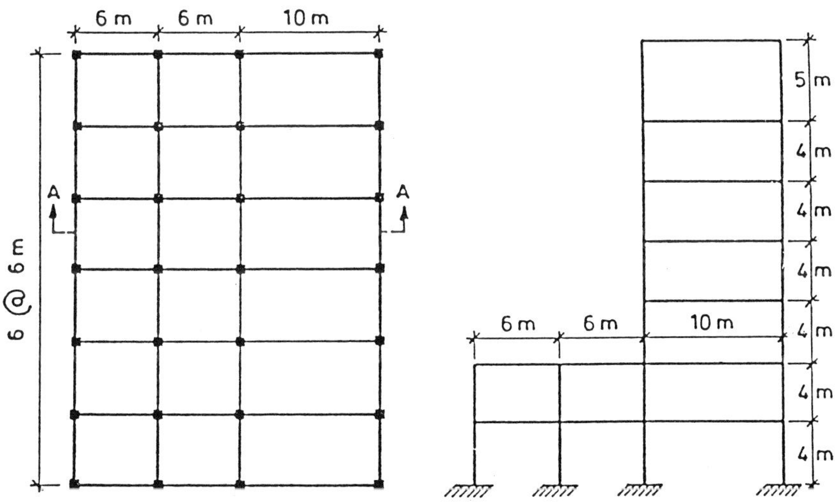

a. Plan

b. Section A-A

Figure 2 Multi-storey frame

a. Elevation                         b. Section A-A

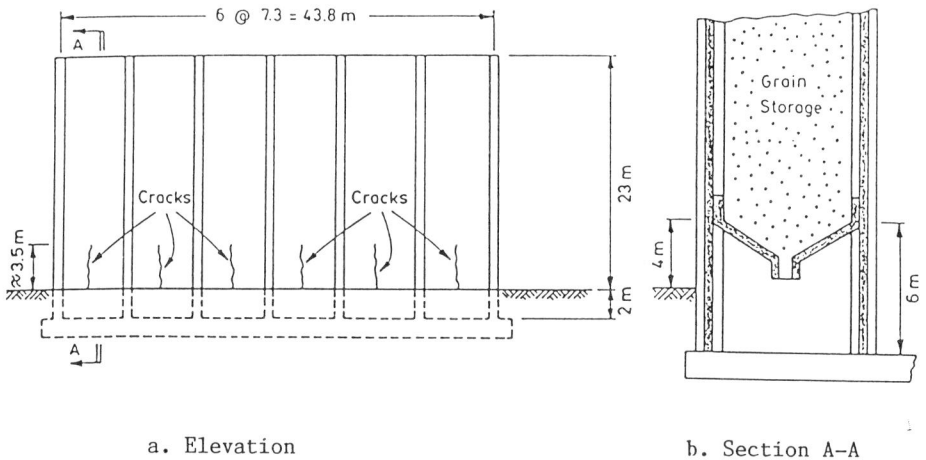

Figure 3   Grain silo

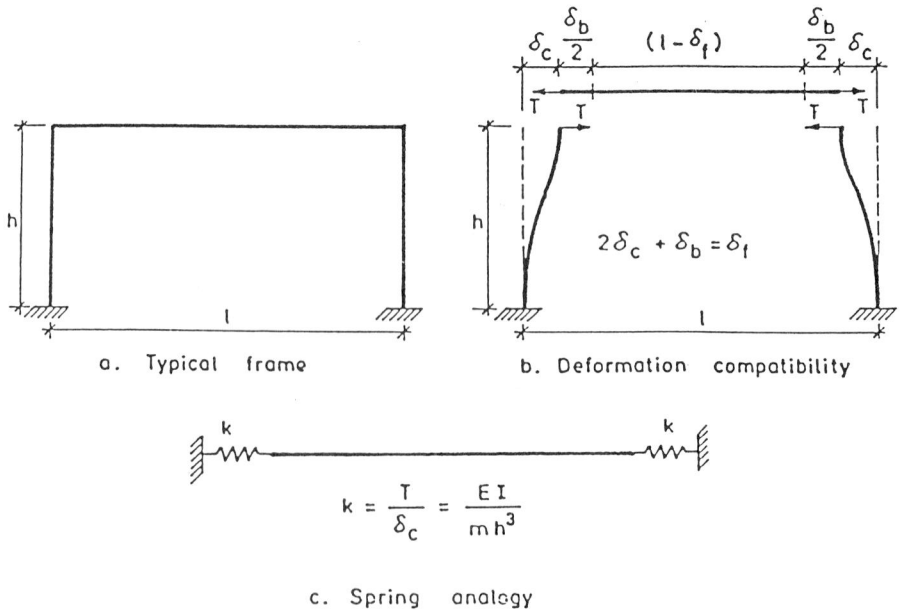

a.  Typical  frame          b.  Deformation  compatibility

$$k = \frac{T}{\delta_c} = \frac{E I}{m h^3}$$

c.  Spring  analogy

Figure 4   Principles of restraint analysis for frames

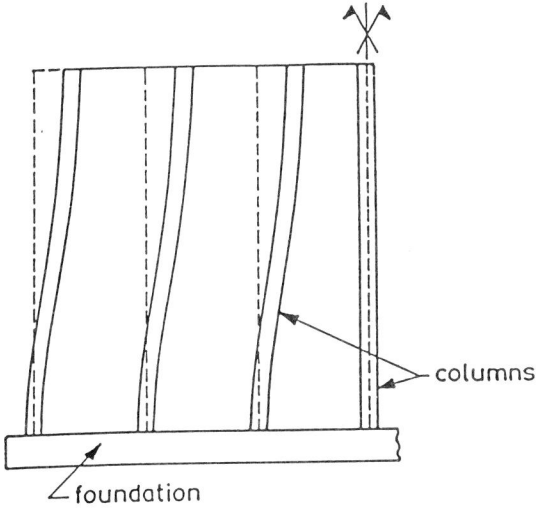

a. Deformed shape                    b. Spring analogy

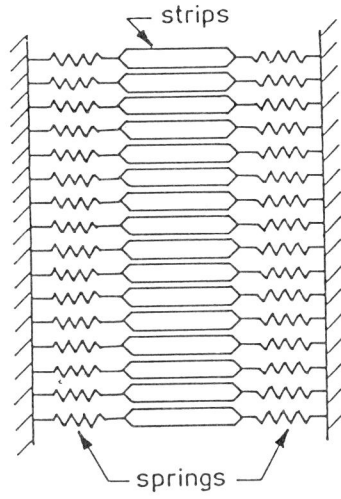

Figure 5   Analytical model for the silo wall

# 30 Lifetime performance and test-induced damage in wood structures

**J J Salinas,** Carleton University, Ottawa, Canada

The basic principles of theory of probabilities
and the concept of damage accumulation are used
to characterise the reliability of wood
structural systems. The concept of load equivalency,
derived from a damage accumulation formulation is
introduced to compare various test schemes used to
assess the reliability of engineered wood products.
The results of studies on the effects of proof-loading
are used to describe the improved reliability resulting
from a shift of the strength distribution.

## INTRODUCTION

Estimation of the structural reliability of wood structures can be
assumed to evolve through various stages each characterised by design
paramters with increasing degrees of confidence and detail. During
the initial design stage, the designer uses building codes prescribing
nominal design loads for occupancy, wind, snow, etc. These prescribed
loads are based on load surveys and regional meteorological information.
Allowable stresses or specified design strengths are also prescribed by
the codes. At this stage the designer determines the loads on the
structure prescribed by the code and selects an appropriate material
strength in order to determine an adequate size or some other geometric
property of the system. At this stage the designer implicitly accepts
the level of safety targeted by the code.

After construction, once the structure starts its service life, there is
improved knowledge of the as built dimensions resulting in a better
estimate of the magnitude of the dead load. If occupancy and weather
records are kept, a better estimation of the actual life loads will also
be possible. Knowledge of the actual material strength, however, may not
necessarily be any better than that originally assumed or adopted from
the code. To improve our confidence on the estimate of the reliability
of the system a better assessment of material strength is required and
often achieved by testing or proof-loading the structure under a well
defined test regime.

The design of this test regime should receive careful consideration.
While a high proof-load will ensure that the surviving structure is

capable of resisting the specified loads, the test load itself may introduce damage resulting in a lower residual strength. A test load of small magnitude may introduce little or no damage into the structure but it may not give a full indication of its potential strength. Sometimes the fact that a structure has survived after many years in service is used as an indication of its implicit level of safety. In general, as damage accumulates with time under load, the probability density function of the strength will shift towards the lower end of the scale increasing its probability of failure.

On the other hand, long-term survival indicates strength above minimum characteristic values and, consequently, impoved levels of safety. This study attempts to incorporate the concept of damage accummulation with time under load for wood structures in order to model their relative levels of safety throughout the life of the structure.

## THE DESIGN ENVIRONMENT

The nominal capacity of a wood structural element may be expressed as

$$C_n = \phi S_c X_p K^* \tag{1}$$

where $\phi$ = Resistance factor; $S_c$ = Specified strength; $X_p$ = Geometric property; $K^*$ = Adjustment factor including moisture content, duration of load, size and other effects. The nominal load effect calculated by the designer is given by:

$$B_n = X_g(\alpha_D D_n + \alpha_L L_n) \tag{2}$$

where $X_g$ = Geometric factor used to convert loads to stresses; $\alpha_D$ = Dead load factor = 1.25; $\alpha_L$ = Live load factor = 1.50; $D_n$ = Nominal dead load; $L_n$ = Nominal live load (prescribed). In a design environment $C_n \geq B_n$ and the required minimum size, $X_p$ is determined from equations (1) and (2)

$$X_p \geq \frac{X_g (\alpha_D D_n + \alpha_L L_n)}{\phi S_c K^*} \tag{3}$$

At this stage, the structural element of size $X_p$ is assumed to have a minimum level of safety close to that targeted by the code. It is possible to estimate this level of safety by making some assumptions regarding the underlying probability density functions for the strength values prescribed by the code and with regards to the acting loads.

## THE PERFORMANCE ENVIRONMENT

Once the structure is in service, its performance is dictated by the actual material strength and the imposed loads. These loads are specific realizations of a broad range of values characterised by the

263

nominal values used in the design environment. Within the performance environment the actual capacity is given by:

$$C_a = S_a X_p \qquad (4)$$

and the actual load effect is given by

$$B_a = X_g (D + L) \qquad (5)$$

where $S_a$ is the actual material strength; $X_p$ is the size determined by equation (3); $X_g$ is the geometric factor used in equation (2); D and L are the acting dead and live loads, respectively.

The material strength, $S_a$, is a random variable and it is possible to estimate the parameters of its probability density function in terms of the characteristic strength specified by the code, $S_c$. Reference (2) gives some recommendations for the selection of parameters for the probability density function for loads.


RELIABILITY CALCULATIONS

The reliability of the system can be formulated in terms of the performance function

$$G = C - B \qquad (6)$$

where C = capacity and B = load effect. The probability of failure can be defined as that of attaining the failure state, $G \leq 0$, obtained from basic theory, by direct integration of the convolution integral of the probability density functions $f_C$ and $f_B$. A commonly used measure of reliability is the reliability index, $\beta$, defined as $\beta = M_G / \sigma_B$. Substituting equations (3), (4) and (5) into (6) the performance function takes the form

$$G = \sum_{i=1}^{3} A_i x_i \qquad (7)$$

where

$$A_1 = \frac{1}{S_c} \qquad\qquad x_1 = S_a$$

$$A_2 = - \frac{\phi K * (\frac{D_n}{L_n})}{\alpha_D (\frac{D_n}{L_n}) + \alpha_L} \qquad x_2 = \frac{D}{D_n}$$

$$A_3 = \frac{A_2}{(\frac{D_n}{D_L})} \qquad\qquad x_3 = \frac{L}{L_n}$$

264

To illustrate these concepts, Reference (1) has used the example of a roof system consisting of 38 by 286 mm joists with 5.5 m span spaced at 400 mm. The lumber is Spruce-Pine-Fir No 2, under dry service conditions. Nominal dead load, $D_n$ = 0.5 kPa; nominal live load (snow), $L_n$ = 2.32 kPa; load factors $\alpha_D$ = 1.25 and $\alpha_L$ = 1.50; resistance factor, $\phi$ = 0.7.

## Uncorrelated gaussian variates

The random variables defining the performance function given by equation (7) are first assumed to be normally distributed and uncorrelated, they take the values given in Table 1 and are based on formulations recommended by reference (2). The reliability index, $\beta$, was calculated and is plotted in Figure 1 for a range of values of the coefficient of variation of the material strength variables, $S_a$.

## Correlated gaussian variates

The material strength, represented by variable $x_1$, as well as the load variables, $x_2$ and $x_3$, are assumed to be normally distributed and perfectly correlated. The reliability index was calculated and the results plotted in Figure 1.

Table 1

### VALUES FOR STRENGTH AND LOAD VARIABLES

| VARIABLE | MEAN | COEFFICIENT OF VARIATION |
|---|---|---|
| $x_1 = S_a$ | $\dfrac{2.1 S_c}{1 - 1.645 V_{x_1}}$ | $V_{x_1}$ |
| $x_2 = \dfrac{D}{D_n}$ | 1.00 | 0.10 |
| $x_3 = \dfrac{L}{D_n}$ | 0.82 | 0.26 |

This example assumes an invariant probability density function for the short-term material strength. The loads, however are assumed to achieve maximum value over the lifetime of the structure. Time under load reduces the strength of structures by a process of damage accumulation. The mathematical formulation of this process has been discussed by Gerhards (3) and Barret and Foschi (4) and used to characterise the effect of a known load history on the strength of wood structures.

## Reliability degradation

Gillard (5) and (6) has used the formulation developed by Barret and Foschi (4) to evaluate a possible degradation of strength and reliabiality of wood joist floor systems in service. The total damage, $\alpha$, accumulated to time $T_2$, is calculated as a function of the load history, $\sigma(t)$, the strength threshold value, $\sigma_o$, and the previously accumulated damage, to time $T_1$ by

$$\alpha(T_2) = e^{\lambda T_2}\{\alpha(T_1)e^{-\lambda T_1} + \int_{T_1}^{T_2} a[\sigma(t) - \sigma_o]^b e^{-\lambda t}dt\} \qquad (8)$$

A Monte Carlo simulation was performed to generate dead and live load components over a lifetime of 30 years and equation (8) was used to evaluate the damage accumulated by the end of the period. The resulting datapoints were used to determine the probability of failure of the system over the lifespan of 30 years. The results of this simulation, for solid wood joist floor systems, are shown in Figure 2.

## PROTOTYPE TESTING

Prototype testing forms the basis for evaluating the performance of some building components. Salina (7) found that engineered wood floor systems tested under standard performance evaluation procedures exhibited reliability levels considerably higher than those observed in traditional timber construction. Figure 3 shows that the reliability index of engineered wood products range from 6 to 9 whereas that of traditional construction is in the neighbourhood of 3.

In a study of testing procedures used to evaluate timber trusses, Salinas and Duchesne (8), questioned the ability of the traditional test methods to identify and reject unacceptable designs. Taking into consideration damage accumulation, rate of loading and variability of material properties they proposed a methodology to assess and compare different testing procedures. Due to the nature of the outcome of a test (pass-fail) the binomial distribution was used to define the testing plan in terms of the maximum applied test load and the sample size (number of specimens required to pass the test). For each testing scheme there is an equation describing the probability of accepting a given design, $P_a$, in terms of the probability of a single specimen surviving the procedure, q. This is known as the Operating Characteristic (OC) curve and it can be used to compare different sampling plans. Figure 4 shows the OC curves for two test procedures at two different test load levels, C = 2.0* Design load and C = 2.4* Design load. This study shows that Test procedure # 2, corresponding to a larger sample size, is better able to identify poor designs.

## Equivalent test procedures

Some early test procedures used to evaluate wood floor designs were believed to be too demanding, time consuming and expensive. Industry required more economical, faster and reliable test methods and in order to compare different proposed alterntives, Salinas and Gillard (9) introduced the concept of equivalence based on damage accumulation theory. Figure 5 shows two load histories and their corresponding damage curves. Under load history # 1, a constant load, $S_1$, is applied to a specimen until it fails at time $t_1$. Under alternate load history # 2, a ramp load is applied until the specimen fails at a load level $S_2$ in time $t_2$. Consider a specimen ramp-loaded following load history # 2 and failing at point B corresponding to a load level $S_B < S_2$ and time to failure $t_B < t_2$. The damage accumulated up to $t_B$ is $\alpha_B$. If we enter the damage curve corresponding to load history # 1 with $\alpha_B$, we can determine the time, $t_A$, required to accumulate this amount of damage and locate point A on load history # 1. We can then say that point A on load history # 1 is equivalent to point B on load history # 2. A weaker specimen C on load history # 2 will have an equivalent point A' on load history # 1 with a lower time to failure $t_{A'} < t_A$.

Salinas and Gillard (9) were able to conclude that equivalent, short-duration, test procedures can be used to determine the reliability of structural systems when a damage accumulation approach is used to take into account load-duration effects in wood components.


## PROOF LOADING

Proof-loading testing is a direct means of eliminating exceptionally weak structural elements from a production line or from a given population. This can improve the strength distribution of lumber or other engineered wood products, thus increasing their utility.

Economical proof-loading schemes generally require to test at a faster rate of loading than that prescribed for strength testing. Lumber appears stronger when tested at a fast rate; therefore, a proof-load test at a fast rate of loading is no guarantee that the surviving lumber has the required "strength" usually determined from standard tests.

While the effect of loading rate in proof-loading systems is generally accepted, whether lumber is damaged during proof-loading, and subsequently reduced in strength, is still a controversial issue having significant commercial implications.

Salinas and Cunliffe (10), based on the work by Cunliffe (11), used a damage accumulation model to account for the difference in the proof-load and strength test histories and any reduction in strength that may occur as as result of damage incurred during the test.

Figure 6 shows the effect of proof-load tests on a population of lumber subject to a 10-second ramp load. The right tail of a

load-effect probability density function is shown together with the left
tail of the resulting distribution after application of proof-loads of
various magnitudes. As can be seen in Figure 5, the resulting
probability density functions are not truncated at the load level
corresponding to the magnitude of the proof-load. Due to damage as
introduced by the test itself, the resulting distributions exhibit smooth
tails which are shifted to the right, resulting in a smaller "overlap"
between the load and strength distributions and, consequently, a smaller
probability of failure or larger reliability index.


DISCUSSION

The reliability of structural systems can be estimated at various stages
of their intended service life. Making some reasonable assumptions it is
possible to estimate the initial reliability of the system "as built".
With time under load, there is a degradation of its strength which can be
interpreted as a "shift" of the strength probability density function to
the left (lower) end of the strength scale. On the other hand, by
proof-loading certain components up to a predetermined load level it is
possible to eliminate exceptionally weak elements essentially shifting
the strength distribution t the right (higher) end of the scale. The
fact that a structure has survived a given load history gives improved
knowledge of its capacity to accumulate damage without failure. At this
stage, the concept of equivalence of the large variability associated
with lumber products this estimate must be within very broad, and perhaps
impractical, limits. In general, the assessment of the reliability of a
structure which does not exhibit creep-rupture characteristics will be
more accurate.

Knowledge that a structure has survived for a given time-span, with the
corresponding improved characterisation of the load and strength
probability density functions, can only be measured in terms of the
confidence associated with the estimate of its residual life, even if the
results indicate a certain degradation of its reliability. In these
terms, it would then be appropriate to consider a "lifetime under
service" as a useful indicator of the level of safety of a structure.


REFERENCES

1) Salinas, J.J. and R.I. Cunliffe. 1987. Lifetime performance and load
   tests of wood structures. Proceedings of the 5th International
   Conference on the Application of Statistics and Probabilities in Soil
   and Structural Engineering. Volume 1. Vancouver, B.C., Canada. May
   24-29, 1987. pp 246-253.

2) National Bureau of Standards. 1980. Special Publication # 577.
   Development of a probability based load criterion for American
   National Standard A58. U.S. Department of Commerce, Superintendent

of Documents, U.S. Government Printing Office, Washington, D.C., U.S.A. 20402, 228 pp.

3) Gerhards, C.C. 1977. Effect of duration and rate of loading on strength of wood and wood-based materials. U.S.D.A. Forest Service Research Paper FPL 283. Forest Products Laboratory, Madison, Wisconsin, U.S.A.

4) Barret, J.D. and R.O. Foschi. 1978. Duration of load and probability of failure in wood. Part I: Modelling creep rupture. Part II: Constant, ramp and cyclic loading. Canadian Journal of Civil Engineering, 5(4), pp 505-532.

5) Gillard, R.G. 1986. Reliability analysis of waferboard I beams. A damage accumulation approach. M. Eng. Thesis. Department of Civil Engineering, Carleton University, Ottawa, Canada. 141 pp.

6) Gillard, R.G. and J.J. Salinas. 1987. Reliability degradation of wood floor systems. Proceedings of the 5th International Conference on the Application of Statistics and Probabilities in Soil and Structural Engineering. Volume I. Vancouver, B.C., Canada. May 24-29, 1987, pp 441-449.

7) Salinas, J.J. 1979. Performance criteria and reliability of floor trusses with metal webs. Proceedings of the Metal Plate Wood Truss Conference, Forest Products Research Society, St. Louis, MO, U.S.A., November 1979. pp 243-247.

8) Salinas, J.J. and D.P.J. Duchesne. 1984. Evaluation of truss testing procedures. A.S.C.E. Specialty Conference on Probabilistic Mechanics and Structural Reliability. Berkley, California. January 1984. pp 135-139.

9) Salinas, J.J. and R.G. Gillard. 1987. Equivalent criteria in acceptance testing. Proceedings of the 5th International Conference on the Application of Statistics and Probabilities in Soil and Structural Engineering. Volume I. Vancouver, B.C. Canada. May 24-29, 1987. pp 325-333.

10) Salinas, J.J. and R.I. Cunliffe. 1988. A probabilistic evalution of the effects of proof-loading on the strength of wood structures. A.S.C.E. Engineering Mechanics Division, Joint Specialty Conference on Probabilistic Methods. Virginia Polytechnic Institut, Blacksburg, VA., May 1988. p. 131.

11) Cunliffe, R.I. 1988. Effect of proof-loading on lumber properties. M. Eng. Thesis, Department of Civil Engineering, Carleton University. Ottawa, Canada. December 1988. 246 pp.

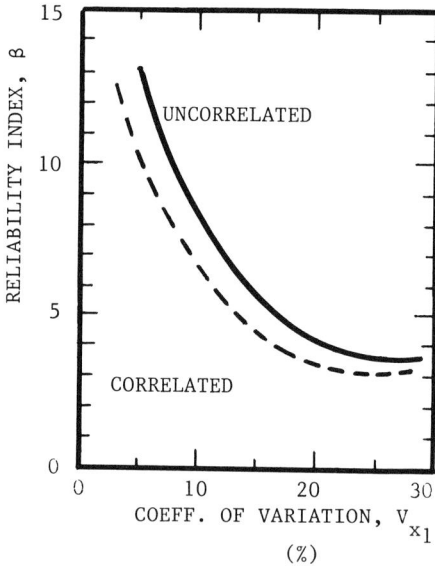

Figure 1   Initial reliability
index

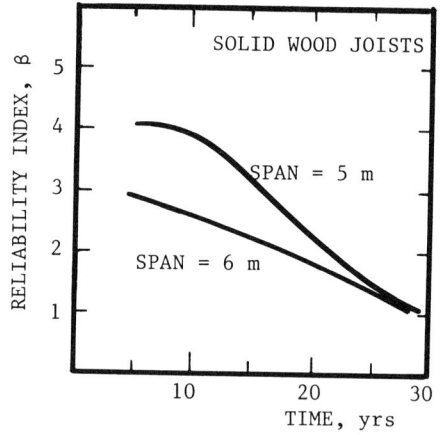

Figure 2   Reliability index
degratation

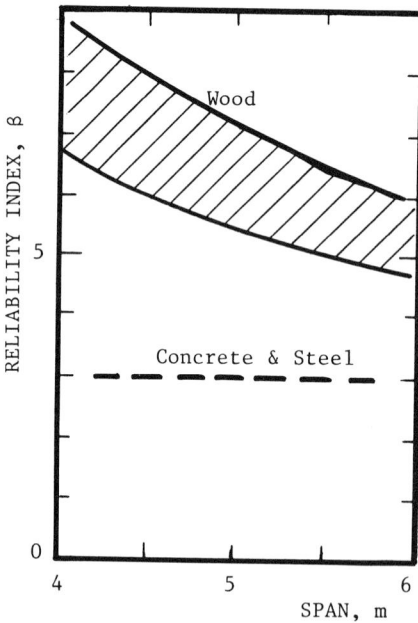

Figure 3   Reliability index of
engineered]wood systems as
standard test procedures

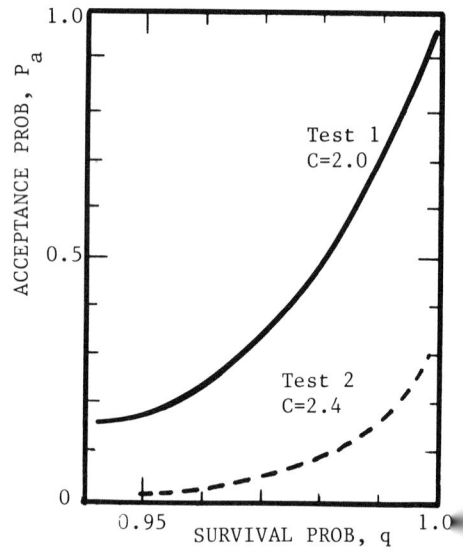

Figure 4   Comparison of test
procedures

270

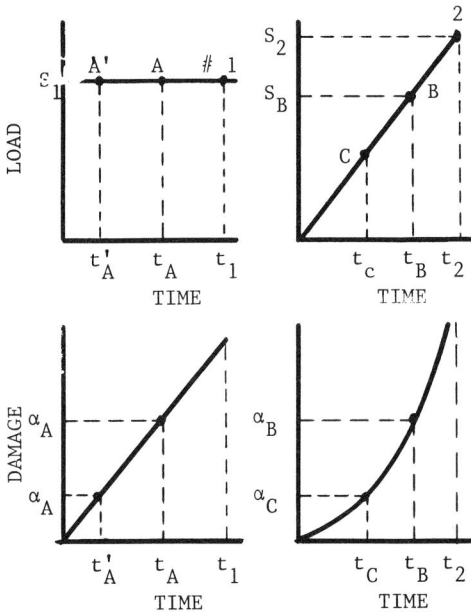

Figure 5    The concept of load
            equivalence

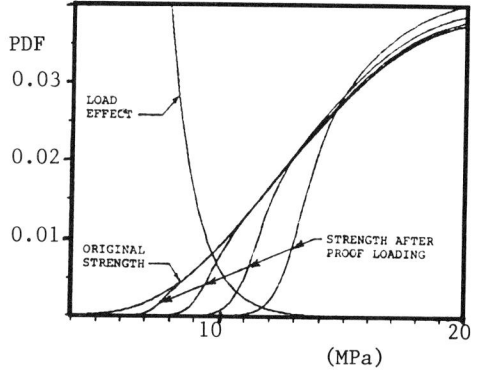

Figure 6    Effect of proof-load tests
            on lumber

# 31 The durability of adhesive bonds in structural applications

**C M Davies,** British Steel Technical, Swinden Laboratories, Rotherham, UK

This paper reviews the techniques which have been used to assess the durability of structural adhesive bonds. The role of non-destructive testing (NDT) is assessed and new developments are examined. It is concluded that suitable NDT techniques should be used to correlate the results of systematic periodic inspection of existing bonded structures with accelerated laboratory tests. Information from such a programme could then be used to assist in the design of structures having the optimum economic balance between performance and durability.

INTRODUCTION

Adhesives have been known since ancient times. In this century synthetic materials have supplanted animal and vegetable glues, and have allowed engineers to consider adhesive bonding techniques as an alternative to bolts, rivets and welds in structural applications. The aerospace industry pioneered such techniques during World War 2, and the success demonstrated in such demanding applications as combat aircraft airframes has led to a desire to transfer this technology into other industries. Automotive engineers see adhesive bonding as contributing towards a solution to the problem of building lightweight, rigid car bodies using a range of modern structural materials such as high strength steel, aluminium alloys and polymeric composites. Adhesives have been widely used in civil engineering applications since the early 1950's(1). Many of these uses have been non-structural in the sense that the adhesive has not been required to transmit significant stresses. In recent years, however, there has been an upsurge of interest in the use of adhesives in truly structural civil engineering applications. These include resin injection for the structural repair of cracked concrete, thin joints between precast prestressed concrete units in segmental construction of bridges and bonding new concrete to old, for the extension of existing structures. Interest is being expressed in the potential for the use of adhesive bonding in structural steelwork connections. In terms of the demands made upon the adhesive and the extent to which the procedure has been employed, one of the most interesting techniques used in civil engineering has been the bonding of steel plates to concrete highway bridges to act as supplementary reinforcement.

BRIDGE STRENGTHENING

The technique of using resin to form the shear connection between steel
plates and concrete dates back to the early 1960's and has been used
widely in Japan and in some European countries.  A considerable amount of
research into the design of externally reinforced concrete beams has been
performed in the UK, notably at the Univesities of Sheffield(2)and
Dundee(3).  The technique has been applied to a numberof bridges in the
UK;

- in 1975 a group of four bridges on the M5 motorway at Quinton, near
  Birmingham, were reinforced when a design check indicated that he
  bridges would not fully meet serviceability requirements under certain
  loading conditions(4),

- in 1977 two bridges at the newly opened M20/M25 motorway interchange at
  Swanley, Kent, were strengthened(1),

- in 1982 a bridge carrying the M1 motorway over a minor road near
  Rotherham, South Yorkshire, was upgraded for the passage of abnormal
  individual loads by the bonding of transverse steel plates to the
  underside of the deck(5),

- more recently, a bridge carrying the M1 motorway in Derbyshire was
  reinforced using steel plates bonded by injected epoxy adhesive, rather
  than the trowelled adhesives used in earlier projects(6).

These are all examples of repair and upgrading of existing structures.
There is at first sight nothing to prevent the technique being extended
to new construction, and prefabricated deck slabs in which the concrete
is poured onto an adhesive coated steel plate are under evaluation(7).
The chief objection to such a technique is that the durability of the
adhesive system is a largely unknown quantity.

The design lives of various types of structure differ greatly.  For cars,
domestic applicances and office furniture a life of approximately10 years
would normally be regarded as acceptable.  Aircraft and offshore
structures typically are expected to last for 30 years.  Structures such
as factories, shops and office blocks may be intended to last for 60
years, with major alterations at 20 year intervals.  In the UK the design
life for bridges is usually 120 years.  If a repair to a bridge is needed
urgently, or if the purpose of strengthening is to permit the passage of
a single heavy load, then the risk involved in using a technique of
un-proven durability may be worth taking; and in any case a design life
for the repair itself of, say, 30 years may well be judged acceptable.
It is clearly quite another matter to specify the use of an adhesive
bonding technique in the fabrication of a main member of a new bridge.  A
number of factorswill influence the behaviour of the adhesive bond over
the design life of a bridge;

- the mechanical and physical properties of the adhesive may be
  significantly affected by the absorption of moisture,

- moisture may attack the chemical bonds between the adhesive and the adherend,

- the adhesive may gradually deform under sustained load, causing a redistribution of stresses within the structure,

- temperature changes (day-night, summer-winter) may cause cumulative damage to the bond due to the differing coefficients of thermal expansion of the adhesive and the adherend or adherends,

- stresses due to live loading may cause cumulative damage to the bond, in a process analogous to fatigue of metal structures,

- structural performance may be affected by localised deterioration of the adherends in the vicinity of the bond (eg corrosion of the bonded surface of a steel plate),

- infrequent occurrences such as vehicle impact loads, spillages of vehicle fuel or tanker cargoes in accidents, and minor fires, may also contribute to long term changes in the behaviour of adhesive bonds.

There is obviously a need to evaluate these factors by experiment before adhesive bonding techniques can be confidently used in new structures.


ASSESSMENT TECHNIQUES

A number of approaches to the experimental evaluation of the durability of adhesive bonds is available. The simplest technique is to make a large number of small adhesively bonded specimens, expose them to an "accelerated" environment (eg 40°C, 95% relative humidity) for various periods (eg 1000 h) and then perform mechanical tests. A slightly more sophisticated technique involves the exposure of stressed small scale testpieces to such an environment and determining the time to failure as a function of applied stress(8). Such techniques have the advantage that their cheapness and simplicity allows a large number of tests to be performed and so statistically valid results can be obtained for a wide range of adhesive formulations, surface pretreatments and curing cycles. The data obtained are, however, suitable only for guidance of designers as the simple specimen geometry does not necessarily reproduce the stress distribution obtained in service, and there is no theoretical basis for relating performance in the test environment to performance in the service environment. A rather more meaningful, but more expensive and less convenient, technique is to fabricate representative structural elements and expose them to a "real" service environment selected for its severity. For example, "open sandwich" steel/concrete composite slabs have been exposed to the elements in the Indian Ocean island of Mauritius prior to destructive testing(9). The slabs were exposed to conditions of temperature and humidity far more severe than would be experienced in the UK, for a period of 4 years (although no attempt could be made to impose service loads during exposure).

The central problem in assessing the durability of an adhesively bonded structure lies in ensuring that the conditions experienced by the testpiece are sufficiently representative of service conditions for the appropriate degradation and failure mechanisms to be reproduced, whilst obtaining data within a convenient period of time. A secondary problem is to relate the degree of degradation experienced during the test period to the degree of degradation to be expected during a given number of years in service. A similar type of problem is experienced by the automotive industry in evaluating the durability of vehicle bodies. The approach used by the automotive industry is to drive the vehicle on a specially designed test track where past experience has an empirical correlation between test track mileage and road mileage/years of service to be established(10). More recently there has been a move towards laboratory simulation of certainaspects of test track performance. The essential requirement for the use of any such technique is service experience with items broadly similar to the item under evaluation. The systematic and periodic inspection of existing bonded structures would therefore be important in three ways;

a) to establish the continued serviceability of these structures and so give an assurance that long term durability is achievable,

b) to confirm that the effects produced in accelerated tests are the same as those produced by service over a number of decades,

c) to enable empirical acceleration factors to be derived for short term tests.

The inspection of traditionally fabricated steel structures is a relatively mature technology. The use of welded construction for critical applications such as ship hulls and nuclear reactor pressure vessels has been made possible by advances in techniques for the detection of discontinuities (particularly crack-like defects) and for the assessment of their significance. Such techniques can be used not only for assessing the quality of fabrication but also for detecting and monitoring in-service defects such as fatigue cracks. One of the challenges presented by adhesive bonding, and fibre-reinforced polymer composites, is the development of suitable NDT methods and techniques for assessing the significance of the defects detected. The nature of the materials, the configuration of the structures and the nature and orientation of the relevant defects conspire to make traditional ultrasonic techniques, radiography, dye penetrant, electro-magnetic methods and fracture mechanics theories of limited use.

Common adhesive bond defects are illustrated in Figure 1. Voids, porosity and poor cure are associated with imperfect mixing and application of the adhesive. The processes of interest in the study of durability are long term changes in cohesive properties of the adhesive, growth of disbonds and weakening of the adhesive-adherend bond.

The simplest technique for checking the adequacy of the plate-to-concrete bond is to lightly tap the plate with a hammer and listen to the sound produced. This technique has been used for the annual inspection of externally reinforced bridges(4). A less subjective version of this technique involves fitting a force transducer within the head of a hammer and monitoring the forces applied during the tap(11).

275

Ultrasonic techniques have been the subject of much development effort in recent years. The technique developed for NDT of metals, echo amplitude measurement at normal incidence, can be used to detect debonds in the same manner as cracks in metal(11). Scanning acoustic microscopy (SAM) is a related technique which differs from conventional ultrasonic inspection in that the frequencies used are much higher and that an acoustic lens is used to give a focused beam (12). Image contrast in the SAM arises from variations in the elastic properties of the sample. The SAM is able to produce a high resolution image from below the surface of an optically opaque material; this characteristic has been utilised to demonstrate the potential for the SAM to detect poor bonding in diffusion bonds(13) and in adhesive bonds between 4 mm thick aluminium plates(14). Ultrasonic spectroscopy can be used to establish the resonant frequencies of the adhesive layer; this information can be used to calculate both the thickness of the adhesive layer and the elastic modulus of the adhesive(15). Ultrasonic phase spectral analysis has been used in experiments with 6 mm thick steel plates bonded using a two part epoxy adhesive. Wide band pulses were analysed to provide data on the variation of velocity and of attenuation with frequency during curing of the adhesive and during subsequent cyclic loading(16). It was foundto be possible to monitor the progress of the cure, and marked changes in the ultrasonic propagation characteristics were observed during the final 20% of specimen fatigue life, providing warning of failure. Although total lack of bonding can be readily detected by ultrasonic techniques, the detection of poor bonding is a problem. A recently published paper reports significant progress, using an ultrasonic oblique incidence technique (17). If transverse waves impinge obliquely upon a bonded interface then the reflection factor depends upon, amongst other things, the rigidity of the bond. Using high frequencies at an angle of incidence of 30°, it was found to be possible to discriminate between different standards of surface preparation for a cured epoxy resin - aluminium sheet bond.

One of the disadvantages with ultrasonic techniques is that contact with the testpiece via a coupling medium is required. Pulse video thermography is a recently developed non-contacting technique for examining sub-surface defects(18). A flash tube is fired to deposit up to 1kJ of light on a selected area of the object during a 10 ms interval. An infra-red camera is focused onto the surface of the object and the images recorded on a video cassette as the heat diffuses into the object. Defects below the surface, producing thermal insulation, are indicated as bright areas.

Acoustic emission (AE) is a technique which detects defects by means of their characteristic responses to applied loads. Since the acoustic events monitored are generally produced by propagation of the defects the method is not, strictly speaking, an NDT technique. An array of piezo-electric transducers mounted on the testpiece respond to acoustic events by producing voltage signals which are then processed. These signals can be sorted according to their amplitude, their total energy or, by comparison of arrival times at different transducers, the locatio of the acoustic event.

Techniques such as those described above are used to detect the presence of defects and measure their size. In the NDT role these results are then used to assess the significance of the defects in terms of structural performance. For the present purpose it may be more convenient to monitor structural performance directly. this can be done using electrical resistance strain gauges bonded to the underside of the external reinforcing plate. Data logging equipment could then be used to monitor the magnitude and distribution of stresses in the plates under traffic loading. By correlating these measurements with traffic patterns on the bridge at intervals over a period of years, any changes in the structural performance of the adhesive bond could be detected. Areas in which bond degradation was indicated could then be more closely scrutinised using other techniques.

The techniques described are at various stages of development. Normal incidence echo amplitude ultrasonic testing and long term electrical resistance strain gauge monitoring, for example,' are well proven techniques which could be readily used. Others such as ultrasonic phase spectral analysis and the ultrasonic oblique incidence technique have only been demonstrated in the laboratory on simple testpieces, and would require some development before they could be confidently used in the field. Acoustic emission, although it now has a track-record in a variety of applications and proven hardward is available off-the-shelf, would need to undergo proving trials to confirm that acoustic events characteristic of the propagation of adhesive bond defects can be detected and distinguished from signals arising elsewhere in the bridge structure. Access to the underside of the bridge is also an important consideration in the selection of testing and monitoring techniques. Acoustic emission and electrical resistance strain gauge monitoring techniques require access only for the initial installation of the gauges and transducers, which can be wired to a more readily accessible monitoring point.

By using such techniques to inspect the existing population of bonded plate bridges at intervals over a 5 year period, it would be possible to build up a picture of the processes acting in the adhesive bonds over a 20 year period. Similarly, test specimens and representative structural elements could be monitored during accelerated ageing trials. With the information obtained from such a programme of research, a series of standard durability tests could be devised which would enable the properties of new designs and adhesive formulations to be evaluated in a relatively short period of time.

Lifecycle costing is the collective title for a range of techniques used to assess the total cost of a structure over its life, including initial cost and the cost of maintenance(19). This branch of accountancy is of use to design engineers when they are presented with a choice between a component which is durable and needs little maintenance and another component which has lower initial cost but requires more maintenance or earlier replacement. The same techniques can be used to schedule planned maintenance eg deciding upon the optimum period for re-painting a steel structure. In order to use such a technique it is necessary to know the planned life of the entire structure, the rates of interest and inflation to apply in the financial calculations, and the durability of the components under consideration. A number of factors contribute to inaccuracy in analysis, and so lifecycle costing is likely to remain a rough tool. The lack of precision inherent in an accelerated durability test should not, therefore, prevent such a test making a worthwhile contribution to achieving the optimum economic balance between performance and cost in new construction.

SUMMARY AND CONCLUSIONS

Adhesive bonding has been successfully used in the UK for repair and upgrading of concrete bridges since 1975. The existence of a series of strengthened bridges of various ages provides an opportunity for establishing a body of data on the durability of adhesive bonds in structural applications, and so remove a major obstacle to the use of such techniques in new construction. A variety of physical techniques are suitable for assessing the condition of bonds, and these can be used to inspect existing bridges and validate accelerated durability tests.

ACKNOWLEDGEMENT

The author wishes to thank Dr. R. Baker, Director of Research and Development, British Steel plc, for permission to publish this paper.

REFERENCE

1) G.C. Mays, Materials Science and Technology, vol. 1, pp 937-943, (1985).

2) R.N. Swamy, R. Jones and J.W. Bloxham, The Structural Engineer, vol. 62A, No. 2, pp 55-68, (1987).

3) G.C. Mays and A.E. Vardy, Int. J. Adhesion and Adhesives, vol. 2, No. 2, pp 103-107, (1982).

4) R.F. Mander, 'Structural bonded reinforcement-theory and practice', Proc. Conf. "Adhesives Sealants and Encapsulants 1985", London, (1985).

5) B.L. Davies and J. Powell, 'Strengthening of Brinsworth Road Bridge, Rotherham', Proc. Conf. IABSE 12th Congress, Vancouver, pp 401-407, (1984).

6) A.D. Leadbeater and C. Russell, 'The practical applications of externally strengthening existing highway bridges using steel', Proc. Symposium "Strengthening and Repair of Bridges", Leamington Spa, (1988).

7) A. Anandarajah and A.E. Vardy, The Structural Engineer, vol. 63B, No. 4, pp 85-92, (1985).

8) P.G. Sheasby, M.J. Wheeler and D. Kewley, 'Aluminium structures in volume car production', Institute of Metals Book 391 "Aluminium Technology '86", (1986).

9) G.C. Mays, 'Structural adhesive applications for bridges', Proc. Conf. "Adhesives, Sealants and Encapsulants 1985", London, (1985).

10) P.G. Selwood, et al., 'The evaluation of an adhesively bonded aluminimum structure in an Austin-Rover Metro vehicle', SAE Technical Paper Series 870149, (1987).

11) R.D. Adams and P Cawley, NDT International, vol. 21, No. 4, pp 208-222, (1988).

12) G.C. Smith, Materials Science and Technology, vol. 2, September 1986, pp 881-887.

13) P. Kapranos and R. Priestner, Metals and Materials, April 1987, pp 194-198.

14) N.J. Burton, IEE Proceedings, vol. 134, pt. A, No. 3, pp 283-289, (1987).

15) C.C.H. Guyott and P. Cawley, NDT International, vol. 21, No. 4, pp 233-240, (1988).

16) R.A. Kline, C.P. Hsiao and M.A. Fidaali, Trans. ASME J. Eng. Mat. Tech., vol. 108, July 1986, pp 214-217.

17) A. Pilarski and J.L. Rose, NDT International, vol. 21, No. 4, pp 241-46, (1988).

18) W.N. Reynolds, NDT International, Vol. 21, No. 4, pp 229-232, (1988).

19) S.B. Tietz, The Structural Engineer, vol. 65A, no. 1, pp 10-11, (1987).

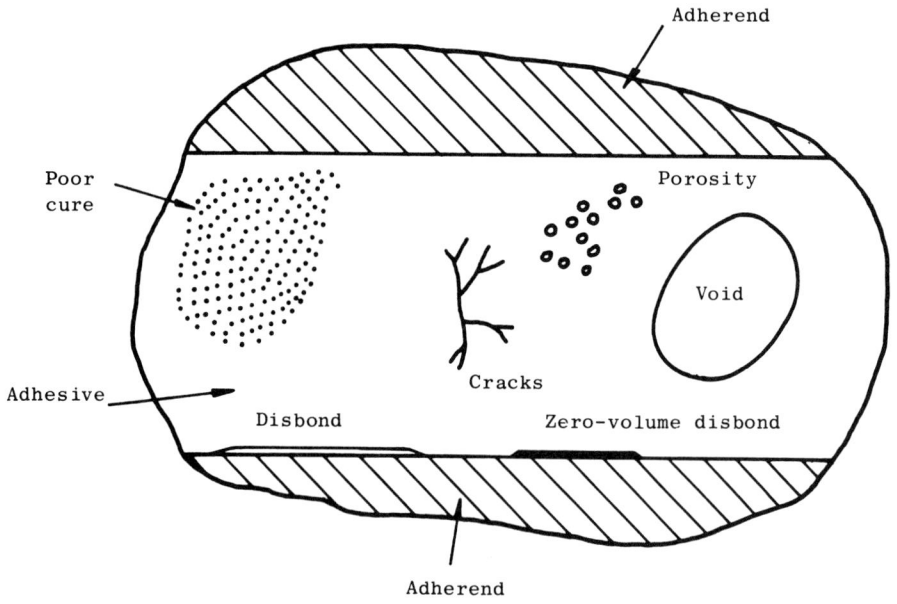

Figure 1　Common Defect Types

E Dore, CIRIA, London, UK

The author has illustrated his paper by a number of examples of bridge strengthening by bonded external plates without the auxiliary provision of mechanical fixings for structural action. Does this mean therefore that the author recommends the method only for the improvement of serviceability performance – or could a designer afford to include the contribution of such reinforcement to a normal factor of safety for ultimate load, bearing in mind the uncertainty of several aspects of durability?

## Author's reply

The question of durability naturally assumes a far greater significance when the adhesive bond is being relied upon for ultimate strength rather than for serviceability considerations. Clearly experience of the long term behaviour of bonds subjected to significant stress in the actual service environment is needed to complement the results of trials involving the weathering of laboratory specimens (such as the 10 year exposure tests recently reported by TRRL (1)).

The ability of externally bonded plates to increase the ultimate moment capacity of under-reinforced concrete beams has been the subject of studies at Sheffield University in recent years. It has been shown that suitably prepared bonded plates can act compositely up to the full theoretical capacity of the strengthened beam (2). Under certain conditions, however, premature failure can occur due to plate separation and horizontal cracking of the concrete at the ends of the plate. A variety of techniques for avoiding this problem were tested, and it was found that glued anchor plates provided the best solution whilst anchor bolts were only partially successful (3). More recently this team has studied the behaviour of structurally damaged beams strengthened by bonded plates whilst still under load, and reported that the repaired beams were able to preserve their structural integrity and maintain composite action until failure (4).

## References

(1) Calder, A J J. "Exposure tests on externally reinforced concrete beams – performance after 10 years", Transport and Road Research Laboratory Report RR 129, 1988.

(2) Swamy, R N., Jones, R. and Bloxham, J W. "Structural behaviour of reinforced concrete beams strengthened by epoxy-bonded steel plates", The Structural Engineer, 65A (2), pp 59-68, February 1987.

(3) Jones, R., Swamy, R N and Charif, A. "Plate separation and anchorage of reinforced concrete beams strengthened by

epoxy-bonded steel plates". The Structural Engineer, 66 (5), pp 85-94, March 1988.

(4) Swamy, R N., Jones, R and Charif, A.   "The effect of external plate reinforcement on the strengthening of structurally damaged RC beams", The Structural Engineer, 67 (3), pp 45-66, February 1989.

**J F A Moore, Building Research Establishment, U.K.**

What is the effective range of penetration of Pulse Video Thermography?

Author's reply

Pulse thermography depends upon changing the temperature of a thin surface layer and then monitoring the temperature distribution at a surface as heat diffuses through the sample.  Clearly the greater the depth at which a given defect is located the less will be the contrast.   The resolvable temperature differences, the intensity of the pulse, the thermal characteristics of the material and the size and nature of the defect will all affect the depth at which a defect can be resolved or its presence indicated.  A recent paper (1) on the application of the technique of laminates and adhesive bonds calculated a theoretical limit of delamination depth to diameter ratio of the order of 5 for an isotropic material, i.e. it would be impossible to detect by thermal transmission a delamination in a sheet of isotropic material if the diameter of the defect were less than approximately 1/5 of its depth below the surface.  The materials ranged from 0.4 to 1.2.

References

(1) Reynolds, W N.   "Inspection of laminates and adhesive bonds by pulse video thermography", NDT International, 21 (4), pp 229-232.

# 32 Establishing and implementing the long term constitutive behaviour of structural plastic pipe linings

**J C Boot and A J Welch,** University of Bradford, UK

This paper identifies the factors affecting the long term structural behaviour of plastics under stress, and the generalised procedures available for their quantification. It is concluded that taking into account the current level of development of such constitutive models, it is preferable to use simple extrapolation techniques based on data obtained from stress paths similar to those requiring prediction. Such a technique suitable for implementation in a finite element analysis is then described.

## 1) INTRODUCTION

It is a well known fact that many of our sewers are in a poor state of repair and that minimum surface disruption techniques for their rehabilitation are required. One such technique is that of spraying a structural plastic lining inside the existing pipe. The prime structural concern with this process (1) is resistance to the external head of ground water that often builds up once hydraulic integrity is restored. If the lining is other than perfectly circular and of constant properties throughout, this loading induces bending deformations which increase with time. Determination of resistance to long term failure due to constrained creep buckling within the confines of the original pipe is thus a major design criterion. A suitable research programme (1) was considered to comprise:

(i)     A series of short term buckling tests in order to assess the imperfection sensitivity of the situation, and provide an upper bound on field strength.

(ii)    Development of a mathematical model capable of simulating both the behaviour under increasing load to failure obtained from (i) and the performance subject to long term creep under constant load.

(iii)   Determination of the material properties required to enable the mathematical model to undertake the calculations described in (ii).

In this paper we describe the work carried out in respect of (iii) above.

## 2) CHOICE OF OPTIMUM CONSTITUTIVE MODEL

It can be demonstrated (1) that the problem posed is capable of solution by considering conditions on an arbitrary cross section. The performance of the lining under representative combinations of hoop bending moment and direct thrust per unit width (M,N respectively) over the required 50 year period (9) must therefore be determined. The load-deformation relationships for polymers are a function of stress history, temperature, chemical environment, shrinkage, and time (2). The generalised solution of these problems requires some form of viscoplastic (3,8) analysis. However, polymers behave in a highly stress path dependent manner; extrapolation of viscoplastic parameters to stress paths significantly different from those used to derive them is known to often give highly erroneous results (2,4).

It would seem therefore, that considering the current level of development of generalised constitutive models for polymers, there is considerable merit in utilising simplified techniques capable of providing an adequate representation of performance under a limited range of stress paths. Accordingly we can note the following points concerning the loading sequence in the case of the problems under scrutiny:

(i)    Initially the total load is applied monotonically.

(ii)   This load remains applied for the life of the structure.

(iii)  Sewers are placed at such a depth that the ambient temperature is independent of surface variations and approximately 15°C.

Under conditions of increasing load there is no essential difference between plastic and nonlinear elastic theories. The latter (5) establish a unique relationship between stress ($\sigma$), creep strain ($\epsilon$), and creep strain rate ($\dot{\epsilon}$). Embraced within this concept the 3 fundamental methods for determining total strain are:

$$\begin{aligned} \text{time hardening} \quad & \dot{\epsilon} = f(\sigma,t) \\ \text{strain hardening} \quad & \dot{\epsilon} = f(\sigma,\epsilon) \\ \text{total strain} \quad & \epsilon = f(\sigma,t) \end{aligned}$$

Although in general there is no simple relationship between $\sigma, \epsilon$, and time (t) the total strain theory is sufficiently accurate provided variations in stress are small. In addition, this theory is conservative in its estimation of total strain and is capable of being instituted directly from the data obtained from creep curves, which for polymers are of the form illustrated in Fig. 1: $\epsilon_o$ represents the quasi-instantaneous strain coinciding with stressing, and in general is comprised of elastic and plastic components. The primary creep stage then follows in which the rate of deformation reduces to the steady state of the secondary creep stage. At large strains the creep rate increases (tertiary creep) and rupture of the material follows. The constant temperature creep curves of Fig. 1 may be represented as a surface in $\sigma, \epsilon, t$ space (see Fig. 2).

As discussed earlier, the stress states of interest in the present

context are those obtained at various ratios of M:N. Furthermore, as
pointed out by Hult (7) it is not valid to simply superimpose bending
and axial behaviour when the creep (or instantaneous) response is
nonlinear. To reduce superposition errors, two sets of curves in $\sigma_i$, $\epsilon_i$,
t space are needed, with the stress resultants N & M in turn plotted on
the stress axis (see Fig. 3).

## 3) LONG TERM PERFORMANCE FROM SHORT TERM TESTING

Struik (6)' establishes that, within the limits of physical ageing, the
long term creep compliance of all polymers is of the same general form,
with specific responses a function of the thermal and chronometric
stability of the material. If these characteristics are favourable, a
plot of $\epsilon$ against ln(t) will tend towards linearity. Under these
conditions, provided test duration is sufficient to establish the linear
portion of the curve, extrapolation from 1% of the period of interest
can be made to +5%, −15% accuracy. This figure is increased slightly
if the linear portion is not quite reached. Thus in order to achieve
valid 50 year predictions a test period of 6 months is required.

## 4) TEST PROCEDURES AND RESULTS

The material obtained for test was the filled polyurethane used by
Subterra Ltd. in their 'Polyspray' process. The validity of using the
semilogarithmic extrapolation technique for obtaining the constitutive
behaviour of this material over the required 5o year period was
established using Dynamic Mechanical Thermal Analysis (11).

Unfortunately, within the time scale available for testing, it was not
possible to devise a viable procedure which would enable the
constitutive behaviour to be established in the detail suggested in Fig.
3. Thus, notwithstanding the arguments put forward in Section 2, it was
decided to treat bending and axial behaviour as fully separable and
develop one surface only in each axis system – pure compression and pure
bending. This assumption is conservative as creep redistribution will
reduce the flexural deformations so calculated, and buckling behaviour
is dominated by the presence of bending with small increases in the
axial deformations being relatively insignificant (1).

The axial compression and 4 point flexural creep tests recommended by
ASTM D2990 (12) were used to establish the required information. Each
test was carried out in a temperature controlled room at 15 ± 2°C using
8 identical samples (1) stressed at between 0% and 66% of the short
term rupture stress. Two strain gauges on each specimen were logged
automatically throughout the required 6 month test period, and the
temperature was similarly recorded using a thermocouple. The
temperature expansion coefficient of the test material was found to be
150 $\mu s$/°C, and thus the strains induced by even a 2°C variation are
significant. The temperature corrected strains of the unstressed
samples established material shrinkage over the test period, which was
therefore subtracted from the remaining temperature corrected axial
results. The results from the two gauges on each axial specimen were
then averaged to yield the curves in Fig. 4. The flexural creep results
are illustrated in Fig. 5 and were obtained by averaging the raw data
from top and bottom gauges on each specimen as this process

automatically eliminates temperature and shrinkage effects from these results.

It can be seen from Fig. 4 that the linear portion of the semi-logarithmic plot has been reached by all specimens. Although the linear part of the flexural creep curves has not been reached, the curvature of these lines is clearly tending towards zero. The 50 year extrapolations in Figs. 4 and 5 are therefore accurate to within the lower and upper margins previously quoted.

Polyurethanes are known to react chemically with water (6). In order to assess the effects of immersion in water on the material under test, 4 nominally identical tests (1) in 3 point bending were monitored using dial gauges. 2 tests were carried out in water, 2 in air. The results of all 4 tests are plotted semilogarithmically in Fig. 6, and indicate that material stiffness is significantly reduced by the presence of water.

5) CONSTRUCTION OF THE STATE SURFACES

According to the philosophy adopted in Sections 2 and 4, the information required to establish the incremental constitutive behaviour is the instantaneous tangent $(\partial\sigma_i/\partial\epsilon_i)$ at any desired point on each of the two surfaces. To achieve this the data presented in Section 4 must first be smoothed and defined mathematically in $\sigma_i$, $\epsilon_i$, t space.
Initially a least squares curve fitting procedure was used to identify the optimum constants A,B,C for the following functions which were used to interpolate the data for each test:

$$\epsilon_i = A + \ln(B.t^C + 1) \qquad ; \ 0 \ \text{sec--6 hrs.}$$

$$\epsilon_i = A.\ln(t + 1)^2 + B.\ln(t + 1) + C; \ 6 \ \text{hrs.--6 mths.} \quad 1$$

$$\epsilon_i = A + B.\ln t \qquad ; \ 6 \ \text{mths.--50 yrs.}$$

Fig. 7 illustrates the typical degree of correlation obtained for the second function.
Using the same curve fitting procedure, a series of isochronous curves at 0.5 intervals on a logarithmic scale in seconds was then established. Quadratic functions were chosen as polynomial curves were found to give the best fit, subject to the constraint that slope decrease with increasing $\epsilon_i$.
Assuming the effect of immersion in water to be independent of stress enabled this factor to be taken into account by re-expressing the above functions in terms of $\epsilon_i'$ (the value of $\epsilon_i$ wet), where

$$\text{for } t < t_1 \qquad \epsilon_i' = \epsilon_i \qquad \qquad \left.\begin{array}{l} \\ \\ \end{array}\right\} \quad 2$$
$$\text{for } t > t_1 \qquad \epsilon_i' = \epsilon_i|_{t_1} + [\epsilon_i - \epsilon_i'|_{t_1}].P \quad$$

and $t_1 = e^{7.5}$ hours, P = 3.09 is the ratio of the linear slopes obtained in Fig. 6.

This procedure numerically identifies the two state surfaces as a series of values for $\sigma_i$ on a regular grid in the $\epsilon_i'$, t planes as illustrated in

Fig. 8.
Each surface can now be represented in parametric form (see Fig. 9) as

$$r(s,n) = \epsilon_i' (n).i + t(s).j + \sigma_i(s,n).k \qquad 3$$

where i, j, k are unit vectors in the $\epsilon_i'$, t, $\sigma_i$ directions respectively,
and n,s are parameters of the surface whose directions are chosen to
have projections parallel to the $\epsilon_i'$ and t axes respectively.   The
tangent to each surface is obtained by vector differentiation as

$$r_i = t = \frac{\partial \epsilon_i'}{\partial n} . i + \frac{\partial \sigma_i}{\partial n} . j \qquad 4$$

To establish Equation 4 at any point x, the point is located as
centrally as possible within a bicubic Lagrangian interpolation (8) put
through 16 points on the numerically defined surface.   Thus

$$\sigma_i = \sum_{j=1}^{16} \sigma_{ij} . \varphi_j(n,s) \;\; ; \;\; \frac{\partial \sigma_i}{\partial n} = \sum_{j=1}^{16} \sigma_{ij}.\frac{\partial \varphi_j}{\partial n}$$

$$\epsilon_i' = a + mn \qquad ; \frac{\partial \epsilon_i'}{\partial n} = m$$

when $\sigma_i = N$ the instantaneous tangent modulus is given by

$$E_a = \frac{\partial \sigma_a}{\partial \epsilon_a} = \frac{\partial \sigma_a}{\partial N} . \frac{\partial N}{\partial n} . \frac{\partial n}{\partial \epsilon_a'} = \frac{1}{A.m} \frac{\partial N}{\partial n} \qquad 5$$

where A is the csa and $\sigma_a$ the extreme fibre stress due to axial
effects.   Using similar notation, when $\sigma_i = M$ there is not necessarily
a unique relationship between M and $\sigma_b$ if the latter is modified by
plasticity or creep.   However, at the stress levels considered   the
curves in the M, $\epsilon_b$ plane are very shallow (1) and it is therefore
reasonable to assume that the stress block due to bending remains
approximately linear throughout.   Thus

$$E_b = \frac{\partial \sigma_b}{\partial \epsilon'_b} = \frac{\partial \sigma_b}{\partial M} . \frac{\partial M}{\partial n} . \frac{\partial n}{\partial \epsilon'_b} = \frac{1}{Z_e.m} \frac{\partial M}{\partial n} \qquad 6$$

where $Z_e$ is the elastic section modulus.   Similarly the secant moduli
relating total stresses and strains are:

$$E_A = \frac{1}{A} \frac{N}{\epsilon'} \;\; ; \;\; E_B = \frac{1}{Z_e} \frac{M}{\epsilon'} \qquad 7$$

Hult (4) suggests that as a simple analysis the tensile strain at
rupture can be considered constant.   From initial tensile testing (1)
the strain at failure was found to be 3%.   From Figs. 4 and 5 the
maximum compressive and flexural strains are 3% and 4%.   Thus, in the
present context, the high axial stresses and redistribution due to
creep can be expected to ensure creep buckling occurs before flexural
tensile rupture, and no attempt has been made to establish the state
surface cut off due to rupture.   If required, the rupture line can be
established in the same manner as the creep data, but testing at stress

287

levels chosen to ensure rupture is reached within the test period.

The plots of shrinkage strain against ln(t) used to obtain (1) Figs. 4 and 5 were almost linear, and a simple cubic function has been used in the constitutive model to relate these two parameters.

## 6) IMPLEMENTATION

The incremental equilibrium equations can be expressed as (1,3)

$$K_T \; . \; \Delta u - \Delta f = 0 \qquad\qquad 7$$

where $K_T$, $\Delta u$, $\Delta f$ are the current structure tangent stiffness matrix, and incremental displacement and loading vectors respectively. In the solution of creep buckling problems using the constitutive model described here, $\Delta f$ contains any increase in external loads and effective terms due to reductions in the tangent moduli $E_a$, $E_b$ over the time increment $\Delta t(1)$. Buckling occurs when deflections increase at constant time.

## 7) CONCLUSION

A technique for representing the constitutive behaviour of structural plastic pipe linings as a series of state curves in stress, strain, time space based directly on observed data has been presented.

The significantly different values obtained for $E_b$, $E_a$ (see e.g. Table 1) at identical extreme fibre strain level and time suggests that the inclusion of a number of intermediate curves would enable a more accurate representation.

Since pipe wall thicknesses are small it is only possible to measure extreme fibre strains. The concept of a single modulus to represent bending therefore relies on the identification of a suitable superposition procedure (4) for bending and axial effects using only extreme fibre strains. If such a procedure cannot be justified in a given application, compression and tension state surfaces can be constructed and combined bending and axial behaviour simulated using a layered thickness approach (3).

The proposed method, therefore, represents a generalised approach for the solution of problems involving the creep behaviour of structural plastics subject to approximately constant loading.

## REFERENCES

1) A.J. Welch, Creep buckling of infinitely long constrained cylinders under hydrostatic loading, Ph.D Thesis, University of Bradford, to be published.

2) R.K. Penny & D.L. Marriott, Design for Creep, McGraw-Hill, (1971).

3) D.R.J. Owen & E. Hinton, Finite Elements in Plasticity,

Pineridge Press, (1980).

4) J.A.H. Hult, Creep in Engineering Structures, Blaisdell, (1966).

5) W.N. Findley & G. Khosla, 'Application of the superposition principle and theories of Mechanical Equation of State, strain and time hardening to creep of plastics under changing loads', Jnl. of Physics, 26, 7, 821-832, (1955).

6) L.C.E. Struik, Physical Ageing in Amorphous Polymers and other Materials, Elsevier, (1978).

7) American Society for Testing and Materials, Standard methods for tensile, compressive, and flexural creep and rupture of plastics. ASTM D2900-77 (1987).

8) O.C. Zienkiewicz, The Finite Element Method, 1977.

9) WRc Engineering, Sewerage Rehabilitation Manual, Water Research Centre, Swindon, (1986).

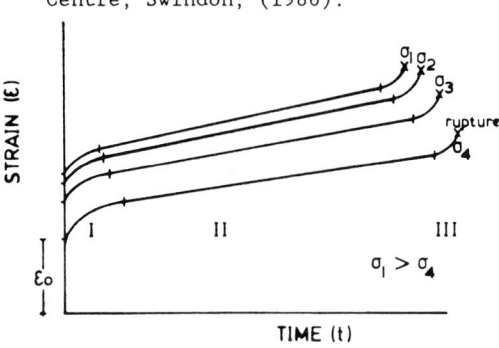

FIGURE 1  Idealised creep curves.

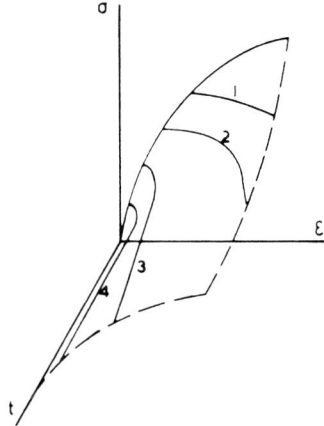

FIGURE 2  Typical creep (state) surfaces.

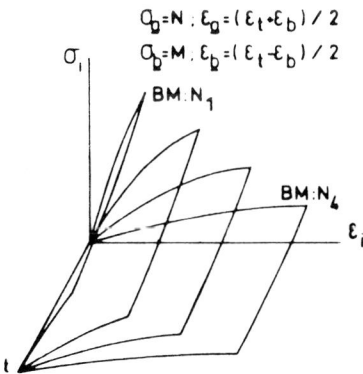

$$\sigma_b = N \; ; \; \varepsilon_a = (\varepsilon_t + \varepsilon_b)/2$$
$$\sigma_b = M \; ; \; \varepsilon_b = (\varepsilon_t - \varepsilon_b)/2$$

FIGURE 3  Range of state surfaces.

| time sec | strain μs | $E_a$ N/mm$^2$ | $E_b$ N/mm$^2$ |
|---|---|---|---|
| 60 | 2000 | 1557 | 2129 |
| 6300 | 2000 | 817 | 1263 |
| 657x10$^5$ | 2000 | 316 | 377 |
| 6300 | 8000 | 532 | 1140 |
| 657x10$^5$ | 8000 | 276 | 360 |

TABLE 1  Typical comparisons of instantaneous moduli

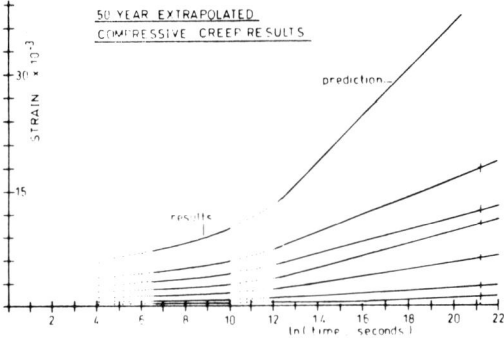

FIGURE 4 Compressive creep results.

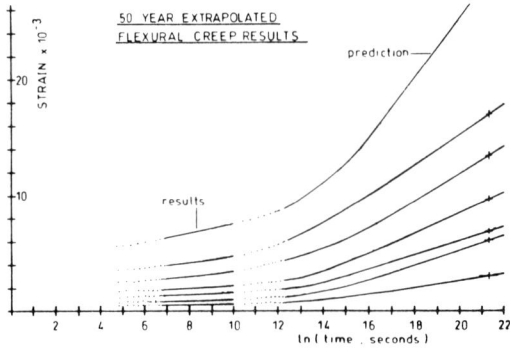

FIGURE 5 Flexural creep results.

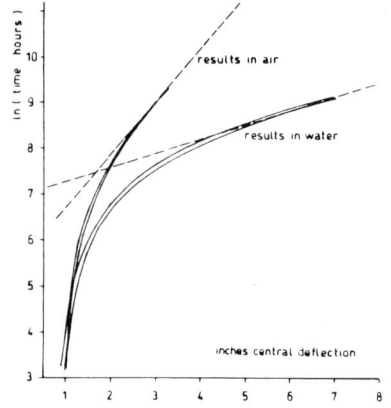

FIGURE 6 Wet/Dry flexural creep tests.

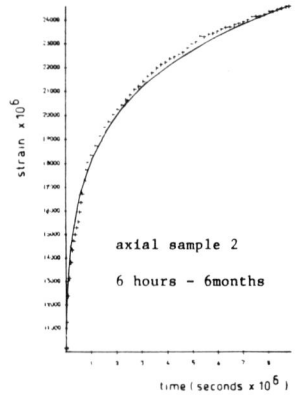

axial sample 2

6 hours - 6months

FIGURE 7 Typical curve fitting results.

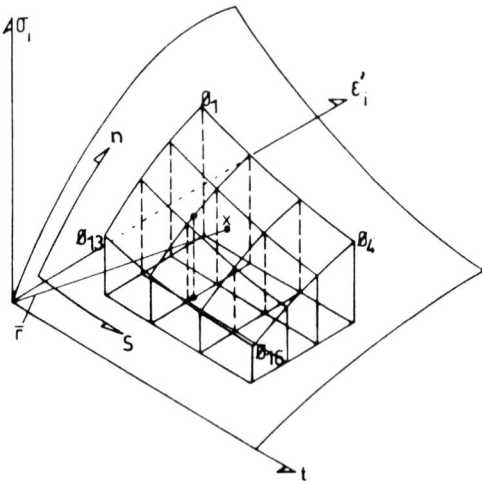

FIGURE 9 Bicubic interpolation of state
surface

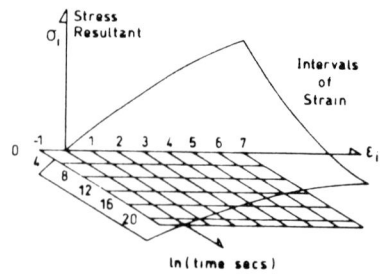

FIGURE 8 Grid defining state surfaces.

# 33 Testing of brick masonry piers at seventy years

**D P Abrams and G S Epperson,** The University of Illinois at Urbana-Champaign, USA

A series of experiments were done to evaluate various types of nondestructive methods that are commonly used for measuring mechanical properties of masonry structures. Test samples consisted of a set of unreinforced brick piers which were extracted from a building that was constructed in 1917. Data is presented related to the velocity of ultrasonic and sonic ' waves, the shear resistance of individual bricks to a horizontal shove force, and estimates of vertical compressive stress from induced stress normal to bed joints. Estimates of shear strength are correlated with the actual strength of test piers loaded to failure in the laboratory.

## INTRODUCTION

If the structure of a building can hold all gravity loadings without excessive deflection or vibration, the structure is deemed serviceable. No testing or evaluation is necessary. However, the capacity of the structure to resist lateral forces such as wind, earthquake or blast is seldom tested because extreme lateral forces rarely occur. It is difficult to surmise by the life of the structure its adequacy under intense lateral forces. If the integrity of the gravity-load system is preserved for several decades, then the proof of its capability is all the more certain. However, for lateral loadings the converse is true because the probability of exceeding the return period of an extreme event increases with the age of the building.

Nondestructive evaluation can play an important role on the life of a masonry building structure. Most older brick buildings have been designed with excessive amounts of conservatism because of uncertainties involved with estimating mechanical properties. If accurate methods of assessment can be developed and verified, then neglected structural strength can be considered effective, and demolision or strengthening may be precluded.

The purpose of the experimental study reported herein was to check the suitability of various nondestructive methods. The overall goal was to develop methodologies for interpretting measurements to evaluate the shear strength of an actual masonry pier or wall. Test samples were piers that were extracted from a building constructed in 1917, and transported to the laboratory. Each pier was placed in a rig which could simulate gravity and lateral stresses. Through correlation of predictions with actual shear strength, the worthiness of each nondestructive method was assessed.

NONDESTRUCTIVE TESTS AND SAMPLE DATA

The types of nondestructive methods used were (a) ultrasonic wave
velocity, (b) pulse-wave velocity, (c) shove test and (d) flat-jack test.
Each method was used on a test pier which was placed in the loading rig
shown in Fig. 1. A set of eight hydraulic jacks were used to apply a
uniformly distributed vertical compressive stress. The vertical stress
was varied in increments of 40psi up to a maximum of 120psi for each of
the NDE tests. Following the nondestructive measurements, test samples
were loaded to failure by applying horizontal forces at the top of a wall
with twin servohydraulic actuators.

## Ultrasonic Wave Velocity Test

The use of ultrasonic measurement is common for flaw detection with a wide
variety of civil engineering structures, however, only recently has its use
been applied to masonry structures (2,4,5). The method consists of
transmitting a wave through a portion of wall and sensing the arrival time
at a remote location. The wave velocity is easily computed as the arrival
time divided by the distance traveled. Equipment is simple, and can be
purchased as a shelf item from a number of manufacturers. Flaws can be
detected because velocities are slower for waves travelling around voids
than for waves travelling through solid material over the same nominal
distance.

Arrival times for ultrasonic waves were measured across the area of a test
wall, and are plotted in Fig. 2 versus distance traveled. It is evident
from the set of curves that waves took longer to travel vertically than
horizontally, which can be attributed to the difference in the number and
spacing of mortar joints in the travel path. The break in slope for the
vertical wave suggests that a flaw of some sort must have been at a
distance of approximately four feet from the base. This is likely to be
attributable to some delamination of the brick and mortar because the break
in slope is eliminated when vertical compressive stress is applied to the
wall (Fig. 3). Whereas the ultrasonic method was found to be sensitive to
vertical compressive stress, a general theory needs to be developed to
predict vertical stress from wave velocity for it to be of general use.
However, the method is suitable for assessing the homogeneity of a wall,
which needs to be defined so that results of other NDE tests may be
extrapolated.

The limitation of the ultrasonic method is that the wave length is too
short relative to the scale of masonry dimensions. Reflection and
refraction of waves tends to be a problem across mortar joints.
Furthermore, waves tend to attenuate within a distance (on the order of
four to five feet) that is shorter than needed to obtain a representative
sample.

## Pulse Velocity Test

This method consists of exciting mechanical pulses by hitting the wall
with a hammer. Sonic waves are transmitted through a wall and detected
with an accelerometer that is secured to the wall at a distant location.
Impact is made normal to the plane of a wall inducing a compression wave.
Accelerations can be measured on the same side of the wall as the impact
force, or through the wall on the opposite side. A typical impluse and

resulting waveforms for accelerations in the horizontal direction as well as directly through the wall are shown in Fig. 6.

Like the ultrasonic test, the arrival time of the wave is of primary concern. Again, through comparison of wave velocities, flaws can be detected as shown in Fig. 4. Waves always took longer to travel vertically than horizontally because of the larger number of mortar joints. Velocities were also found to be sensitive to the amount of vertical compressive stress as shown in Fig. 5, but again it would be dubious to obtain a quantitative estimate of the ratio between vertical stress and wave velocity for a general class of walls. Attenuation of sonic waves was observed to occur at longer distances than for the ultrasonic waves, on the order of seven feet or greater.

The frequency content of the wave and the crosscorrelation between the impulse and response appear to contain salient characteristics but require further study.

Shove Test

The shove test is a name given to a procedure that measures the shear strength of the brick-mortar interface along the length of a single brick. The method is slightly destructive in that a brick has to be removed so that a hydraulic jack and load cell can be placed in the brickwork (Fig. 7). In addition, the head joint needs to be removed on the opposite side of the brick so that the full applied load can be transferred through shear. The ultimate shear stress is then extrapolated to give the shear strength of a wall or pier by multiplying it by the shear area of the element. A sample set of test data is shown in Fig. 7. Load applied to a single brick is plotted versus the horizontal movement of the brick. Each loading and unloading curve in the figure' represents a different vertical compressive stress. Although the previous loading history should influence the shear strength for a subsequent loading, the strength measured during the final loading nearly replicated that at the first loading. Two separate tests were performed on either side of the test wall. From these data, ultimate shear stress is plotted versus vertical compressive stress in Fig. 8. The slope of a best-fit line suggest a coefficient of friction in the range of 0.7.

Flat-Jack Test

Knowledge of the axial compressive stress in a particular portion of wall is tantamount to knowing the shear strength of unreinforced brick masonry. Flat jacks have been used in rock mechanics to detect the amount of vertical compressive stress at a joint. A thin, hydraulic jack is placed in a void created by cutting out a joint. The amount of jack pressure (with correction factors for the stiffness of the flat jack and its area) needed to reposition the joint to its original position is equivalent to the axial stress across the joint before testing. The method has been extended to masonry structures by a number of investigators (3,6). A standard has been proposed as well (1). The technique was explored by cutting two joints on either side of a wall, and subjecting the wall to different amounts of vertical compressive stress. A typical distribution of vertical deflection along a stressed joint was monitored and is plotted in Fig. 9 for varying applied stress levels. Predictions of vertical compressive stress are plotted versus actual stress in Fig. 10 where it is apparent that the flat-jack method provides reasonably reliable stress estimates.

MEASURED BEHAVIOR OF TEST PIER

To date, one test wall has been loaded to failure. Failure began as a flexural failure, with cracking initiating at the region of net tensile stress. Cracking then halted, and the pier again resisted an increasing load. Ultimate load occurred following the extension of diagonal cracking from the top central region to the lower compression region. The ultimate load was reached at 121 kips (Fig. 11), resulting in a gross shear stress of approximately 76psi. This is a lower shear strength value than determined from the shove test. However, values of shear strength measured from the shove test are likely to vary considerably over the specimen cross section. A reduction factor should be considered due to probable flaws and inhomogeneous regions discovered during ultrasonic and sonic testing. Given this, the measured ultimate strength of the test wall is not unreasonable.

SUMMARY AND CONCLUSIONS

Four nondestructive methods were evaluated in an experimental study. Wave velocity tests were found to be acceptable for flaw detection, and were found to be sensitive to amounts of vertical compressive stress. Mechanical pulse waves were found to be a more practical means for measuring flaws in masonry than ultrasonic waves because of their longer wavelength and attenutation distance. Shove tests provided unique data on shear strength of brick-mortar interfaces which could be extrapolated to provide a reliable estimate of the ultimate shear resistance of a brick pier. Vertical compressive stress could be estimated by measuring the hydraulic pressure needed to displace brickwork back to its undisturbed position.

ACKNOWLEDGMENTS

Research was funded by the Army Research Office through the Advanced Construction Technology Center at the University of Illinois at Urbana-Champaign. Opinions and findings presented in this paper are those of the authors and may not reflect those of the Army or the Center. Appreciation is extended to Mr. Joseph Vieceli and Mr. Lyndon Lawyer for their help with the laboratory investigations.

REFERENCES

1. "General recommendations for methods of testing load-bearing masonry," RILEM Draft Recommendations (1988).

2. Kingsley, G R, J L Noland, and R H Atkinson, "Nondestructive evaluation of masonry structures using sonic and ultrasonic pulse velocity techniques," Proceedings of Fourth North American Masonry Conference, UCLA (1987).

3. Noland, J L, G R Kingsley, R H Atkinson, "Utilization of nondestructive techniques into the evaluation of masonry," Proceedings of the 8th International Brick/Block Masonry Conference, Dublin, Ireland, (1988).

4.  Bocca, P, "A study of microcracking in masonry construction: the use of pulse velocity measurements," Proceedings of the 8th International Brick/Block Masonry Conference, Dublin, Ireland, (1988).

5.  Calvi, G M, "Correlation between ultrasonic and load tests on old masonry structures," Proceedings of the 8th International Brick/Block Masonry Conference, Dublin, Ireland, (1988).

6.  De Vekey, R C, "Nondestructive test methods for masonry structures," Proceedings of the 8th International Brick/Block Masonry Conference, Dublin, Ireland, (1988).

Figure 1   Test Rig

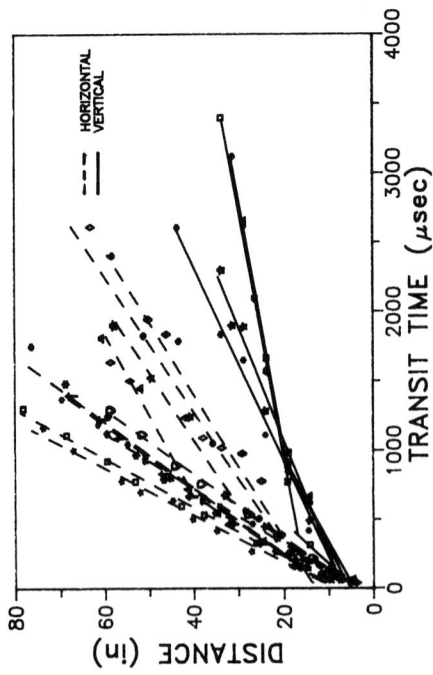

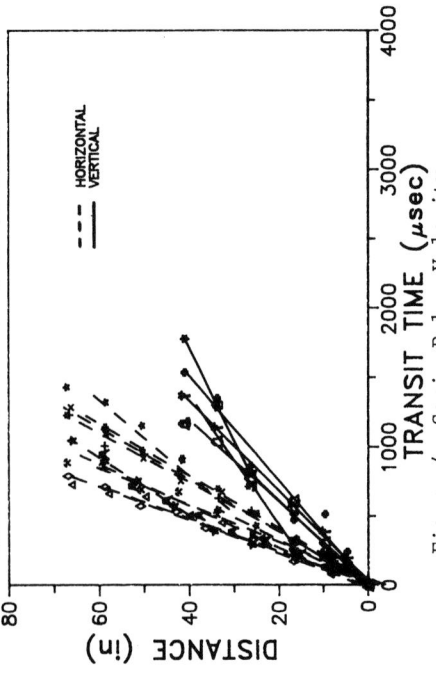

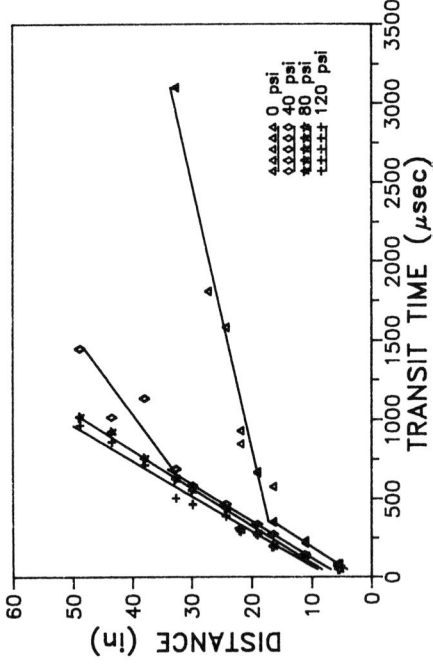

Figure 3  Ultrasonic Test – Varying Axial Stress

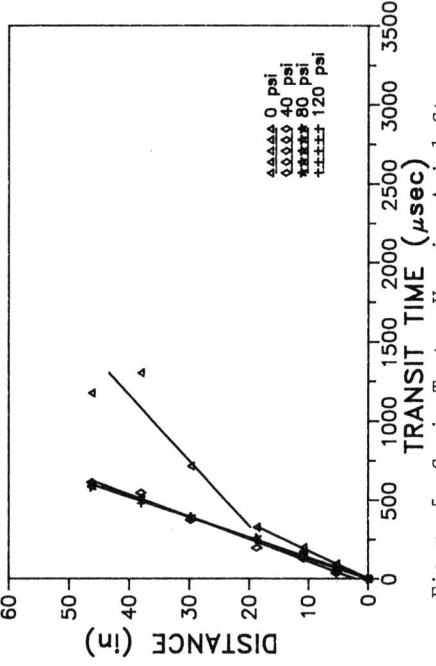

Figure 5  Sonic Test – Varying Axial Stress

Figure 2  Ultrasonic Pulse Velocity

Figure 4  Sonic Pulse Velocity

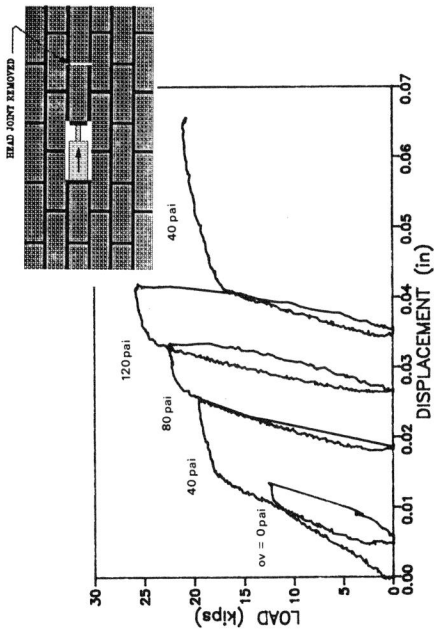

Figure 7  Shove Test – Varying Axial Stress

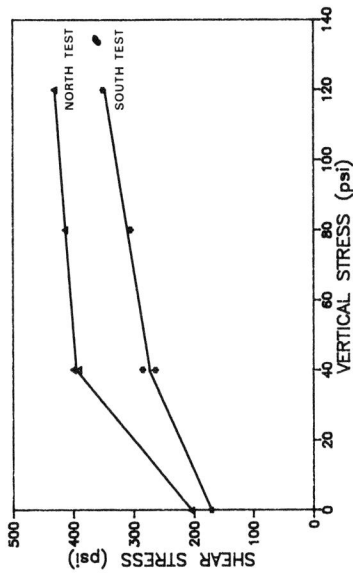

Figure 8  Ultimate Shear Stress

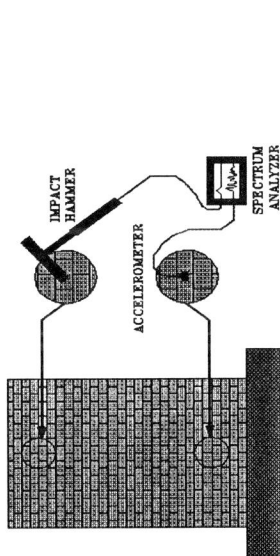

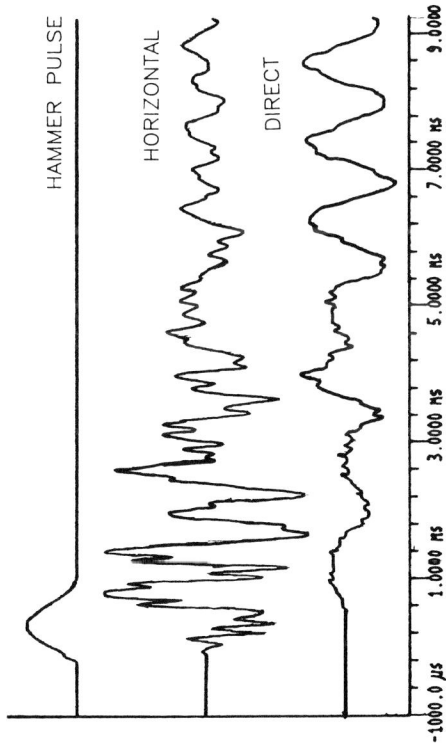

Figure 6  Typical Waveforms from Sonic Test

297

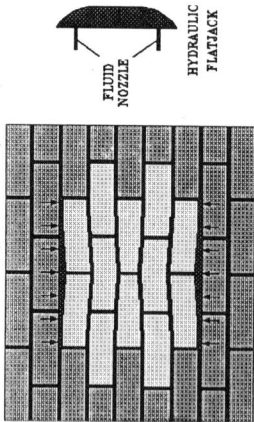

Figure 10 Evaluation of Flatjack
Axial Stress Predictions

MEASURED STRESS (psi)

ACTUAL STRESS (psi)

BEST FIT
1:1

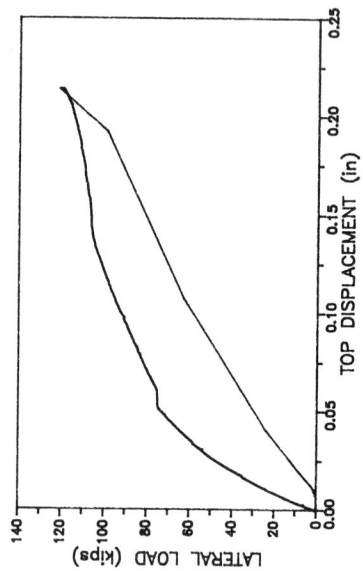

Figure 11 Ultimate Strength of Test Wall

LATERAL LOAD (kips)

TOP DISPLACEMENT (in)

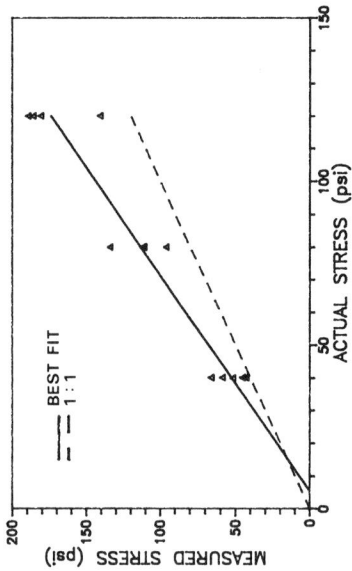

FLUID
NOZZLE

HYDRAULIC
FLATJACK

40 psi APPLIED VERTICAL STRESS

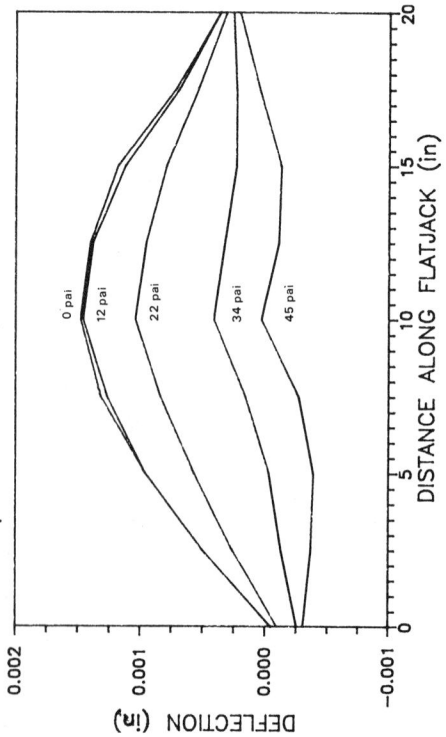

Figure 9 Typical Distribution of
Deflection - Flatjack Test

DEFLECTION (in)

DISTANCE ALONG FLATJACK (in)

0 pai
12 pai
22 pai
34 pai
45 pai

298

# 34 Determination of in-situ stress in masonry structures

**D Lenczner,** The University of Wales College of Cardiff, UK

This paper describes a new method for determining the
in-situ stresses in masonry walls or piers.  It has been
used successfully in a number of buildings and other
masonry structures.  The method is based on measuring the
elastic recovery strain in a member which is then cut out
and reloaded to give the same strain recovery.  The
corresponding stress with a creep correction factor
applied to it gives the in-situ stress in the structure.

INTRODUCTION

In the regeneration of the inner cities the engineer is often faced with
the problem of having to assess the existing stresses in the more heavily
loaded parts of the building which is being refurbished or extended and
where additional loads are imposed on the structure.  Alternatively,
where cracking or other symptoms of distress are detected, it is
essential to determine the stresses in the vicinity of the affected area
for a proper evaluation of the problem.

The author has developed a method for determining the in-situ stresses in
brick/block masonry and has used it successfully in a number of buildings
and other masonry structures.  The method and its applications are
discussed in the following sections.  It has been recently validated in
the laboratory by applying a known load to a masonry wall which was then
tested in the same manner as in the in-situ tests.  The results indicated
that an accuracy of 5-10% can be expected using this method.

DETERMINATION OF IN-SITU STRESSES IN MASONRY STRUCTURES

Basically, the method used for determining the in-situ stresses in
brick/block masonry structures is quite simple.  A strain measuring
device is stuck to the surface of the wall, column, pier or other element
to be tested.  A demec gauge is suitable for measuring the strain.  It is
simple to use, generally it has sufficient sensitivity and one can use it
even in the most difficult environment where the more sophisticated
devices would be difficult to use.  After initial readings have been
taken the stress in the area under examination is released by removing a
small panel or prism from the wall or pier by cutting.  After the release

of stress the elastic recovery of the specimen is measured. One then applies a known load to the specimen such that, on release, the elastic recovery is the same. As a first approximation, the stress so determined is the in-situ stress. For a more exact assessment of stress one must take into account the loading history of the member. During the time under load a certain amount of creep will have occurred and the elastic recovery will not be the same as when the member was loaded for only a brief duration. It is outside the scope of this paper to explain in detail the effect of creep on the elastic strain recovery. From previous findings, it is reasonable to assume that, on the whole, the effect of creep and the associated stress relaxation is to reduce the initial elastic recovery by some 10 percent. To allow for this, the stress as determined by the short duration loading test needs to be multiplied by a factor of 1.1 to give the correct in-situ stress.

A certain amount of expertise is required to determine the in-situ stresses in brick/block masonry. The area to be investigated must be thoroughly examined and any rendering or coating must be removed before the demec pips are stuck to the surface of the loadbearing member. A number of gauge points must be secured to allow for the variation of stress from point to point. It is not uncommon for the elastic recovery in the small area under investigation to vary by 15-20%. Some older masonry crumbles easily on removal and great care must be taken when removing a panel or a prism from a wall or pier. Cutting by saw causes least disturbance to the masonry but when a member is over 200 mm in thickness it may be more practicable to cut out a specimen by a stitch drilling technique. For this a supply of water is necessary. Wherever possible the cut should be such that in addition to the removal of the panel or prism a free standing plinth should be left on it original base so that an in-situ loading test can be performed (see Figure 1). The cut out specimen is best tested in a laboratory.

It is useful to have some idea about the magnitude of the in-situ stress. This will govern the load to be applied to give the required elastic strain recovery. Trial and error loading cycles and some degree of extrapolation are often necessary to get the stress corresponding to the elastic recovery. A further problem arises when the wall or pier carries an eccentric load. For this reason, whenever possible strain readings should be taken on both faces. Where this is not possible the test can still serve a useful purpose.


LIVE EXAMPLE

A large structure in London is supported on a grid of brickwork vaults and arches spanning between massive brickwork piers. The space under the arches had been subsequently filled in with thick brick walls. As part of a new development on this site it was necessary to remove the infill walls. Calculations had shown that some of the piers were already highly stressed and there was some concern that by removing the walls, any additional load, which they were carrying over and above their self weight as a result of differential settlement between the pier and wall foundations, would be transmitted to the piers. In order to establish whether the walls were in fact carrying additional load and what was the in-situ stress in the piers the author was called in to carry out an investigation on the state of stress in the structure. Below is a brief outline of the procedure which was adopted.

## EXPERIMENTAL PROCEDURE

After careful examination a number of suitable locations were selected in the walls and piers to determine the in-situ stresses. For reasons given previously it was decided to use a demec gauge to measure strain in the walls and piers before and after the removal of the load. A typical layout of demec pips in the walls and piers is shown in Figure 2. Because of the thickness of the walls and piers it was decided to use a stitch drilling technique for cutting out the samples. In each wall selected for testing two vertical cuts were made about a metre long and 400 mm apart. A horizontal cut at the top and at mid-height enabled the top sections to be subsequently removed, leaving the bottom sections freely supported on their original base (see Figure 1). A similar procedure was used for the piers but here a corner section was cut instead. The sequence of operations was as follows:

1. Demec pips were stuck by epoxy to the wall and pier surfaces.

2. Initial gauge readings were recorded.

3. Sections were cut out from walls and piers leaving free standing plinths.

4. Elastic recovery strain was measured in the cut sections and free standing plinths.

5. Load/deformation tests were carried out on the free standing plinths and on the cut out sections in the laboratory to determine the stress which gave the same elastic recovery as the initial values on the release of stress.

6. A creep factor was applied to the calculated stresses to give the in-situ stresses.

## EXAMPLE

### Pier A

$$\text{Mean elastic recovery strain} = 25 \times 1.07 \times 10^{-5}$$
$$\text{Maximum elastic recovery strain} = 30 \times 1.07 \times 10^{-5}$$
$$\text{where } 1.07 \times 10^{-5} = \text{demec gauge factor}$$

### Site loading test

Cross-sectional area of free standing prism = 44,306 mm$^2$

From Figure 3, the load corresponding to the mean elastic recovery of $25 \times 1.07 \times 10^{-5} = 4.65$ tons

$$\sigma_{mean} \text{ (before correction)} = \frac{4.65 \times 2240 \times 4.45}{44306} = 1.05 \text{ N/mm}^2$$

The load corresponding to the maximum elastic recovery of $30 \times 1.07 \times 10^{-5}$
(by extrapolation) = 5.58 tons.

$$\sigma_{max} \text{ (before correction)} = \frac{5.58 \times 2240 \times 4.45}{44306} = 1.26 \text{ N/mm}^2$$

Allowing for creep effect

$$\sigma_{mean} = 1.05 \times 1.1 = \underline{1.16 \text{ N/mm}^2}$$

$$\sigma_{max} = 1.26 \times 1.1 = \underline{1.39 \text{ N/mm}^2}$$

Laboratory loading test

Cross-sectional area of prism removed for laboratory testing = 44,306 mm$^2$

From Figure 4, the load corresponding to the mean elastic recovery of
$25 \times 1.07 \times 10^{-5}$ = 4.65 tons.

$$\sigma_{mean} \text{ (before correction)} = \frac{4.65 \times 2240 \times 4.45}{44306} = 1.05 \text{ N/mm}^2$$

The load corresponding to the maximum elastic recovery of
$30 \times 1.07 \times 10^{-5}$ = 5.925 tons

$$\sigma_{max} \text{ (before correction)} = \frac{5.925 \times 2240 \times 4.45}{44306} = 1.33 \text{ N/mm}^2$$

Allowing for creep

$$\sigma_{mean} = 1.05 \times 1.1 = \underline{1.16 \text{ N/mm}^2}$$

$$\sigma_{max} = 1.33 \times 1.1 = \underline{1.46 \text{ N/mm}^2}$$

We therefore have the following values of in-situ stress in Pier A

| Type of test | In-situ stress (N/mm$^2$) | |
|---|---|---|
| | Mean | Max. |
| Laboratory | 1.16 | 1.46 |
| Site | 1.16 | 1.39 |

These stresses compared well with the client's calculated stresses and
were high for this type of brickwork.

Using similar procedure, the in-situ stresses in an adjoining infill wall
which was to be removed was

| Type of test | In-situ stress ($N/mm^2$) | |
|---|---|---|
| | Mean | Max. |
| Laboratory | 0.16 | 0.23 |
| Site | 0.12 | 0.18 |

Stress in wall due to self-weight was 0.06 $N/mm^2$.

It is clear that the wall had in fact picked up some additional loading during its life, probably as a result of differential settlement. On the removal of the wall the extra load would be transferred to the adjacent piers but calculations have shown that as a result, the load in piers would increase by only 3.7% which in this case was not considered serious.

VALIDATION OF THE METHOD

Recently a laboratory test was carried out to validate the method described above. A single leaf wall 1.17 m high by 1.4 m wide was built and stressed to a level of 4.83 $N/mm^2$ The wall was subjected to the same test routine as previously described, with strain readings taken before and after the release of the load. The cut out panel was then loaded in a testing machine and the stress which gave the same elastic recovery as that observed in the wall was recorded. Three load cycles gave stresses of 5.4 $N/mm^2$, 4.6 $N/mm^2$ and 5.50 $N/mm^2$ respectively with an average of 5.17 $N/mm^2$, a percentage discrepancy of
$(\frac{5.17 - 4.83}{4.83})$ x 100 = 7%.

CONCLUSIONS

A method is decribed for determining the in-situ stresses in loadbearing brick/block masonry members. It is based on measuring the elastic recovery strain in a panel or a prism which is cut out by sawing or stich drilling from the loadbearing member. The panel or prism is then loaded to give the same strain recovery and the stress is recorded. A creep correction factor is to allow for previous loading history of the member. A laboratory test to validate the method gave an accuracy of 7 percent. More tests are planned, this time on site, to confirm the reliability of the method.

REFERENCES

Rossi, P. 'Flat-jack test for the analysis of mechanical behaviour of brick masonry structures'. Proc. of the 7th Int. Brick Masonry Conf. Melbourne, Australia (1985).

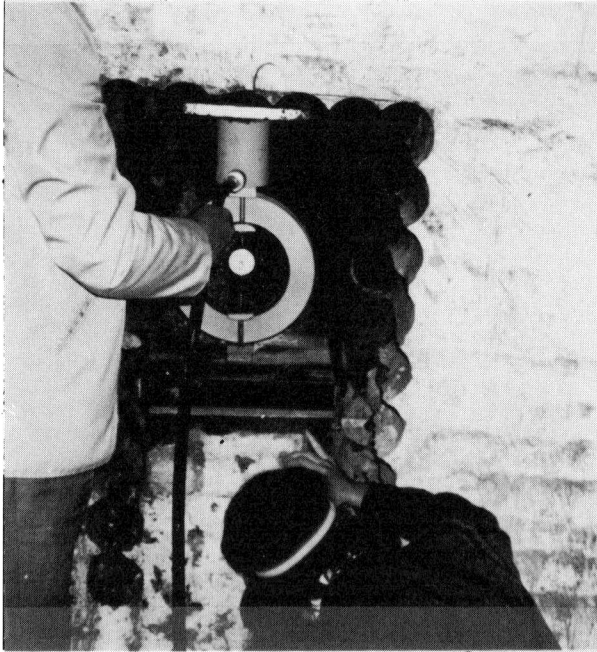

Figure 1   Typical section cut out from a wall by stitch drilling technique, leaving a free standing plinth.

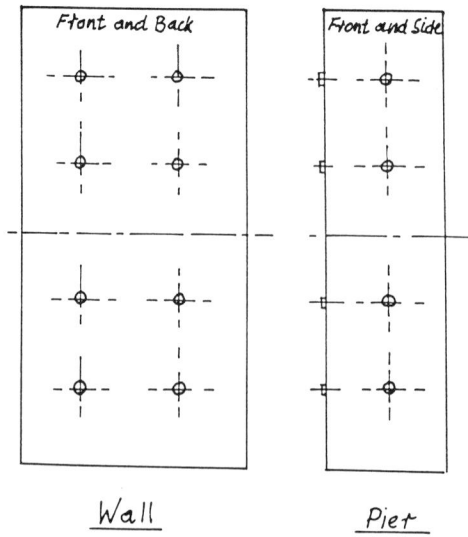

Figure 2   Layout of demec pips in wall and pier specimens.

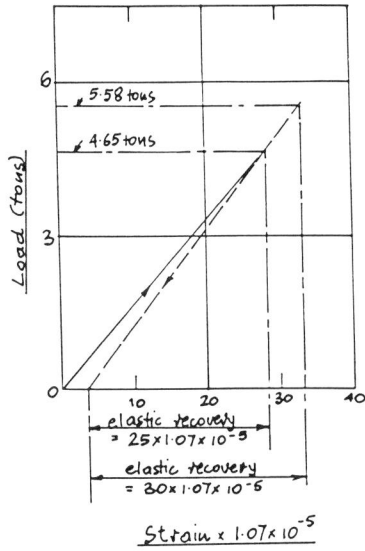

Figure 3    Load/deflection cycle carried out in-situ on a pier
            specimen.

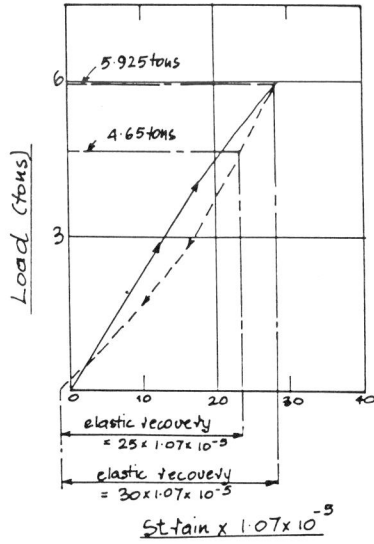

Figure 4    Load/deflection cycle carried out on a pier specimen in
            the laboratory.

A Kennedy, British Rail Research, U.K.

Dr Lenczner stated that he is concerned that flat jacks could provide misleading data because they make no allowance for shear lag between the area under test and the rest of the structure. Although this could be a problem when testing the elasticity of brickwork in-situ, I submit that it is irrelevant when trying to assess the force being transmitted through a particular joint. This is because the flat jacks are used to restore the strain regime, in the area of the joint, which existed prior to the test.

## Author's reply

The force used by the flat jack to restore the strain regime prior to the cut does not necessarily indicate the force which was transmitted through the joint before the cut. This is because of the bridging effect which occurs after the cut. In an extreme case this bridging, or arching could be so effective that no force is required by the jack to restore the strain regime. It does not necessarily follow that the force across the joint was zero before the cut.

I would agree that the degree of accuracy of my method is largely independent of the stress level and that the percentage error would therefore appear smaller when higher stress levels are monitored.

The reason why I have used a higher stress level when checking the accuracy of the method is that I checked it in the first instance on a stronger brick and I used the working stress level for that type of brickwork. I would therefore agree that with a weaker brickwork which I described in my paper the percentage error could have been higher.

I certainly intend to carry out more validation tests, especially for weaker brickwork, at lower stress level. Furthermore, I intend to compare the accuracy of my method with the flat jack technique which I believe tends to underestimate the in-situ stress.

I would be pleased to receive more information about the single core method of testing masonry which have been developed.

P Jackson, Gifford and Partners, U.K.

I would like to question the claimed accuracy of the method, 7%, which appears to be based on the results of a single test.

Colleagues at Gifford's have developed a related stress assessment method for concrete and they are now considering extending it to masonry. They have found that more sophisticated (and more extensive) instrumentation is needed to achieve the required accuracy although, because they use a single core, their test is still more convenient.

They have also found that a large component of the errors (expressed in N/mm$^2$) is independent of the level of the stress being assessed. Against this, errors due to material non-linearity might be expected to increase with stress. In view of this, I would like to ask why Dr Lenczner chose to perform his validation test at a stress which was three times greater than the stress he was measuring in his 'live example', even though he described the later stress as 'high'. I would also like to know if he has performed, or plans, any further validation tests; particularly tests at lower stresses.

## D W Begg, Portsmouth Polytechnic, U.K.

Can this paper claim to describe 'A new method for determining stresses in masonry' when N Davey of the Building Research Station published a similar paper entitled 'Measuring the existing stresses in masonry structures' in The Engineer, April 26, 1935. Davey's use of a vibrating wire gauge gave measurements an .order of magnitude more accurate than a demec gauge.

## Author's reply

I must confess I was unaware of the paper published by Davey in 1935 on "Measuring the existing stresses in masonry structures" and I will try to do this as soon as possible. I would, however, like to add that in the past I have used vibrating wire gauges as well as demec gauge to measure strains in masonry and I could not go along with the statement that using a vibrating wire gauge would give an accuracy an order of magnitude higher. Indeed, those tests which I have carried out in the past showed that, by and large, the degree of accuracy was about the same using both methods. I also found that where site conditions were difficult especially where there was frequent exposure to rain, the demec gauge method gave more consistent results.

# 35 Calibration of the British Ceramic Research Limited (BCRL) panel freezing test against exposure site results

**F Peake and R W Ford,** British Ceramic Research Ltd, Staffordshire, UK

Damage resulting from nine Winters on an exposure site has been compared with the results obtained using the British Ceramic Research Limited panel freezing test, for 23 different types of clay bricks. Excellent concordance has been demonstrated, with 10 freeze/thaw cycles of the laboratory test being roughly equivalent to exposure for one Winter when bricks are situated in a coping course. It is suggested that 100 cycles of the panel freezing test, without damage, is more than adequate to demonstrate that a brick is Frost Resistant as defined in BS 3921. Further work is required to establish the corresponding number of cycles for Moderately Frost Resistant bricks.

## INTRODUCTION

Most national standards concerned with the properties of bricks and brickwork make reference to the resistance to freeze/thaw conditions when saturated with water. Ever since the development of the mechanical refrigerator which enabled freezing conditions to be readily obtained in the laboratory, many simulative test procedures for assessing this property have been suggested, and some have been adopted as standard test methods. It is now generally accepted that uni-directional freezing and thawing is a prime requirement for this type of test and various methods have been described based on this principle(1-3).

Before the results obtained with an accelerated test can be accepted, it is necessary to demonstrate that they bear a relation to results obtained under natural exposure conditions. However, an accurate quantitative relationship which can be used to define required levels of durability is not easy to establish. This is mainly because of the wide range of conditions of exposure which can occur both in terms of location of the brick in the structure, and the variation in wetting and freezing conditions which occur during different Winters in different geographical locations.

It must be accepted that there is a natural variation in the durability of a sample batch of bricks which, even after allowing for the normal variation in firing treatment during manufacture, can result in individual bricks within the sample batch being damaged after different intervals of exposure. For these reasons, it can be misleading to judge any particular freezing test method based on a restricted amount of data e.g. one Winter's exposure and a small number of bricks covering a limited range of durability(4).

BS 3921:1985(5) classifies bricks into one of three categories:

1. Frost resistant (F)
   Bricks durable in all building situations including those where they are in a saturated condition and subjected to repeated freezing and thawing.

2. Moderately frost resistant (M)
   Bricks durable except when in a saturated condition and subjected to repeated freezing and thawing.

3. Not frost resistant (O)
   Bricks liable to be damaged by freezing and thawing if not protected.

At present, the classification (F,M or O) of a particular brick is based on the evidence that the brick has been in service for at least three years and has performed satisfactorily under the appropriate conditions of exposure. Where this evidence is not available, the manufacturer can only assume a classification based on his knowledge of how similar bricks have behaved in service.

Reference is made in BS 3921:1985 to the Panel Freezing Test(1) which is a simulative accelerated procedure for assessing the resistance of brickwork to cyclic freeze-thaw conditions. This test has been used since 1972 by both manufacturers and users and over 600 panels have been tested covering the whole range of bricks available in the UK. Experience with the test over this period has demonstrated that the results relate closely to the predicted durability of the brick based on its practical performance. However, in order to establish a more definitive correlation between the laboratory results and this practical performance a test site was established in 1979 where 23 different types of bricks covering a wide range of durability were exposed to natural weathering conditions over the next nine Winters and any damage was closely monitored. Samples of the same bricks were tested using the panel freezing test.

## 1. BRICK SAMPLES

The bricks were chosen in consultation with manufacturers to be representative of the different manufacturing methods and raw materials used by the industry, and also to cover a wide range of durabilities. Each sample consisted of 150 bricks. Ten bricks were taken for the determination of initial rate of suction, water absorption and compressive strength. A general description of each brick type, and its physical properties is given in Table 1.

The remaining bricks in each sample were used to construct the panel for testing in the laboratory (30 bricks) and the test wall on the exposure site (110 bricks).

## 2. THE CERAM RESEARCH PANEL FREEZING TEST

A panel (3-bricks wide × 10-courses high) was constructed from each type of brick being tested using a conventional 1:1:6 mortar. After an adequate curing time, the panel was saturated by complete immersion in water for 7 days. One face of the panel was then subjected to 100 freeze/thaw cycles and any damage noted during periodic examination throughout the test.

Table 1. Properties of sample bricks

| Brick | Suction rate Kg/m² /min | 24 h WA % | 7 day WA % | 5 h boil WA % | Saturation coefficient | Compressive strength MN/m² |
|---|---|---|---|---|---|---|
| 1. W/C 3-hole red common | 2.40 | 13.0 | 15.0 | 19.3 | 0.67 | 18.4 |
| 2. W/C solid brown facing | 3.11 | 13.6 | 17.9 | 21.5 | 0.63 | 44.5 |
| 3. Stiff plastic red facing | 2.03 | 16.1 | 16.8 | 18.0 | 0.89 | 35.9 |
| 4. W/C 3-hole brown facing | 0.40 | 6.2 | 6.1 | 6.3 | 0.99 | 73.1 |
| 5. W/C hole red facing | 1.46 | 12.7 | 15.4 | 17.1 | 0.74 | 29.3 |
| 6. W/C 3-hole brown facing | 0.92 | 8.7 | 11.4 | 14.4 | 0.60 | 40.9 |
| 7. Soft mud multi red | 1.41 | 9.1 | 9.7 | 13.1 | 0.69 | 32.8 |
| 8. Soft mud multi red | 1.78 | 20.7 | 21.5 | 29.1 | 0.71 | 6.8 |
| 9. Repressed W/C | 0.30 | 3.2 | 3.8 | 4.7 | 0.68 | 64.0 |
| 10. W/C multi perf brown | 0.12 | 1.0 | 1.5 | 1.6 | 0.64 | 96.4 |
| 11. Soft mud buff | 5.37 | 24.8 | 25.9 | 33.6 | 0.74 | 18.6 |
| 12. Soft mud multi red | 2.75 | 20.5 | 22.1 | 27.8 | 0.74 | 18.6 |
| 13. Soft mud light multi | 3.11 | 23.5 | 23 7 | 30.1 | 0.78 | 17.2 |
| 14. W/C multi perf rustic | 1.35 | 10.0 | 10 9 | 12.6 | 0 79 | 38.8 |
| 15. W/C multi perf smooth red | 0.53 | 4.8 | 5.5 | 6.1 | 0.79 | 56.8 |
| 16. W/C Multi perf | 0.30 | 4.7 | 5.1 | 6.0 | 0.78 | 61.9 |
| 17. W/C multi perf | 0.23 | 3.6 | 3.8 | 4.0 | 0.91 | 79.9 |
| 18. W/C multi perf | 0.23 | 4.1 | 4.4 | 4.7 | 0.87 | 54.2 |
| 19. W/C multi perf rustic | 0.97 | 6.1 | 7.6 | 12.0 | 0.51 | 34.0 |
| 20. Semi-dry pressed fletton | 1.83 | 12.5 | 13.6 | 15.9 | 0.78 | 22.1 |
| 21. Soft mud | 0.59 | 8.8 | 9.6 | 16.3 | 0.54 | 26.5 |
| 22. Soft mud | 0.60 | 8.6 | 9.8 | 16.4 | 0.53 | 26.1 |
| 23. W/C 3-hole smooth red | 0.70 | 8.0 | 8.6 | 10.5 | 0.76 | 61.0 |

The test equipment and the procedure have been described in detail(1). At the time of writing the method is being produced as a draft for development by British Standards Institution (BSI). This provides an opportunity for interested parties to comment or to carry out further development work before the method goes forward as a British Standard Test. There is the additional possibility of using the method to establish a performance specification which will enable bricks to be classified F, M or O, based on the test results.


## 3. EXPOSURE SITE TESTING

The site was situated about 5 miles to the west of Newcastle-u-Lyme where an existing concrete-paved area was made available. The site was partly protected to the south and east by trees, but was open to the west and the north where a large building stood at a distance of about 70 m.

The test walls were built in the Spring of 1979 to the suggested design given in BS 3921:1969 (see Figure 1). A 1:1:6 cement/lime/sand mortar was used. The walls were arranged in five rows, 2 m apart, with 1m between each wall. The walls in adjacent rows were staggered to give the maximum exposure. A general view of the site is given in Figure 2.

The walls were examined at intervals during the Winter months and any damage noted. In the Autumn of 1981, before the third Winter, it was decided to remove the top dpc beneath the coping course of each wall to allow greater penetration of water into the bricks between the bottom dpc and the coping. Subsequently it was noted that the damage to bricks in the lower levels of the wall was largely determined by the effectiveness of the coping which varied from wall to wall. It was therefore decided that only damage to the bricks in the coping course should be monitored. In fact, as expected, these were the first bricks to exhibit damage (Figure 3).

A summary of the weather records during each of the nine Winters is given in Table 2. The total rainfall and the total number of air frosts which occurred during the six months October-March are reported, together with the mean values over the last 30 years at Manchester (the nearest principal metrological station). Also included in this Table is the Freezing Index for each winter. This index is evaluated from the total rainfall and the number of air frosts occurring during the winter and gives an indication of the severity of the weather conditions in relation to potential frost damage(6). Damage was observed in walls 8 and 20 during the first Winter. The worst damage was sustained by Wall 20 where 11 of the 12 coping bricks had failed after the second Winter. Walls 12 and 13 started to show signs of sulphate attack during the second Winter with cracking occurring along the centre of the mortar joints.

Chemical analysis of this mortar confirmed the sulphate attack.

After the second Winter, the weather conditions were fairly typical in terms of total rainfall and air-frosts and no particular year could be said to be worse than any other so far as frost damage was concerned.

## 4. RESULTS

The results obtained with the panel freezing test and from the exposure site are summarised in Table 3 and are compared in different ways in Figures 4 and 5. Figure 4 compares the number of freeze/thaw cycles required in the laboratory to initiate damage, with the number of natural freeze/thaw cycles on site before damage occurred. This plot clearly demonstrates the congruity between the two sets of results. Out of 23 sets of bricks, 10 showed no damage in either situation

Table 2. Weather records at exposure site

| Winter | No. of air frosts | Total rainfall (mm) | Freezing index |
|---|---|---|---|
| 79-80 | 54 | 613 | 1135 |
| 80-81 | 51 | 610 | 773 |
| 81-82 | 48 | 564 | 623 |
| 82-83 | 58 | 493 | 677 |
| 83-84 | 43 | 532 | 602 |
| 84-85 | 42 | 518 | 416 |
| 85-86 | 50 | 637 | 437 |
| 86-87 | 42 | 368 | 442 |
| 87-88 | 28 | 442 | 227 |
| Mean | 46 | 509 | 592 |
| 30 year mean | 39 | 408 | |

311

## Table 3. Summary of laboratory and exposure site

| Sample no. | Laboratory | | Exposure site | | Comments |
|---|---|---|---|---|---|
| | No. of cycles to initial damage | % of bricks damaged after 100 cycles | No. of winters to initial damage | % of bricks damaged after 9 winters | |
| 1 | 34 | 29 | 2 | 25 | |
| 2 | 71 | 3 | 7 | 25 | |
| 3 | 18 | 100 | 2 | 42 | |
| 4 | ND | Nil | ND | Nil | |
| 5 | 28 | 100 | 3 | 100 | |
| 6 | 36 | 66 | 6 | 60 | Slight damage in both tests |
| 7 | 100 | 6 | 9 | 8 | |
| 8 | 18 | 31 | 1 | 50 | |
| 9 | 100 | 11 | ND | Nil | |
| 10 | 50 | 100 | 5 | 100 | Pitting due to ironstone particles |
| 11 | ND | Nil | ND | Nil | |
| 12 | ND | Nil | ND | Nil | } Sulphate attack but no |
| 13 | ND | Nil | ND | Nil | } damage to coping bricks |
| 14 | 35 | 100 | 4 | 100 | |
| 15 | ND | Nil | ND | Nil | |
| 16 | ND | Nil | ND | Nil | |
| 17 | ND | Nil | ND | Nil | |
| 18 | ND | Nil | ND | Nil | |
| 19 | 50 | 29 | 6 | 58 | |
| 20 | 19 | 100 | 1 | 100 | |
| 21 | ND | Nil | ND | Nil | |
| 22 | ND | Nil | ND | Nil | |
| 23 | ND | Nil | 2 | 25 | Suspected sample variation |

ND = Not Determined, i.e. no failure after 100 cycles in the laboratory test or no failure after 9 winters on the exposure site.

while the remaining results, with one exception, fall about a line which indicates that as a broad generalisation, each Winter of exposure is equivalent to about 10 freeze/thaw cycles in the laboratory test in terms of the stresses produced within the bricks. Although, on average, during each Winter there were 46 air frosts, i.e. freeze/thaw cycles, each cycle must be considered to be considerably less aggressive than the artificially imposed laboratory cyclic conditions, both in terms of the depth of frost penetration into the exposed brick, and the level of saturation of the bricks at that time.

Brick No. 23 does not fall within the general pattern of behaviour depicted by Figure 4. Bricks from this sample safely withstood 100 cycles in the laboratory test, but some damage occurred after two Winters on the exposure site. It may be significant that processing problems were being experienced in the manufacture of this brick at the time of sampling which could have resulted in some variation in the quality of the product.

Figure 5 shows the correlation between the proportion of coping bricks damaged after nine Winters on site, and the proportion of bricks damaged in the laboratory test after 100 cycles. Again, as in Figure 4, the concordance between 100 cycles of the panel freezing test and nine Winters natural exposure is indicated. A good correlation would not be expected because of the inherent variation in the durability present within any sample batch of bricks.

## 5. CONCLUSION

The work has clearly demonstrated that the cyclic freeze/thaw regime imposed on a brick in the laboratory panel freezing test gives results that are directly related to those obtained when the same brick is subjected to weather in an exposed situation over a number of Winters. As a broad generalisation, for bricks prone to frost damage in these situations, the rate at which damage becomes apparent, as measured by the number of freeze/thaw cycles, is about five times faster in the laboratory test than under natural exposure conditions.

In the laboratory test the brick is maintained in a continuously saturated condition so that the mechanical stresses imposed by freezing are maximized. In naturally exposed situations, this combination of freezing under saturated conditions will only occur occasionally. Additionally, the artificially imposed conditions in the laboratory produce a greater depth of penetration of freezing than normally occurs in typical Winter weather.

All the bricks used in the investigation would be categorized as either moderately frost resistant (M) or frost resistant (F).

Bricks used in a coping course which are undamaged after nine Winters would certainly be classified as frost resistant. This suggests that 100 cycles of the panel freezing test without damage is more than adequate to demonstrate this level of durability.

A test requirement to differentiate between O and M category of frost resistance is less obvious. The present results indicate that the ability to withstand 10-15 cycles in the panel freezing test without damage is a reasonable criterion but further work is required to confirm this.

## REFERENCES

1. WEST, H. W. H., FORD, R. W. and PEAKE, F., 'A Panel Freezing Test for Brickwork'. Br Ceram Trans J (83) 112, 1984.

2. VAN DER KLUGT, L. J. A. R., 'Frost Testing by Uni-Directional Freezing'. Br Ceram Trans J (87), 8, 1988.

3. GERMAN STANDARDS COMMITTEE, 'Test Methods for the Determination of the Frost Resistance of Fair Faced Clay Bricks and Clinker Bricks; One-side Freezing of Test Walls'. DIN 52252. Part 3, 1986.

4. BRUNING, H., 'Frost Resistance of Masonry Bricks'. ZI Int 40 (11), 570, 1987.

5. BRITISH STANDARDS INSTITUTION, 'Clay Bricks'. BS 3921: 1985.

6. BEARDMORE, C. and FORD, R. W., 'Winter Weather Records Relating to Potential Frost Failure of Brickwork'. Br Ceram Trans J (86) 7, 1987.

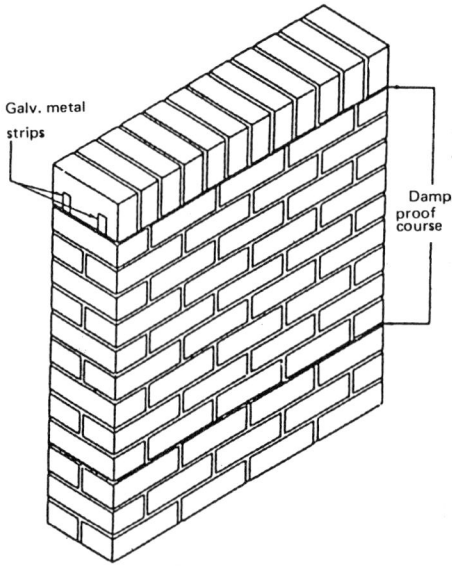

Figure 1. Exposure-site test wall

Figure 2. General view of exposure site

Figure 3. Typical failure of coping course

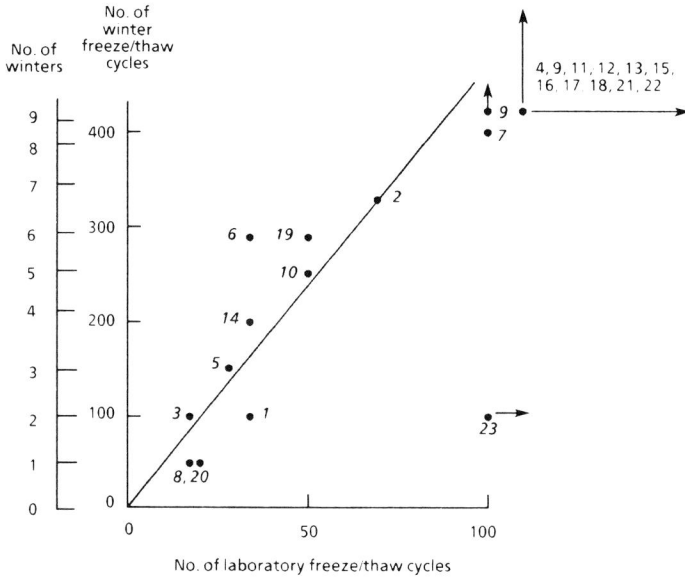

Figure 4. Relation between laboratory and natural freeze/thaw cycles to produce failure

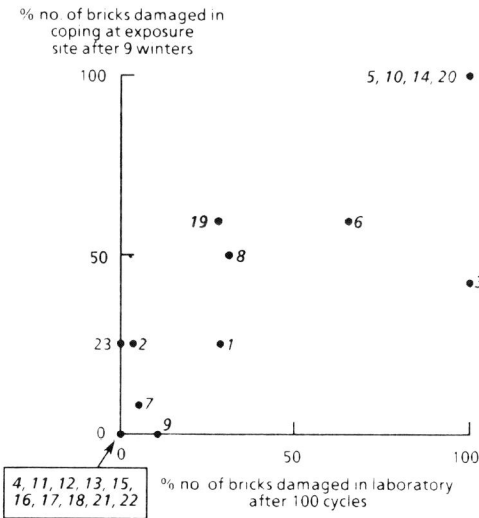

Figure 5. Relation between frost damage in the laboratory and exposure site

### R C de Vekey, Building Research Establishment, U.K.

BS3921, the clay brick standard contains a class M - 'moderately frost resistant', defined as 'not resistant to repeated freezing and thawing while saturated with water, but suitable for many external situations!'

The test described does not apply a test regime appropriate to these units - is it therefore suitable?

### Author's reply

The various types of brick which have failed in less than 100 cycles can be placed in a loose ranking order based on the number of cycles before initial failure occurs. This ranking order, as far as we can judge, does relate to the way these different bricks stand up to exposure in practical building situations and this suggests that, although the test regime is extremely severe, it can be used to assess the durability of Moderately Frost Resistance Bricks (M). However, as stated in the paper, more data is required before a test criterion can be defined (in terms of the minimum number of cycles) to differentiate between 'M' and 'O' classes of brick.

Dr J F A Moore, Building Research Establishment, UK

This lengthy session of 10 papers has struck at the heart of the issue of the "Life of Structures". The theme of "Materials in Service" is concerned first with what happens in reality in the lifetime environment. Then we need to know how to model behaviour so that practicable and reliable tools can be developed for design and assessment. This heterogeneous collection of papers has not only touched on the full range of issues but done so for concrete, reinforcing steel, plastics and masonry.

Starting with a paper of an essentially mathematical approach using damage accumulation and then one on electrochemical aspects of corrosion of steel in concrete, the focus moved to technology (concrete cover) and a tool for assessment of performance in situ (concrete permeability). Full-scale problems with some concrete in the Middle East were frankly revealed to complete a run of papers from overseas.

Papers on adhesives and plastic pipe linings emphasised the considerations particular to structures other than buildings and the need to characterise the environment to which the material in a component is subjected. The remaining papers were on masonry. By treating the properties of 70 year old brick piers, stress under load in service and freeze-thaw action they emphasised that time is the the essence of life and of the changes which may occur in properties and performance. Despite the potential aggressivity of freeze-thaw action the final paper appeared to be the only one in the conference on this topic.

The diversity of papers has provided more useful pieces for the jig-saw which will facilitate the estimation of life at the design stage. They will also help assessment, made at some stage during service, of the history of performance of the materials in a structure and of its likely remaining life.

# 36 The performance of masonry arch bridges

**J Page,** Transport and Road Research Laboratory, Crowthorne, UK

This paper describes the full scale load tests being
done by the Transport and Road Research Laboratory as
part of work to revise the assessment of the traffic
load carrying capacity of masonry arch bridges.
Techniques for strengthening bridges with insufficient
capacity are described. The remaining life of the
structure is taken into account in deciding whether to
strengthen or replace; quantified guidelines would be
desirable but may be difficult to develop.

## 1. INTRODUCTION

Forty per cent of the UK bridge stock, or about forty thousand bridges,
are brick or stone masonry arches. They therefore represent a vital part
of the nation's transport system. They come in a vast range of shapes
and sizes; the longest span in the UK is 61 m (Grosvenor Bridge, Chester
Many were built in mediaeval times, but the great era of building began
with the construction of the canals in the second half of the eighteenth
century and ended when the railway network was substantially complete at
the beginning of the twentieth century.

From time to time the structural adequacy of arch bridges must be assessed
An assessment may show that:

1) the bridge simply needs routine maintenance.

2) the bridge needs to be widened because traffic flows have increased.

3) the bridge needs to be strengthened because traffic loads have
   increased or because of structural deterioration (alternatively a
   traffic weight limit may be applied).

4) the bridge needs to be replaced because it cannot be upgraded.

An important part of the assessment procedure is to assess traffic load
carrying capacity and this paper briefly describes the presently used
method and the Transport and Road Research Laboratory (TRRL) programme of
analysis and full scale load testing to revise this method.

Cost will be an important factor in the decision whether to retain or
replace a bridge; in general it is cheaper to repair or strengthen an
arch bridge than to replace it. The main repair and strengthening

techniques are described here.  The bridge may be a listed structure in which case the options are more limited.  The remaining life of the structure will be taken into account in an assessment, relying on the experience and training of the engineer responsible.


## 2.  THE MEXE METHOD

The present method of assessing the traffic load capacity of arch bridges – the "MEXE" method (1, 2) – was devised during the second world war to provide a quick way of assessing whether a bridge could carry military loads.  It was adapted for civil use after the war.  The method is based on an elastic analysis.

First, a "provisional axle load" is calculated from measurements of arch span and thickness and of the fill thickness at the arch crown.  This is then adjusted by factors which take into account the rise and shape of the arch barrel, the arch barrel and fill materials, the state of the mortar joints, and the general condition.  This adjusted value, the "modified axle load" represents the permissible axle load for a two axle bogie;  a further calculation gives the allowable axle load for a single axle.  A typical assessment is given in Table 1.

Table 1.  MEXE Assessment

| | | |
|---|---|---|
| Span | $L$  (m) | 4.902 |
| Rise at mid-span | $r_c$  (m) | 1.154 |
| Rise at quarter-span | $r_q$  (m) | 0.904 |
| Thickness of arch barrel | $d$  (m) | 0.343 |
| Depth of fill at crown | $h$  (m) | 0.246 |
| Provisional axle load PAL  $=$  $740(d+h)^2/L^{1\cdot3}$          $=$ | | 32.5 tonnes |
| Modifying factors: | | |
| Span/rise | $F_{sr}$ | 1.0 |
| Profile | $F_p$ | 0.92 |
| Material | $F_m$ | 0.87 |
| Joint | $F_j$ | 0.73 |
| Condition | $F_c$ | 0.7 |
| Modified axle load $= 1.0 \times 0.92 \times 0.87 \times 0.73 \times 0.7 \times 32.5$  $=$ | | 13.3 tonnes |
| Allowable axle load for single axle                              $=$ | | 14.9 tonnes |

The calculated axle loads exceed those allowed in the UK and so the assessment suggests that no weight restriction or strengthening is necessary.  However the bridge was replaced because its spandrel walls had moved outwards and were becoming unsafe;  the MEXE assessment does not directly take any structural contribution of the spandrel walls into account.  Three of the modifying factors – material, joint and condition, take into account the deterioration of the structure;  the condition factor $F_c$ relies considerably on "engineering judgement".

319

## 3. THE TRRL RESEARCH PROGRAMME

The MEXE method is very quick and easy to use but many simplifying
assumptions were made in its derivation, it is thought to be conservative
it cannot deal with distorted arches and it is limited to spans of less
than 18 m. For these reasons the TRRL is doing research with the aim of
providing a new or revised assessment method which will resolve the weak-
nesses of the existing method. The research involves the development of
analytical models, and full and model scale load tests to calibrate them.

### 3.1 Analysis

Two methods of analysis have been developed. The simpler is a "mechanism
method (3) which assumes that the arch collapses as a four hinged
mechanism, in which state it is statically determinate and the load
required for collapse can readily be computed from the equations of
equilibrium. The analysis is simple enough to be performed on a personal
computer and at the time of writing is expected to be used as the basis
of a revised assessment technique. An example of a mechanism analysis is
given in Figure 1. The other analysis uses finite elements (4). Both
analyses are essentially two dimensional so that for instance they cannot
take into account the stiffening effect of spandrel walls (it can be
argued that this should not be taken into account in an assessment in
case the joint between arch and wall is weak, as it often is).

### 3.2 Full scale load tests

A series of full scale tests is being carried out on redundant arch
bridges with a range of spans, arch shapes and construction materials to
provide data to calibrate analytical models. So far six tests have been
completed (5-8) and two more are planned in the near future. Suitable
bridges have become more difficult to find as the test programme has
progressed and the alternative of building full scale models is being use
for two bridges; one test was completed recently and another is due to
take place in the near future.

The method of loading adopted for these tests is to apply a transverse
"line" load above the arch at the road surface. The load required for
collapse depends on the longitudinal position of the line and it is
positioned where the minimum load is calculated to be required; usually
it is between quarter and third span. The "line" is made sufficiently
wide to prevent a premature failure of the fill (usually 750 mm) and is
the full width of the bridge between parapets. Load is applied by
hydraulic jacks via the load distribution frame shown in Plate 1. The
jacks, of the prestressing type to provide as much stroke as may be
required, react against ground anchors which pass through holes drilled
through the bridge deck. The load being applied to the bridge is measure
by load cells incorporated in the distribution frame.

Surveying has been adopted as the standard technique for measuring
structural displacements because the targets which are attached to the
bridge (about twenty are used) and which will be lost when the bridge
collapses, are cheap.

Other measurement techniques may be used in appropriate circumstances,
for instance displacement transducers have been used to measure the
change in width of cracks in the arch barrel, and strain gauges to measure

strain in tie rods.

Photography (still, cine and video) is used to provide a detailed visual record of damage to the structure as the test progresses.

The test procedure is to apply an increment of load (chosen so that about twenty increments will be required for collapse) and, when the required load level has been reached, to cut off the oil supply to the jacks until the various measurements are completed. The applied load usually decreases during this period due to relaxation of the structure and so the load at the end is also recorded, just before the next increment is applied.

Table 2 describes the six tests completed so far. There is not space in this paper to discuss the results in detail, but it will be seen from the table that quite good agreement is being achieved between the mechanism analysis and the experimental results. The table also provides a measure of the factor of safety of the MEXE assessment by comparing the collapse load with two MEXE single axles side by side. Plates 2 and 3 show two of the bridges tested and Plates 4-6 show three of the bridges shortly before or at the moment of collapse.

## 4. REPAIR AND STRENGTHENING TECHNIQUES

If an assessment shows that an arch bridge is inadequate to carry the loads required of it then a decision has to be made as to whether to strengthen or replace it. There are a number of techniques available which will increase its life and its load carrying capacity and which will cost less than replacement; they are briefly listed here.

### 4.1 Abutment, piers and foundations

4.1.1 Underpinning. The material beneath a pier or abutment is dug out a short length at a time and replaced with concrete.

4.1.2 Curtain walls. A curtain wall is built to about 300 mm below the footing and 300-500 mm thick, and is tied to the existing masonry with tie rods. The pier is then pressure grouted.

4.1.3 Piles. Small diameter bored piles will limit settlement or provide additional load capacity; to provide continuity they are bored through and cast into the existing pier or abutment.

4.1.4 Invert slabs. Invert slabs are used to prevent scouring; they may also be used to tie the two abutments together to prevent movement.

### 4.2 Arch barrels

4.2.1 Pressure grouting. Pressure grouting is quick and easy to apply and it can increase the capacity of the barrel by improving the joints in the ring. If the fill can be grouted its strength will be increased and pressure on spandrel walls will be relieved.

4.2.2 Saddling. A reinforced concrete ring is cast on top of the existing arch barrel. It will relieve the existing barrel of all but its self weight.

Table 2. Details of load tests to collapse on six bridges

| | Bridgemill | Bargower | Preston | Prestwood | Torksey | Shinafoot |
|---|---|---|---|---|---|---|
| Span (square) mm | 18300 | 10000 | 4950 | 6550 | 4900 | 6160 |
| Span (skew) (mm) | - | 10360 | 5180 | - | - | - |
| Angle of skew (degrees) | 0 | 16 | 17 | 0 | 0 | 0 |
| Rise at midspan (mm) | 2850 | 5180 | 1636 | 1428 | 1155 | 1185 |
| Arch thickness at crown (mm) | 711 | 558 | 360 | 220 | 343 | 540 |
| Arch shape | parabolic | segmental | elliptical | originally segmental | segmental | segmental |
| Arch material | sandstone | sandstone | sandstone | brick | brick | stone |
| Arch density (kN/m³) | 2.1 | 2.7 | 2.3 | 2.0 | 2.1 | 2.5 |
| Fill density (kN/m³) | 2.2 | 2.1 | 2.2 | 2.0 | 2.0 | 2.0 |
| Spandrel wall thickness (mm) | - | 1400 | 610 | 380 | 380 | 365 |
| Parapet thickness (mm) | - | 400 | 370 | - | 380 | 365 |
| Spandrel/parapet material | sandstone | sandstone | brick | brick | brick | stone |
| Total width (mm) | 8300 | 8680 | 5700 | 3800 | 7805 | 7030 |
| Fill depth at crown (mm) | 203 | 1200 | 380 | 165 | 246 | 215 |
| Width of line load (mm) | 750 | 750 | 750 | 300 | 750 | 750 |
| Maximum applied load (kN) (L) | 3100 | 5600 | 2100 | 228 | 1080 | 2500 |
| Collapse mode | - | crushing of arch material | crushing of arch material | 4 hinge mechanism | 3 hinge "snap-through" | 4 hinge mechanism |
| MEXE single axle load (kN) ($L_{MEXE}$) | 181 | 642 | 298 | 18 | 146 | 376 |
| $L/2*L_{MEXE}$ | 8.6 | 4.4 | 3.5 | 12.7 (1) | 3.7 | 3.3 |
| Mechanism best estimate (kN) ($L_{MECH}$) | 2550 | 5750 | 1760 | 177 | 1415 | 2330 |
| $L/L_{MECH}$ | 1.22 | 0.97 | 1.19 | 1.29 | 0.76 | 1.07 |

(1) this bridge is not wide enough for two axles side by side

4.2.3 Lining with gunite. A layer of concrete about 100-150 mm thick and usually reinforced with fine mesh reinforcement is sprayed onto the soffit of the arch from a gun.

4.2.4 Prefabricated liners. A liner usually of corrugated steel is tightly fitted beneath the arch and the gap between it and the arch filled with sand/cement grout. Lining with gunite or corrugated steel is unlikely to improve the appearance of the structure.

## 4.3 Spandrel and wing walls

The traditional means of restraining walls which are moving outwards is to tie both walls together with rods and spreader plates. Another solution is to expose the walls and to backfill them with concrete; if the barrel is being saddled at the same time this is the most appropriate method.

## 5. DISCUSSION

Arch bridges are capable of remaining in service for a very long time (there are Roman bridges still in use). There are many techniques available to repair or strengthen them and they are usually cheaper than replacement. Arch bridges are usually only replaced if they have been allowed to deteriorate beyond economic repair or if they are not capable of being strengthened or widened to meet traffic needs. As an example, replacement would take place if the arch barrel or spandrel walls had become grossly distorted. Judgement of remaining life of the structure both before and after repair and strengthening rests with the experience and training of the engineer responsible for the assessment. Guidelines to help him would be desirable and could be developed by pooling the experience of practitioners. Such guidelines would be qualitative; making them quantitative may be difficult and not cost effective.

## 6. REFERENCES

1. Department of Transport. The assessment of highway bridges and structures. Departmental Standard BD 21/84, DTp, London (1984).

2. Department of Transport. The assessment of highway bridges and structures. Advice Note BA 16/84, DTp, London (1984).

3. Crisfield, M.A. and Packham, A.J. A mechanism program for computing the strength of masonry arch bridges. TRRL Report RR 124, Transport and Road Research Laboratory, Crowthorne (1987).

4. Crisfield, M.A. Finite element and mechanism methods for the analysis of masonry and brickwork arches. TRRL Report RR 19, Transport and Road Research Laboratory, Crowthorne (1985).

5. Hendry, A.W. et al. Test on stone masonry arch at Bridgemill - Girvan. TRRL Report CR 7, Transport and Road Research Laboratory, Crowthorne (1985).

6.  Hendry, A.W. et al. Load test to collapse on a masonry arch bridge at Bargower, Strathclyde, TRRL Report CR 26, Transport and Road Research Laboratory, Crowthorne (1986).

7.  Page, J. Load tests to collapse on two arch bridges at Preston, Shropshire and Prestwood, Staffordshire. TRRL Report RR 110, Transport and Road Research Laboratory, Crowthorne (1987).

8.  Page, J. Load tests to collapse on two arch bridges at Torksey and Shinafoot. TRRL Report RR 159, Transport and Road Research Laboratory, Crowthorne (1988).

The work described in this paper forms part of the programme of the Bridges Division of the Structures Group of TRRL and the paper is published by permission of the Director.

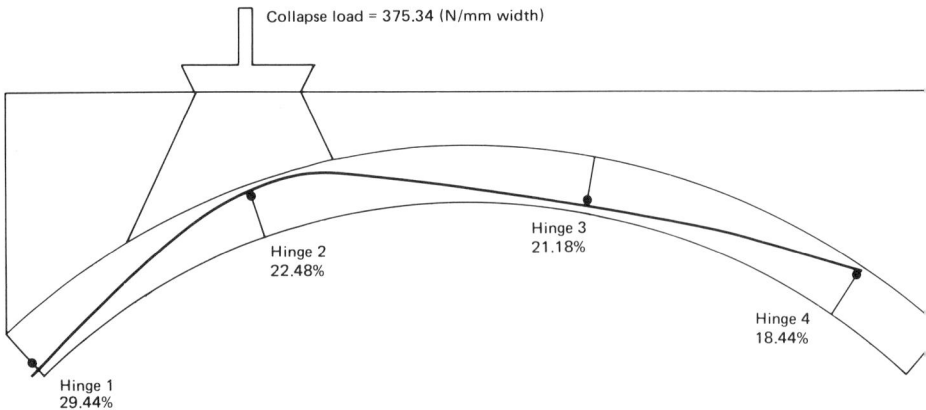

**Fig. 1 Example of a mechanism analysis**

Plate 1 Loading system

Plate 2 Torksey: North face

Plate 3 Shinafoot: West face

Plate 4 Prestwood shortly before collapse

Plate 5 Torksey at moment of collapse

Plate 6 Shinafoot shortly before collapse

R Lavender, Tower Hamlets Borough Engineers Service, UK

Has any research been carried out on masonry brick arches that have arch rings angled across the width of a bridge? I ask this question because many brick arch bridges currently in service are constructed in this manner.

Author's reply

I know of no published research on the behaviour of skewed brick arches. As far as I know the brick courses are skewed relative to the springings only when the bridge itself is skewed. The TRRL research has deliberately concentrated on bridges with little or no skew. They may suffer from ring separation which is common with multi-ring brick arches whether square or skewed, and which is referred to in paper 37. They may all suffer from problems because of their skew but at the present state of our research I can offer no comments.

# 37 The performance of masonry arch bridges with ring separation

**Dr P J Walker, Dr A N Qazzaz and Dr C Melbourne,** Department of Civil Engineering and Building, Bolton Institute of Higher Education, UK

The paper discusses the results of a series of tests carried out on small and full-scale arch bridges to study the effects of ring separation. Tests showed that the defect can be initiated by a relatively small live loading and that fully developed ring separation can significantly reduce the ultimate load carrying capacity of masonry arch bridges.

## INTRODUCTION

Many of the existing stock of over 40,000 masonry arch bridges in Britain suffer from defects. Consequently, it is essential that the effects of these defects on the load carrying capacity is quantified. One such defect from which brickwork masonry arches are prone is ring separation. The barrel of a brickwork masonry arch generally comprises of more than one ring of brickwork. Ring separation is the process by which these originally bonded rings of brickwork become partially or wholly separated. The probable causes of ring separation are a combination of weathering of the mortar and the stress history of the bridge.

As part of a preliminary study for a full-scale testing programme seven model arches were tested to study effect of ring separation on ultimate carrying capacity. A 6000mm span segmental brickwork arch was constructed and load tested to collapse. The latter bridge represented the eighth to be tested as part of a testing programme co-ordinated by TRRL to be used to update the current assessment techniques for masonry arch bridges (4).

## STUDIES OF RING SEPARATION

The models built were of a parabolic profile with a 1000mm span and span/rise ratio of 3 : 1. The arch ring comprised two rings or courses of brickwork either bonded or de-bonded around the full arc of the arch, the total ring thickness was approximately 110mm. The models were built using half-scale fletton bricks and constructed without spandrel walls. Prior to laying, the bed-face of each brick was coated with oil in order to minimise the effects of the bond strength at small-scale. Sand was used as the backfill material. Arches 1 - 3, Table 1, were built such that the two courses of brickwork were fully bonded together using a cement : sand mortar. Initially, ring separation was achieved using two oil coated polythene sheets sandwiched between the two courses of brickwork and the cement : sand mortar bed-joint, Arch 4. However, due to the lack of friction

Table 1.   Results for ring separation test

| Arch No. | Inter-ring Bed-joint material | Loading position | Experimental ultimate load (kN) |
|----------|-------------------------------|------------------|----------------------------------|
| 1 | mortar | 1/4 | 13.2 |
| 2 | mortar | 1/4 | 18.6 |
| 3 | mortar | crown | 31.0 |
| 4 | polythene/mortar | 1/4 | 4.8 |
| 5 | sand | 1/4 | 6.9 |
| 6 | sand | 1/4 | 8.1 |
| 7 | sand | crown | 8.8 |

between the two courses this form of de-bonding was considered unrepresentative and inappropriate thus the bed-joint material was replaced by a damp building sand, Arches 5-7 in Table 1. A knife edge load (KEL) was applied incrementally up to failure at either the 1/4 or crown points. Backfill pressures around the extrados, radial deflections and brickwork strains were monitored throughout the loading history of each model.

The present discussion is limited to a consideration of the failure modes and ultimate loads of each model, ultimate load for each of the seven arches is reported in Table 1. All of the models loaded at the 1/4 span point failed due to the formation of a four hinge mechanism, a typical failure is shown in Figure 1. For those models loaded at the crown failure was due to the development of a 'classical' (1) five hinge mechanism, Figure 2. At failure in Arches 1 - 3 no or little ring separation was reported and formation of the hinges was as a monolithic 'single' ring. In Arches 4 - 7, other than in the region of the applied load, the de-bonded rings were measured by embeddment strain gauges as separating with increasing load. Development of the hinges in the models with initial ring separation was such that two thrust lines developed, one in each of the inner and outer rings. Hinges in the two rings were coincident and developed simultaneously, Figures 1 and 2. Collapse of the Arches 4 - 7 was sudden and caused by almost total physical separation of the two rings and full development of the hinges. Transfer of the load between the separated rings would seem to have occurred through contact between the rings underneath the loading point and friction between the two rings.

The effect of ring separation on the ultimate load carrying capacity is clearly illustrated in Table 1. For the arches loaded at the quarter span the average reduction in ultimate load, compared to Arches 1 and 2, was 53%, for a sand bed-joint (Arches 5 and 6), and 70% for rings separated by oiled polythene (Arch 4). The greater reduction in the ultimate load for the Arch 4 would seem to have been due to the absence of friction between the rings. Reduction in the ultimate load carrying capacity due to ring separation for crown point loading was on average 72%, comparison of Arches 3 and 7. The effect of ring separation is therefore dependent upon the loading position since the resultant thrust line will be directly related to the loading position and it is the magnitude and formation of the thrust line that will influence the subsequent development of the ring separation.

Reduction in ultimate load carrying capacity due to ring separation around the total arc of the arch has been demonstrated by the small-scale tests and as previously stated this work is now being continued at large-scale. However, total separation represents only an extreme case of ring separation and it is more likely the case is one of partial separation. Consequently, a further series of small-scale tests are planned to study partial ring separation and the effects of cyclic loading.

## FULL-SCALE TESTING OF A MASONRY ARCH

Details of the full-scale arch bridge built and loaded to failure at the Institute's large-scale testing laboratory are illustrated in Figures 3 and 4. A segmental profile was selected for the arch ring and when completed the total width of the bridge was 6000mm. Concrete engineering bricks were used throughout and bonded together using a 1 : 2 : 9 by volume (cement : lime : sand) mortar mix, average 28 day compressive strength of 2.3 N/mm$^2$. A 50mm graded limestone 'crusher run' was used for the backfill material, compacted in 100mm layers to 200mm above the crown. The completed arch was surfaced with a 100mm thick bituminous layer.

Throughout the construction process instrumentation was provided to monitor all aspects of the bridges behaviour. Thirty-four vibrating wire type earth pressure cells, placed around the extrados of the arch barrel and spandrel walls, were used to monitor the backfill pressure at the brickwork/backfill interface. A total of fifty-six strain gauges were installed around the masonry to measure both the brickwork surface strains and also to detect formation of ring separation. Movement of the bridge during testing was measured using sixty linear voltage displacement transducers (LVDT's). Two separate series of load tests were conducted on the bridge. The first comprised of the application of a 50kN point load at various points across the width and span of the bridge, these tests were carried out in simulation of vehicular loading. Following the point loading tests, the bridge was subjected to a KEL at the quarter span, loading was applied incrementally through to collapse.

The response of the bridge under the 50kN point load may be described as being a localised deformation in the region of the loading position. Only whilst the load was directly above or adjacent to a pressure cell was any increase in pressure recorded. Similarly, negligible movement of both the arch barrel and spandrel walls was recorded by the LVDT's (reading to 0.05mm). However, during the programme of point loading localised cracking was observed and recorded when the load was applied adjacent to the spandrel wall at the quarter span. Loading caused separation of the spandrel wall and arch ring and the cracking manifested itself inside the arch barrel as ring separation along the line of the quarter point, detected by strain gauges embedded in the brickwork. Further testing across the bridge caused the cracking to open and close with load, this is illustrated in Figure 5 as an 'influence line' diagram for the ring separation and spandrel wall/ring defects. The 'influence lines' represent crack opening due to a 50 kN 'point' load applied at different positions adjacent to the spandrel wall. Figure 5 shows the localised nature of the deformation as cracking was nearest when loading was nearest to the position of the crack. The 50 kN load tests highlighted a fatigue problem that many arches may suffer from when subjected to a large number of vehicular loadings.

The arch failed under the KEL due to the formation of a four hinge mechanism at a total applied load of 1173 kN. At collapse the initial ring separation had propagated to cover the full arc of the ring arch barrel. The spandrel walls were lifted and rotated by the deformation of the arch ring, similar to that observed in small-scale tests which have been reported elsewhere (2). A 'mechanism' analysis (3) was used to calculate the ultimate load of the arch, the method assumes a four hinge mechanism and the ultimate load is calculated from an equilibrium with the vertical dead load of the backfill acting on the arch. However, it is clear that both the lateral pressures of the backfill and spandrel wall stiffening acting on the arch have a significant influence on the ultimate load carrying capacity, hence, these variables were incorporated into a modified analysis proposed by the authors (2). Using the modified mechanism method an ultimate load of 906 kN was predicted for the arch, this compares with a predicted ultimate load of only 271 kN if lateral backfill pressure and spandrel wall effects are ignored. The improved agreement between the experimental and theoretical ultimate loads illustrate the need to consider the effects of the lateral restraint of the arch due to the backfill and spandrels.

## CONCLUSIONS

The following conclusions can be made:

1       A reduction of 72% in the ultimate load carrying capacity of model brickwork masonry arches was recorded due to ring separation.

2       A modified mechanism type analysis, developed from small-scale tests to allow for the effects of spandrel wall stiffening and backfill lateral pressures, can be used to calculate an improved estimate of the ultimate load carrying capacity of full scale arch bridges.

3       A relatively small point time . live load (approximately 5% of the ultimate load) applied to the road surface of the full-scale arch was sufficient to cause localised damage to the structure.

## ACKNOWLEDGEMENTS

The authors wish to acknowledge the support of SERC, TRRL, British Rail, NAB and the Department of Civil Engineering and Building, Bolton Institute of Higher Education.

**REFERENCES**

(1)     Heyman, J.; <u>'The Masonry Arch'</u>, Ellis Horwood, Chichester, (1982).

(2)     Melbourne, C. and Walker, P.J.; 'Load tests to collapse of model brickwork masonry arches', Proceedings of the Eighth International Brick and Masonry Conference, Dublin (1988), pp 991-1002.

(3)     Pippard, A.J.S. and Baker, J.F.; <u>'The Analysis of engineering structures'</u>, 2nd Edition, London, (1943), Edward Arnold.

(4)     Department of Transport Roads and Local Transport Directorate; 'The assessment of highway bridges and structures', Departmental standard BD21/84.

Figure 1    Typical failure mode of ring separation model test – quarter point loading

Figure 2    Typical failure mode of ring separation model test –
crown point loading

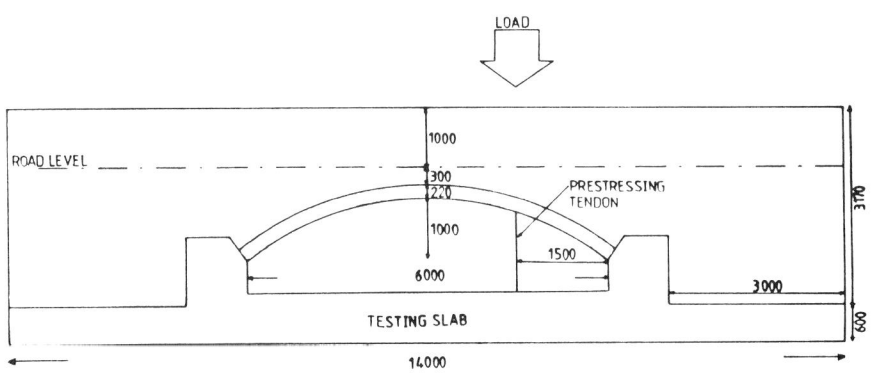

Figure 3    Details of full-scale test

Figure 4    Full-scale masonry arch test

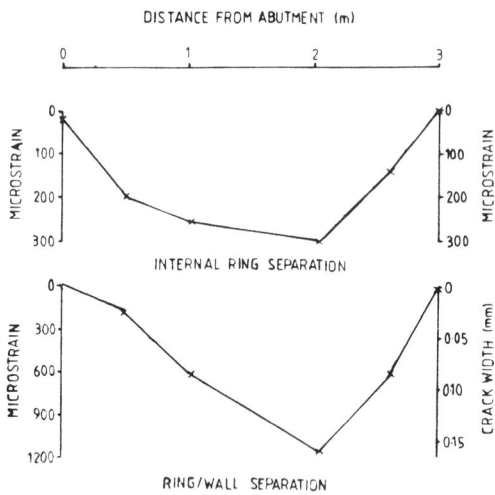

Figure 5    Influence of load position on ring and spandrel
            wall separation

R C de Vekey, Building Research Establishment, U.K.


Dr Melbourne mentioned that the trial structure was fully bonded. Did he mean by headers or metal ties between the arch-rings or simply by a continuous mortar collar joint?

Would the structure perform better if the spandrel walls were tied on positively to the arch rings by, for example, stainless steel ties?

Author's reply

In answer to the first question, the arch rings were held together simply by the mortar joint. The short answer to your second question is yes. However, the complexity of the interaction between the arch barrel and the spandrel walls is not fully understood. Certainly, many masonry arch bridges suffer from spandrel wall separation but it is usual with this defect for the barrel to develop a longitudinal crack, thus allowing the spandrel wall to move and take part of the arch barrel with it. Consequently, connecting the spandrel wall to the barrel may not enhance the bridge performance with respect to this type of defect. However, introducing stainless steel mesh transversely into the mortar arch barrel bed joints, may prove more successful.

# 38 A discussion comparing the performance of bridges in Dorset from 1066 to present day

**P Jones and J Brownlow,** Dorset County Council, UK

The bridges built in Dorset over a period of 900 years have been influenced by several factors. The earlier bridges were constructed by local labour using local materials. In those days there were several small brickworks and quarries convenient to the sites and so the forms of construction were usually arch bridges in brick or stone. The age of the iron, cast iron and steel are all identified in bridges in the County including the rolling lifting bascule bridges serving Poole and Weymouth harbours. Overall there are some 910 highway bridges of which half are pre 1922, and the opportunity has been taken in this paper to examine the performance of the bridges and the associated maintenance works and costs.

## 1. THE OLDER BRIDGES

1.1 The earliest bridge constructed within Dorset dates from around the eleventh century. This is located in Lyme Regis and use to provide a crossing place over the river Lime. Whether this bridge was constructed for local or strategic reasons is not clear but at the time Lyme Regis formed an important port on the south coast of England. The materials used in the bridge construction were local won stone which largely remains in position after the passage of some nine centuries.

1.2 Following on from the eleventh century other bridges were constructed at a number of other important river crossings as indicated on Appendix 1 including:-

A - Julians

8 span stone arch bridge built in 17th century widened in brick and stone when it was reconstructed in 1844. All the arch voussoir stones have been replaced during the last decade.

B - Canford

3 span stone arch built in the 18th century, the 6 stone and brick flood arches were built in 1675. A cantilevered footbridge was added in 1964.

## C - Whitemill

8 span stone arch bridge built in the 12th century, founded on timber piles which are still in position.

## D - Blandford

6 span stone arch bridge built in the 16th century and widened in 1782. There are also 5 stone flood arches all of which remain standing today.

## E - Sturminster Newton

6 span stone arch main bridge built in the 16th century. More work was done on the bridge in 1820 and 1827 as indicated by date stones in the structure. 10 span stone flood arches built in 1828.

## F - Wool

6 span stone arch bridge built in 15th century repaired in 1667.

1.3   A feature of all these early bridges was that they were mainly constructed from locally won material and constructed by local craftsmen.   Furthermore cost would have played an important role as it does today but the intended design life expected of these early bridges was probably put at no more than a generation and it is of great credit to the builders that they are still standing and in the main still carrying their design loads.   It is interesting to note that some of the earlier designers protected themselves by placing weight restriction notices on some of the bridges such as:-

NOTICE

To owners and drivers of traction engines this bridge is insufficient to carry weights beyond the ordinary traffic of the district.

1.4   This is apparently quite simple but what constituted the 'ordinary traffic of the district?'.   Another notice repeats the same message but at much greater length.   Perhaps they were meant for the more educated gentry.   It read:-

County Council of Dorset

Take notice that this Bridge (which is a County Bridge) is insufficient to carry weights beyond the ordinary traffic of the District, and that the owners and persons in charge of Locomotive Traction Engines and other ponderous Carriages are warned against using the Bridge for the passage of such Engine or Carriage.

E Archdall Ffooks (1888)
Clerk of the County Council of Dorset*

1.5   The situation regarding early bridge building is not unique to Dorset, each county having its own characteristics regarding locally won materials and craftsmen.   However, a pattern of ideas regarding shape and size of bridges emerges and this was no doubt due to the influence both of state and religion as new communities

337

were created in the countryside.  There must have been a great
deal of trial and error in the formation stages of the new bridges
and particular problems were most likely encountered at the
foundation building stage.  Investigation shows that in some cases
several types of foundations were tried before a final selection
was made.

2.    THE BALANCE SHEET OF COSTING

2.1   The present balance sheet for bridges in Dorset reads as follows:-

|          | Pre 1922 | 1922-47 | Post 1947 | Total |
|----------|----------|---------|-----------|-------|
| Stone    | 199      | 1       | 0         | 200   |
| Brick    | 229      | 3       | 0         | 232   |
| Concrete | 3        | 51      | 339       | 393   |
| Steel    | 31       | 6       | 10        | 47    |
| Other    | 0        | 9       | 1         | 10    |
|          | 462      | 70      | 450       | 882   |

2.2   The above list is not complete because there are also the Rights
      of Way bridges to take account of and these can vary from the
      simple timber planking arrangement to a suspension type of bridge.
      In addition there are the private bridges and those belonging to
      other authorities such as British Rail which were built mainly
      under various Railway Acts of Parliament.

2.3   In order to assess the merits or otherwise of a particular bridge
      or groups of bridges it is important to have a good well kept up
      to date recording system detailing any maintenance and/or
      reconstruction work on any particular bridge which can be
      identified and hopefully costed in real terms.  The recording
      system in use within Dorset has been operational since late 1950's
      and more recently the records have been computerized.  Secondly
      the system is designed to identify each bridge in detail together
      with a historical mode of maintenance works.

2.4   The make up of the costs can be identified into particular groups
      of bridges such as:-

Figure  1    Pre  1992 bridges,  1922 - 1947  bridges  post  1947  bridges

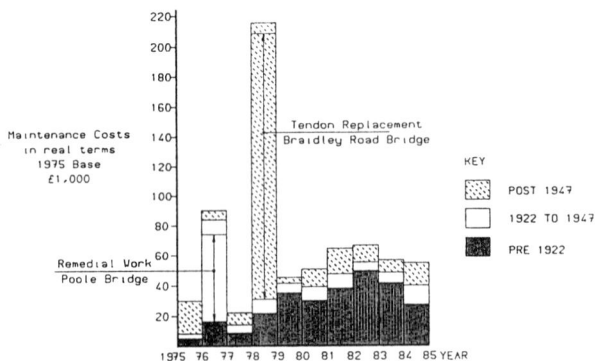

2.5 Quite a large number of the bridges are listed structures totalling some 71 under the Town and Country Planning Acts. These structures are all pre 1922 and include the more ancient bridges going back to the eleventh century. However, the annual maintenance costs for these type of bridges averages at some £15000 per annum is about £7.2 per sq metre of the deck area.

2.6 A further examination of the maintenance cost figures can identify the type of construction material involved as indicated below.

Figure 2  Maintenance for differing materials in bridges

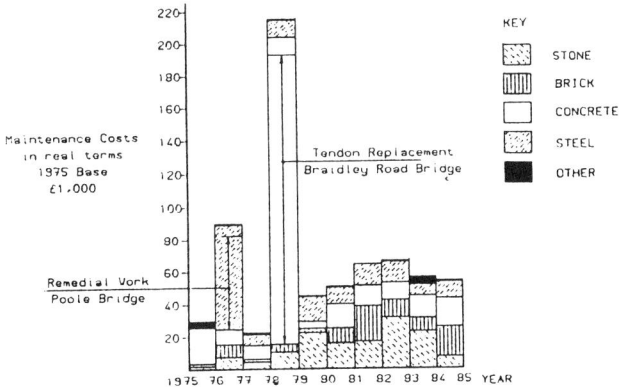

2.7 The above cost figures need to be further broken down to terms of square metre of deck per structure and this exercise indicates the following.

| Stone | £4.41 per square metre) | Average over ten |
| Brick | £4.37 per square metre) | years.  1975 base. |
| Concrete | £2.16 per square metre) | |
| Steel | £6.48 per square metre) | |
| Other | £2.35 per square metre) | |

2.8 The conclusion which can be drawn from the above costing information indicates that there is a relatively high level of cost relating to the more recent bridges ie the post 1947. A further high cost area concerns the steel bridges and the remaining of costs relate to the pre 1922 bridges.

3.    MAINTAINING SOME OF THE BRIDGES

3.1 The earliest concrete bridge is Tuckton Bridge built in 1905 over the river Stour on the eastern fringes of Bournemouth. It is a 12 span structure and the present day strength of the concrete is about $26KN/mm^2$. A survey in the early 1970s showed several areas of distress and during 1975 defective areas were cut out and made good including the provision of replacing corroded reinforcement. At present the bridge has a 13 ton max excluding PSVs weight limit and on an average day carries 20000 vehicles per day.

3.2 Since the days when the Tuckton bridge was constructed improved design techniques have been brought forward including more latterly BS 5400 requirements. The advent of BS 5400 in particular has highlighted the need for greater attention to be paid to the ultimate limit state of the structure and the serviceability throughout its design life. In this respect fatigue and deflection are given more emphasis than hitherto. It is of interest that a design life of 120 years is now assumed for all bridges designed to BS 5400. Nevertheless high maintenance expenditure is still being encountered with regards to:-

i)   Waterproofing bridge decks
ii)  Replacing movement joints

3.3   A further high cost maintenance was related to the replacement of the exposed tendons in the Braidley Road bridge in 1979 at a cost of over £0.5 million when the bridge was 10 years old.

3.4   In connection with steel and iron bridges the main problems relate to:-

i)   Regular painting
ii)  Waterproofing bridge deck

3.5   The advantages of steel structures are that generally the defective areas are clearly visible and can often be cut out and made good followed by painting. This has very much been the case at the lifting bascule bridges both at Poole and Weymouth where current annual maintenance costs are in the order of £95000 per annum. Lifting bridges require servicing and the manning levels are often determined at the formation stage when permission was sought usually through an act of Parliament is being granted in the case of Poole the lifting bridge is constantly manned for 24 hours a day, 365 days and the annual costs are in the order of £80,000. The previous iron Ferrybridge built in 1893 had to be completely replaced because corrosion was very extensive in the critical sections and the cylindrical pile supports had also failed. The cost of the replacement was £2.0 million which was of reinforced concrete construction.

3.6   The remainder of the maintenance costs mainly concerns making good stone and brick work including items such as pointing, removing vegetation growth etc. In some cases traffic growth both in weight and number of vehicles has lead to the need to reconstruct some 15 arches within Dorset. In most cases the arch had failed due to either foundation failure or the lack of maintenance within the arch.

3.7   All structures suffer from road accident damage but most problems of this nature arise from modern large heavy goods vehicles having to negotiate the narrow older structures. Measures to alleviate these problems have included the introduction of one way working controlled by traffic signals and the installation of large kerbs to deflect the wheels of these vehicles away from the fabric of the bridge.

3.8   Unfortunately it has not been possible to complete the whole historical picture because a large number of bridge assessments still have to be carried out on the pre 1922 bridges in order to evaluate their up to date load carrying characteristic. This work should be completed over the next two to three years but allowance is being made for maintenance levels to increase by some 15% per year.

Figure 3   Bridge Maintenance Expenditure Levels

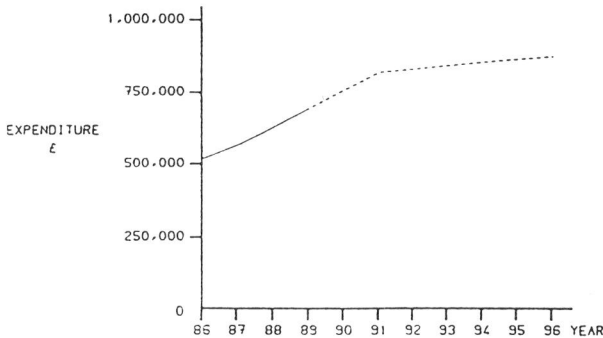

3.9   The majority of the increased expenditure is likely to be spent on:-

i)     Foundation strengthening
ii)    Parapet strengthening
iii)   Generally making good

3.10  The completion of the works should allow adequate strength for 11.5 tonne axle loading up to the 40 tonne loads in accordance with EEC directive. A recent report[0] issued by the Commission of the European Communities [com (87) 34 final] indicated that the Council's Directive 85/3/EEC suggested that modern bridges built after 1961 in the United Kingdom could be judged to be sufficiently strong and those built after 1922 with spans over 20 metres could also be considered strong enough according to the design standards used at that time.   As mentioned above bridge assessment work is continuing and Central Government has already set aside £27 million for strengthening Local Authority bridges and Dorset will be making a bid for some of these funds in the usual manner.

4.    A LOOK AT FUTURE REQUIREMENTS

4.1   It is only from a proper assessment of the historical situation that parameters can be drawn with regards to the durability of materials and research in Dorset shows:-

i)     The bridge must be properly constructed ie with good drainage, waterproofing and movement joints.

ii)    Due account must be taken of the loading intensity type and width of vehicles likely to use the bridge.

341

iii) Regular maintenance needs to be undertaken to ensure the durability of the material.

4.2 Provided proper regard is taken of these parameters the choice and selection of materials for bridges should be taken on the grounds of:-

a) Aesthetic appearance
b) Load carrying strength
c) Ease of construction
d) Cost of construction
e) Time taken for construction
f) Maintenance requirements

4.3 Bridges form an important part of the highway network within the County and it is desirable that well designed examples should continue to be provided. There are many earlier bridges that have toned in with their surroundings largely due to the judicious use of local materials and the construction of more modern bridges whether in steel or concrete needs to fully reflect the atmosphere of their setting. A significant contribution can therefore be made by all bridge engineers not only to improve transport communications but at the same time enhance the beauty and durability of the environment without incurring the need for future high maintenance costs.

BIBLIOGRAPHY

\*       Dorset Bridges. A history and guide. A.J. Wallis.

0       Report by the Commission to the Council on the development of the circumstances which have justified the derogation accorded to Ireland and the United Kingdom as regards certain provisions of Directive 85/3 on the weights and dimensions of commercial vehicles. Commission of the European Communities.

ACKNOWLEDGEMENT

The writers wish to acknowledge the support given by Mr D A Hutchinson, the County Surveyor, and members of his staff who have helped in the preparation of this paper.

# County of Dorset

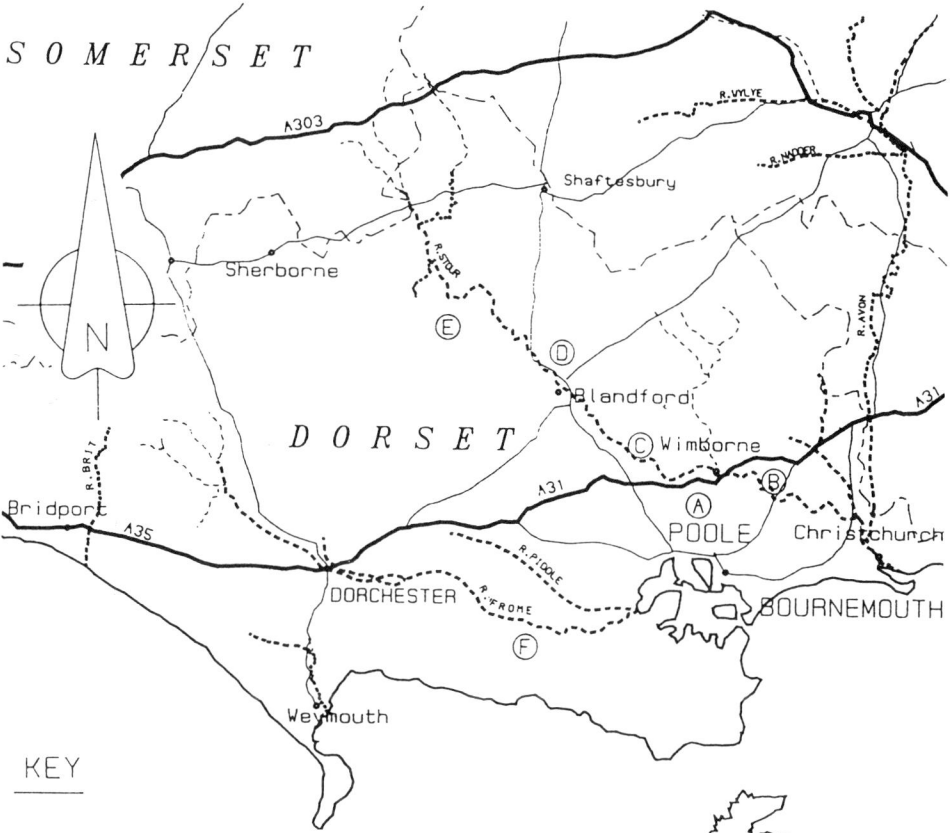

## KEY

—— TRUNK ROADS
—— PRINCIPAL ROADS
······ MAJOR RIVERS
—·—·— COUNTY BOUNDARY

Ⓐ Bridges refered to in paper.

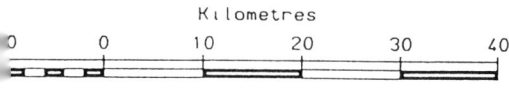

Kilometres

0    0    10    20    30    40

APPENDIX 1

# 39 A concrete ancient monument

**N D Waine,** BSc, CEng, MICE, FIHT, Devon County Council, UK

This paper describes the planning for construction and
subsequent history of what is thought to be the oldest
surviving concrete bridge in the U.K. This stretches
over a period of some 115 years, giving insights into
the optimism which all Civil Engineers share. Recent
investigations into the behaviour and composition of
the structure have led to a programme of repair and
refurbishment, which is still in progress.

INTRODUCTION

Although the concept of design life is increasingly under discussion,
there is no guidance on the parameters which should be used when making
predictions. It is only in the past few years that Terotechnology has
provided any guidance so that the data which we are beginning to collect
will be of more use to future generations than to ourselves. The best
that we can do at present is study and learn from the examples around us.

An interesting case in point is Axmouth Bridge at Seaton (Fig. 1) in the
south east corner of Devonshire, U.K. This was built in 1877 and is
thought to be the oldest surviving all-concrete bridge in the country.
Because of this it has been designated an Ancient Monument and the County
Council as owners are responsible for ensuring that it is preserved.
Design faults, foundation failure, construction defects and the ravages of
time and tide have left a structure which is nearing the end of its useful
life. There are no details of the concrete mix design, but the result is
very poor by modern standards. Construction too, left much to be desired.
Salt water and sea spray are always present, but fortunately there is no
reinforcement to corrode.

Cracking was first reported in 1930, and continued to increase, but it was
not until 1956 that steel relieving beams were installed to remove traffic
loads from the centre span. Continuing deterioration led to a thorough
reappraisal to ensure that the old bridge could be preserved and still
carry traffic. Detailed site investigations and vibration tests have been
carried out to predict the effect of piling operations during proposed
repairs and the construction of a new bridge. The factors which have
affected the choice of remedial works will be discussed, together with the
way in which they will influence the future life of the existing bridge
and the construction of a new one.

# HISTORY

Axmouth Bridge was built in 1877 at the expense of the then Lord of the Manor of Seaton, Sir William C. Trevelyan. He paid the Lord of the Manor of Axmouth, William Hallett, £400 for the rights of the Axe Passage and could therefore charge tolls from those who crossed his bridge. Sir William wished, according to the Seaton Guide of 1885, to promote the use of concrete and so he employed Phillip Brannon, self styled Architect and Civil Engineer from London to carry out the work. Brannon was a prolific inventor but to put it as politely as possible some of his ideas were ahead of his time. He too was a keen proponent of the newly introduced concrete, and was therefore ideally matched with his client from Seaton.

The original proposal was for a clear span of 100 ft., but apparently the Board of Trade were unhappy with the idea and accepted a three-span alternative. With the advantage of hindsight we can now see that this caution probably saved Brannon's reputation. A casual look at the bridge which he built reveals that the western pier has settled, in fact by about 600mm. It is generally believed that this happened while the arch centering was being removed; recent site investigations have revealed that the foundation to the pier rests on a thin gravel stratum overlying soft organic clayey silt. The eastern abutment and pier are founded on the same gravel, but this extends down to the underlying marl so that any settlement was comparatively small.

Brannon's description of three of his inventions which he used at Axmouth show that he believed in the impossible. With flawed logic and without a detailed soil survey, his chances of avoiding failure were small. To quote his own words:-

"CUP BASE OR CONCAVE CENTRICATING FOUNDATIONS for Quick Sands, Mud and other yielding subsoil superseding the use of piling, cylinders and other expensive artificial foundations. The forming to any solid, open, or caissoon structure of an unyielding concave soffit in which the thrust of the arch is received by rigid tension arrangement, the two continually reacting and rendering impossible any spreading abroad or bursting upward. The entire weight of the superstructure thus acts with a concentrating pressure, irresistibly pushing inwards the subjacent sand or mud, instead of continually squeezing them out, in the manner which takes place, without exception, in all other foundations; and there is consequently formed thereby a solid column of resistance, extending downward until it blends with the equivalent density or gravitation force below, or rests on the solid rock or strata underlying the soft soil".

"BALANCE OR COUNTER THRUST OR COMPOUND ARCHING. The thorough arching of all bridging work, in an eccentric arrangement of arch rings, and counter arching, so as that the varying thrusts shall neutralise or utilise each other, and approximate the bridging work to beam effect".

"IMPROVEMENTS IN THE CONSTRUCTION OF BRIDGES AND CONSTRUCTION OF WIDE SPANS OF FREE BEARING, and in which all dead weight is·eliminated. All parts of the structure are made reciprocally supporting, and all strains are converted into supporting elements."(1).

Tolls continued to be collected until 1907, but the bridge remained in private ownership until it was acquired by Devon County Council in 1930.

It was from that time that cracks in the concrete were first reported, but there are no records of any major repairs until 1956, when the fill over the centre arch was removed and replaced by timber decking supported by steel beams spanning between the piers.

Sprayed concrete and masonry repairs were carried out to the cutwaters and piers in 1971 but nothing was done to repair the damage and deterioration of the arches and spandrels. Five years later the bridge was designated an Ancient Monument, as by that time it was the oldest surviving all concrete bridge in the country.

When, in the late 1970's, the first assessment was carried out in accordance with Department of Transport Technical Memorandum BD 31/74, the true state of the bridge became apparent. A weight restriction of 7.5T was imposed, and this highlighted the fact that major structural repairs would be needed.


RECENT INVESTIGATIONS

Gradual deterioration of the bridge is recorded in successive inspection reports, and is matched by continuing repairs. The first strength assessment had confirmed the subjective view that all was not well and it was at this time that the construction of a new bridge was first mooted. This marked the start of a long sequence of detailed investigations. Two soil surveys revealed that the alluvial deposits in the mouth of the River Axe are very varied. On the east bank there are gravels overlying weak Keuper Marl bedrock at a depth of 12m. Under the west pier and west abutment, there is a thin layer of gravel but below this there is very soft to soft clayey silt ($c_u$ = 12 to $26N/mm^2$, $\phi = 0$) some 5m thick overlying gravel and marl. Brannon clearly did not have enough information to avoid the problems which beset him.

During the detailed design for a new bridge, it became clear that piled foundations would be needed, but that this might adversely affect the old bridge during construction. The two main difficulties which were foreseen were settlement brought about by inflow of silt into the pile bores and, more seriously, vibration during driving. Particular concern was felt about the latter problem, as the western approach embankment, abutment and pier are all founded on the same weak silt. Both temporary cofferdams and permanent piling could give rise to unacceptable vibrations as they would be very close to sensitive parts of the old structure.

The first step in trying to assess the effect of construction operations was to measure the vibration induced in the bridge by road traffic. Seismometer readings were taken over a period of 4 hours, with subjective notes of vehicle sizes. The intention was that the results obtained would enable a limit to be set on the vibration caused by future piling operations. The results were reassuring in one sense, but of little use in another. The recorded peak particle velocities proved to be so low that it could be argued that vibration was not really the problem which had at first been feared. However, the poor response of the bridge meant that it was not possible to set limiting values on the vibration levels which could be accepted during pile driving.

346

With plans for a new bridge in the course of preparation it was considered that careful monitoring of vibration during construction would be essential. For these to have any meaning, base levels were required and a new series of measurements were taken using a loaded lorry at the maximum permitted weight of 7.5 tonnes.

The results obtained showed that different parts of the structure responded in very different ways. The centre of the bridge is atypical, as it is made up of simply supported steel beams, and these are comparatively lively. Particular interest centred on the western pier in view of its apparent delicate condition, and this proved to be amply justified. The highest particle velocities were recorded at this location, and the vibrations tended to continue longer than elsewhere. These results, coupled with those obtained from a long-term deflection survey currently in hand, give the impression that the western pier is floating in a sea of silt. Comparative seismometer traces for the two piers are shown in Fig. 2 and Fig. 3.

There are, however, some features of the vibration measurements which have not been fully explained. Particle velocities in the vertical direction were only 0.39mm/s, but in the orthogonal horizontal directions were 2.3mm/s and 1.6mm/s longitudinally and transversely respectively. Similar results, although of a lesser magnitude, were recorded throughout the structure. Frequency analysis of the same results failed to produce any consistent vibration modes.

As the main concern is the effect of piling operations nearby, an attempt was made to simulate a shock wave by driving the same 7.5T lorry over a 50mm piece of timber. This produced such dramatic results, with peak particle velocities of up to 16mm/s in the steel relieving span, that tests were abandoned after the second run.


EXTENDING THE LIFE OF AXMOUTH BRIDGE

There are several features of the bridge which give serious cause for concern. The most obvious is that the western pier, and to a lesser extent the adjacent abutment and approach embankment are founded on weak compressible silts, which probably failed during construction.

The concrete in general, and in the piers in particular, was made with a badly graded aggregate, lacking in fines. This can be remedied, but access in the tidal estuary is not easy. The eastern abutment and pier are founded on gravels, but the s.p.t. results from the two soil surveys provide conflicting evidence about their density.

At present, the centre arch does not carry any traffic loads, but it is very badly cracked. The side spans are in better, but far from perfect, condition. The spandrels, too, are very extensively cracked and as has already been mentioned, the western pier is 600mm lower than Brannon intended, resting on weak silt.

Ever since these shortcomings were recognised, various proposals for building a new bridge and strengthening the old one have been developed. Throughout this period, maintenance has been kept to a minimum as it always seemed probable that all the problems were about to be overcome.

However, the financial and political vicissitudes to which we have all become accustomed have conspired constantly to frustrate the main objective. Early in 1988, tenders were received for the construction of a new bridge which were far higher than had been estimated and this forced one further postponement in the long-awaited relief.

The only way forward was to review and revise earlier proposals for the strengthening and repair of the Ancient Monument. The lengthy catalogue of defects given above dictated what was needed and tenders were invited for the most important items at the earliest opportunity. Four different operations can be identified, each of a specialist nature so that the remedial works will have to be spread over a period to avoid different contractors affecting each other on a very restricted site.

Of prime concern is the integrity of the foundations, particularly the western pier and abutment and of the arches. Advice was sought from specialists in underpinning and, based on their recommendations, a system of micro-piling was designed which could be tendered for by firms offering different proprietary methods. The same firms were all asked to stitch the major cracks in the arches.

The self weight of the bridge has been estimated at 2,580 tonnes, of which 520 is carried by each abutment and 770 by each pier. Compared with this, live loads are insignificant, so the micro-piling is designed to carry dead loads only by end bearing in the underlying marl. Both vertical and raking piles of 170mm nominal diameter are included in the layout to distribute the loads as widely as possible and to cater for any horizontal forces (Fig. 4).

The eastern pier and abutment are far less vulnerable than those to the west but on the evidence from one soil survey, there is the possibility that the gravels are less compact than is desirable. It is therefore planned to grout them, although it is difficult to predict the likely take-up in the absence of any permeability tests.

Concrete in the tidal zone of the abutments and piers is continuing to deteriorate in spite of repairs carried out seventeen years ago. Further repairs are about to start using stainless steel mesh and sprayed concrete on the surface followed by pressure grouting of the core.

In recent years the road surfacing over the centre span has needed repair on several occasions, which has been attributed to rotting of the timber decking installed in 1956. It has always been assumed that the steel beams supporting the timber were rusting as they are in a moist marine environment, and have received no maintenance in 32 years. The anticipated rotting timbers and rusting beams were duly revealed when work started recently on the underpinning contract. Various methods have been investigated for replacing the beams and decking and current proposals are for curved Universal Beams with an insitu concrete deck. The use of high-tensile steel and a compact section will reduce the depth of the beams sufficiently for the road to be reinstated at its original level, with a marked improvement in appearance.

The worst cracking in the arches occurs in the centre span, but this only has to carry its own weight, because of the steel relieving beams. No measurement of these cracks has been undertaken but they are significant

Figure 1. Axmouth Bridge Seaton from the south-west

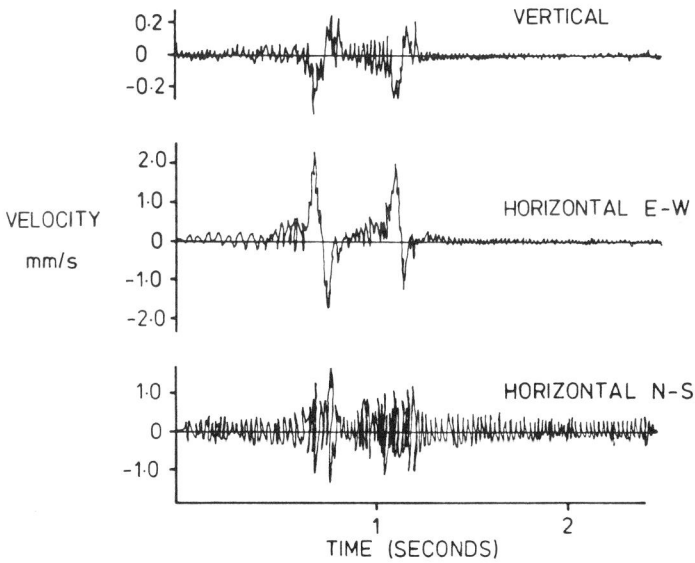

Figure 2. East pier seismometer trace: 7½ T. lorry

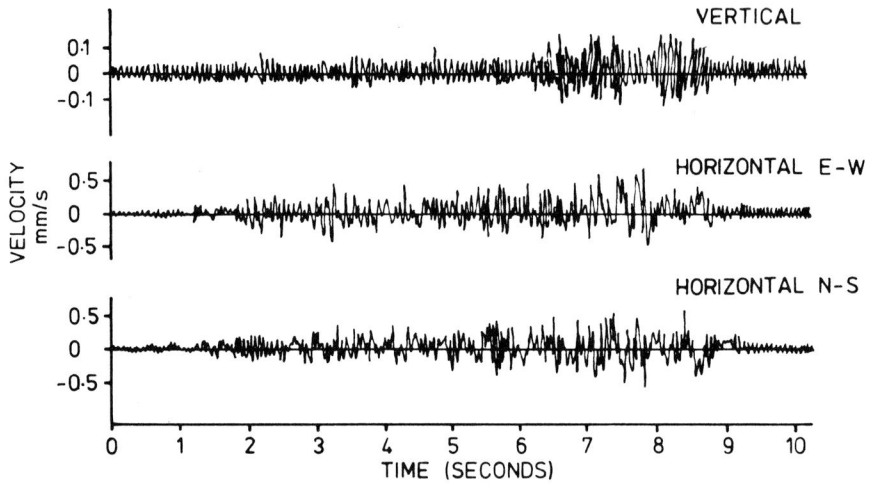

Figure 3.  West pier seismometer trace: 7½ T. lorry

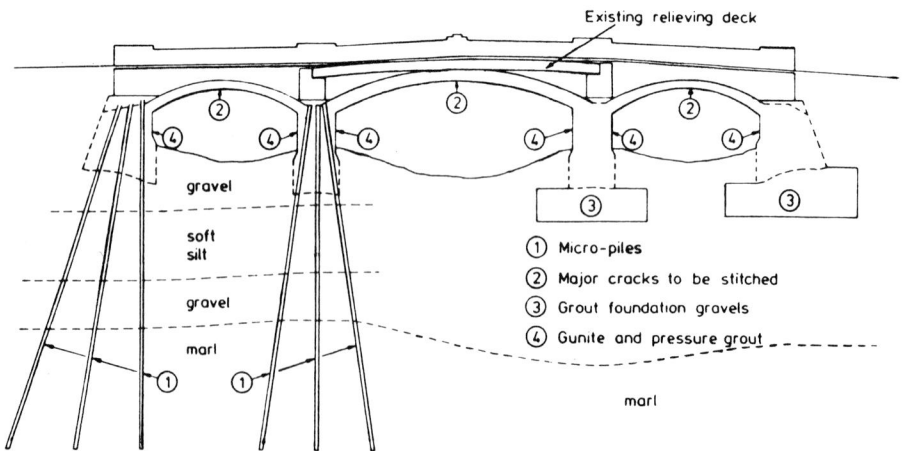

Figure 4.  Remedial measures carried out in 1988

enough to give cause for concern. To ensure the long-term integrity of the structure, it is necessary to reduce their effect, and they are being strengthened by a series of stainless steel dowels set in cement grout. The dowels will be 16mm in diameter 1.0m long in 32mm diameter holes drilled at an angle to the concrete face alternately from opposite sides of the crack.

THE FUTURE

When all the works described above are completed, Axmouth Bridge will have a new lease of life. To provide a quantifiable assessment of their effectiveness, further vibration measurements will be carried out, using the same 7.5 tonne lorry. This will be of great value during construction of a new bridge.

Once the underpinning has been completed displacement and vibration will be much less of a threat to the marginally stable foundations. When a new bridge is built and vehicles are replaced by pedestrians, the old bridge should gradually decline to a venerability in virtual retirement.

Our oldest remaining concrete bridge was built almost at the same time as motor cars made their first appearance. Although it would appear that the increasing weight of vehicles and volume of traffic since that time has been responsible for the obvious deterioration, it is more likely to have been caused by unsatisfactory design, ignorance of soil conditions and inadequate construction. An attempt to remedy these shortcomings is well under way, and this should extend the life of the structure for a considerable time to come.

The most vulnerable aspect of the structure is, unfortunately, the very concrete which makes it so interesting. The exterior is rough, and the interior even rougher. There is a strong temptation to inject cracks, grout porous areas and render or coat everything. This would certainly slow the rate of deterioration and might enhance the appearance, but it would surely destroy the curious charm of a small part of our national heritage.

REFERENCES

(1)   BRANNON P.   The Arcustat (1879)

# 40 Durability assessment of concrete bridges by in-situ testing, early results

**F R Montgomery and M Basheer,** Civil Engineering Department, The Queen's University of Belfast, UK

This paper describes the survey and results of the first of a series of concrete bridges to be tested as a means to develop techniques to assess the durability of major concrete bridges on the motorway and trunk road system in Northern Ireland. The bridge was constructed in 1961 and is, in parts, in poor condition.

## INTRODUCTION

The Civil Engineering Department at Queen's has commenced a research programme in collaboration with the Roads Service of the Department of the Environment for Northern Ireland. The object of the work is to develop methods to assess the state of concrete Bridges on Northern Ireland's motorway and trunk road system to help a planned maintenance scheme. It is hoped to be able to predict time remaining before repair works may be necessary for any chosen bridge element.

A number of bridges have been chosen for detailed study, representing the general bridge population in terms of age, form of construction, environment, traffic volume and so on. It was decided to commence the trials on the three piers of a motorway bridge which were due to undergo major repair. Large areas of the pier surfaces were delaminated and in some places had fallen off to reveal rusted reinforcement. It was known that chloride ion content was generally much higher than normal limits and this was believed to be the main cause of the corrosion and delamination. It was hoped that starting with a structure showing very obvious signs of distress would provide a better opportunity to learn quickly than would a good structure and should also help to provide a benchmark for the test techniques.

## THE BRIDGE

The subject of the investigation is a reinforced concrete, simply supported, four span bridge carrying a single carriageway road over a motorway of two lanes in each direction. The spans are supported on two abutments, one at each end of the bridge, on each side of the motorway, and on three similar basically rectangular piers, one on each side of the motorway adjacent to the hard shoulder and one on the central reservation. The specification for the bridge was drawn up in 1960 and construction was done in 1961 and 1962. The concrete for the piers was specified to have a minimum works compression strength of $21N/mm^2$ and an average of $31N/mm^2$, both at 28 days. Aggregate to cement ratio and water to cement ratios were also both specified as 5.5 and 0.55 respectively.

## EARLY OBSERVATIONS

The face of the west pier adjacent to the motorway was extensively spalled, mainly in two bands, one at low level to a height of approximately 2 m and one at high level below the deck extending downwards about 1 m to 1.5 m. There were other areas of spalling and delamination on this face, in one or two places the top and bottom bands were joined by vertical strips of delamination. The opposite face of this west pier, protected from the motorway, exhibited almost no problems: such as existed were very localised. The central pier had a deliminated, partly spalled band at ground level, on one side to a height of about 2 m and a less extensively spalled band at ground level on the opposite face. The third, or east, pier had only a little delamination at ground level on the face towards the motorway and almost none on the protected face. The deck was sound and played no part in the investigation. It was decided to concentrate the investigation on the west pier since it contained the most damaged face and one of the least damaged faces.

## TEST METHODS

A covermeter survey of the poor, or outer, surface of the pier had been completed over 3 years ago. The results are included here as Figure 5.

A half cell[1] survey was done for the same outer face of the pier before any repairs were carried out and the results are included as Figures 1 and 2.

To measure the strength of the concrete in-situ the pull-off test[2] was used. This measures surface tensile strength which is converted through known relationships to yield compression strength as reported in Tables 1 and 2.

To measure surface permeability of the concrete in-situ the Clam[3] developed in the Civil Engineering Department of Queen's University was used. The tests for permeability were performed at a working pressure of 20 p.s.i. for a test duration of 60 minutes. The permeability index quoted in Tables 1 to 3 is the mean flow rate of water into the surface from the 40th minute to the 60th minute of the test. The rate of flow is fairly steady in most cases during this 20 minute period. Results gained previously on good reference specimens are quoted as Table 2, for comparison.

Chemical tests performed on samples as listed in Table 3 were as detailed in BS4551:1980 for cement content and chlorides content.

DISCUSSION

The bridge deck has a movement joint over the top of each of the three piers. All leak to some extent but that over the west pier leaks more that the other two. Almost all the water getting onto the top of the pier runs down that side facing the motorway because of the general slope of the deck. The deck is supported on rubber bearings which are surrounded in sand which was used during construction as permanent formwork. The sand holds a lot of salt washed in during winter and releases it continually throughout the year when more water enters. This results in a salt wash down the face every time there is rain.

From the chemical analysis in Table 3 it is seen that there are some very high chloride contents. This is reflected in the half cell readings of Figures 1 and 2. Generally the areas where delamination had occured corresponded with the areas of high half cell potential.

There were two areas of the front face of the pier where there was very little delamination, mainly at the two ends, represented by test sites A1, A7, A8 A9, C1 and C2. Examination of the summary of results in Figure 7 shows some variation of strength results overall but generally in agreement with the specification of 1960. On the other hand the permeability results show an interesting grouping of low permeabilities where delamination has not occurred and generally higher results where it has occurred. Some of the deliminated concrete was also extensively honeycombed.

An examination of the results for the protected, and generally drier, face of the pier shows a very great spread of permeabilities but no delamination. This good performance is due, of course, to the fact that it is not subjected to salt from either above or below. See too Figure 4 for half cell results.

After extensive jack hammering to remove all delaminated concrete, the front face of the pier was reinstated using large quantities of gunnite. It is interesting to note from Figure 3 that the high half cell readings still existed after repair. This was taken care of by installing an impressed current cathodic protection system to the pier after repair.


CONCLUSIONS

It is possible to measure corrosion processes, in a bridge structures, likely to lead to deterioration. Strength of the concrete is not a reliable index to deterioration but it seems that permeability results allied to a knowledge of cover and of the exposure to salt, may provide a useful guide to durability. From observations on site it seems probable that the great variability of permeability readings on this pier resulted from poor compaction.


REFERENCES

(1) ASTM C876-80 "Standard test method for 'half cell' potentials of reinforcing steel in concrete".

(2) Long, A.E. and Murray, A.McC. "The pull-off partially destructive test for concrete", Proc. 1984 CANMET/ACI Intl. Conf. on In-situ/nondestructive Testing of Concrete, SP-82, ACI, Detroit, pp327-350

(3) Montgomery, F.R. and Adams, A. "Early experience with a new concrete permeability apparatus, "Proc. Structural Faults 85 Conf. ICE London, U.K., April/May 1985, pp359-363.

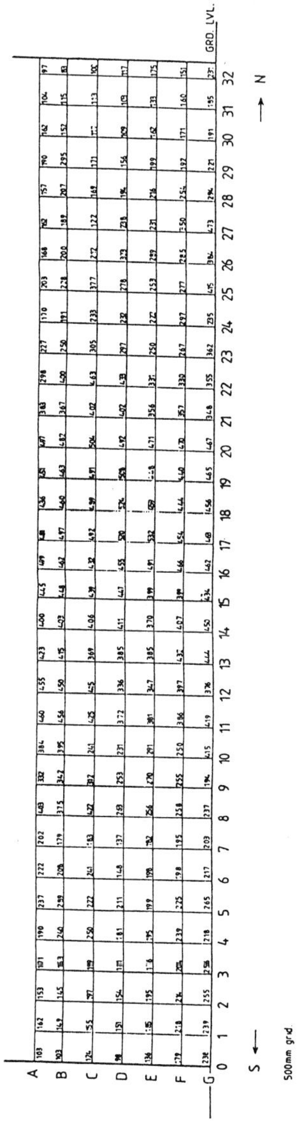

FIG 1 HALF CELL POTENTIALS OF SIDE FACING ROAD PRIOR TO REPAIR

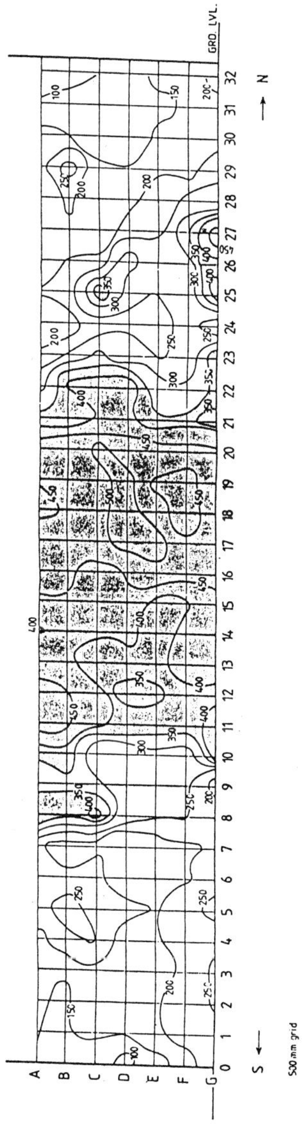

FIG 2 HALF CELL CONTOUR OF SIDE FACING ROAD PRIOR TO REPAIR

355

## TABLE 1 PERMEABILITY, PULL-OFF AND HALF CELL RESULTS ON SITE

| Location | Permeability Index (mL/m²/s) | Average Surface strength (N/mm²) | Half-cell Reading (mV) |
|---|---|---|---|
| A1 | 0.37 | 19.0 | Not available |
| A2 | 0.60 | 37.0 | 458 |
| A3 | 0.72 | 42.0 | 168 |
| A4 | 0.49 | 30.5 | 443 |
| A5 | 0.81 | 42.6 | 455 |
| A6 | 1.99 | 47.0 | 456 |
| A7 | 0.04 | 52.0 | 195 |
| A8 | 0.03 | 36.6 | 215 |
| A9 | 0.04 | Not available | 261 |
| B1 | 0.07 | 39.6 | --- |
| B2 | 5.17 | 24.5 | --- |
| B3 | 15.84 | Not available | --- |
| B4 | 10.81 | 19.2 | --- |
| B5 | 4.21 | 39.9 | --- |
| B6 | 0.58 | 45.4 | --- |
| B7 | 0.04 | 38.4 | --- |
| C1 | 0.06 | 53.1 | --- |
| C2 | 0.06 | 35.3 | --- |
| Maximum value | 15.84 | 53.1 | 458 |
| Minimum value | 0.03 | 19.0 | 168 |
| Average | 2.30 | 37.6 | 331 |
| Std. Deviation | 4.34 | 10.2 | 133 |
| CV % | 186.5 | 27.1 | 40 |

## TABLE 2 TYPICAL PERMEABILITY INDICES FOR REFERENCE SPECIMENS

| Mix & W/C Ratio | Age of Specimens (Days) | Average Cube strength (N/mm²) | Average Surface Strength (N/mm²) | Average Permeability Index (mL/m²/s) |
|---|---|---|---|---|
| 1:2:4-0.4 | 35 | 50.1 | 47.4 | 0.02 |
| 1:2:4-0.5 | 68 | 56.7 | 55.0 | 0.03 |
| 1:2:4-0.8 | 50 | 21.4 | 17.5 | 0.11 |

## TABLE 3 CHLORIDE CONTENT, CEMENT CONTENT & PERMEABILITY INDEX OF DELAMINATED CONCRETE

| Sample No. | Percentage chloride in cement | Percentage cement in concrete | Permeability index (mL/m²/s) |
|---|---|---|---|
| L1 | 2.17 | 14.8 | |
| L2 | 2.79 | 13.8 | 0.29 |
| L3 | 1.40 | 14.8 | |
| L4 | 3.29 | 19.0 | |
| L5 | 4.19 | 12.6 | 13.14 |
| L6 | 6.91 | 14.9 | |
| L7 | 2.12 | 18.0 | 1.18 |
| L8 | 4.43 | 17.2 | |
| L9 | 3.58 | 17.1 | |
| Average | 3.43 | 15.8 | 4.87 |
| Std. Deviation | 1.64 | 2.11 | 7.18 |
| CV (%) | 47.85 | 13.38 | 147.35 |

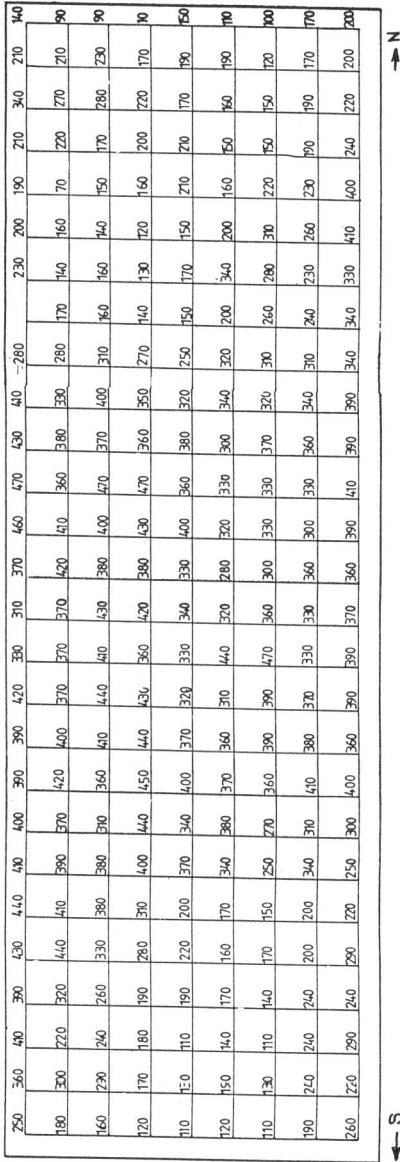

FIG 3 HALF CELL POTENTIALS OF FRONT FACE AFTER REPAIR

600 mm grid

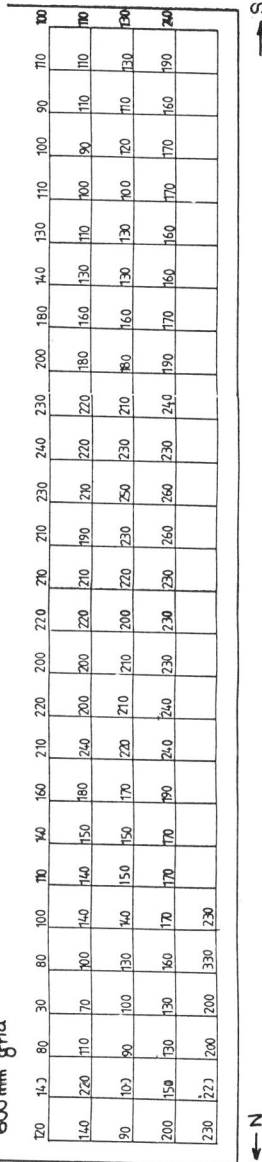

FIG 4 HALF CELL POTENTIALS OF REAR FACE

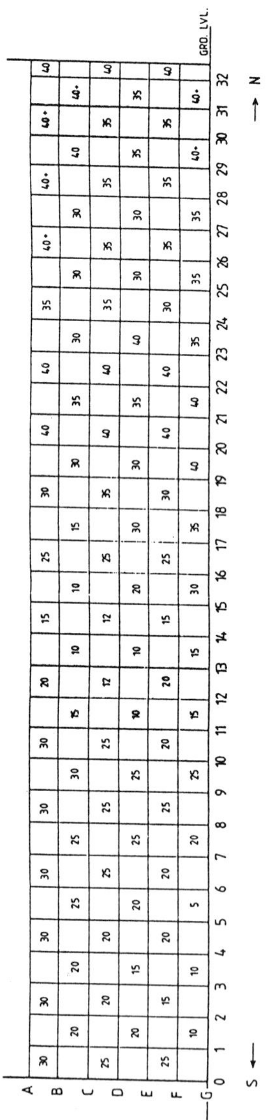

FIG 5 COVERMETER SURVEY

300mm Grid

S ↓     N ↑     GRD. LVL.

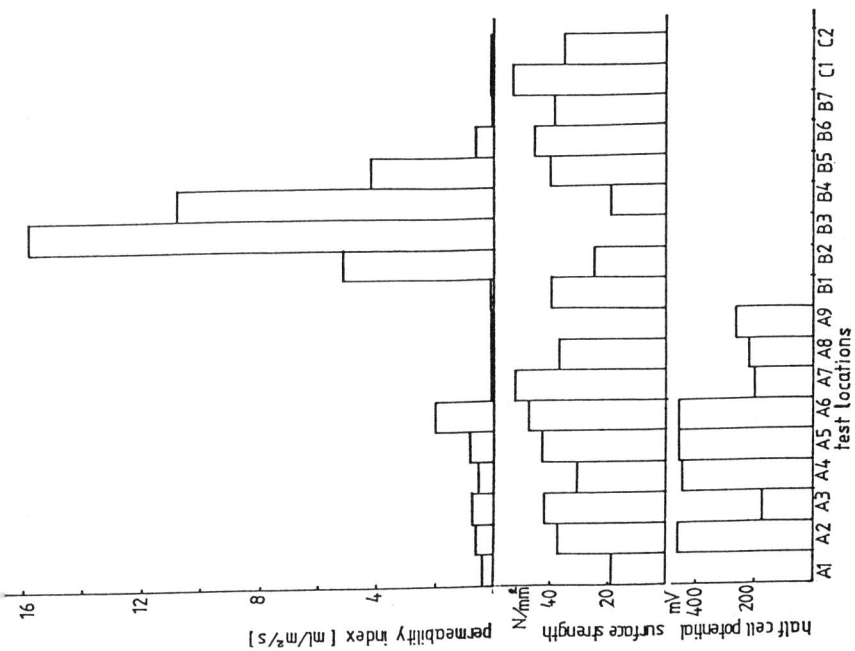

FIG 7 PERMEABILITY, SURFACE STRENGTH & HALF CELL POTENTIALS AT SELECTED POINTS

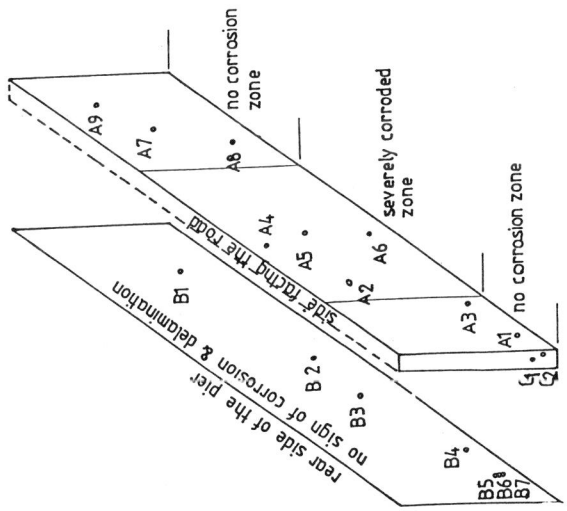

FIG 6. LOCATION OF TEST POINTS

359

# 41 The effect of local corrosion on the life of bridge deck slabs

**P A Jackson,** Gifford and Partners, Southampton, UK

Severe local reinforcement corrosion was simulated by cutting right through eight adjacent bars of the deck slab of a half scale model bridge. Tests on the bridge indicated that this had little effect on strength and that the fatigue life of the remaining reinforcement would be virtually infinite, even if it was suffering from local corrosion. It therefore seems unlikely that local reinforcement corrosion will be the critical factor in determining the life of bridge deck slabs.

## INTRODUCTION

In Britain it has been normal practice for some years to waterproof bridge deck slabs. This has been very effective in avoiding the extensive problems of general reinforcement corrosion which have arisen in North America. However, because the waterproofing excludes the air, it does mean that any corrosion which does occur is likely to be of the local variety. This is generally considered to be more dangerous because it can lead to significant loss of section before it becomes visible. However, it has been found that (because of local strain hardening) significant section loss causes remarkably little reduction in effective yield stress (1). In contrast, because of stress concentration, local section loss has a disproportionate effect on fatigue life.

The Department of Transport have recently drafted a guidance note (2) to enable the effect of local corrosion on fatigue life to be assessed. This gives fatigue curves based on tests on corroded reinforcement, but it is intended for use in assessing the fatigue life of reinforced concrete structures. To do this, a relationship between applied load and reinforcement stress is needed and the fatigue life of the bars can then be obtained using the draft document in combination with BS 5400, Part 10 (3). In the case of bridge deck slabs, the steel stress-wheel load relationship would normally be obtained using elastic plate theory and a fatigue failure in a bar would presumably be taken to constitute failure of the deck slab. As the wheel load stresses calculated in this way are likely to be high and because bridges are subjected to many cycles of wheel loads, this procedure will suggest that the life of bridge deck slabs will be greatly reduced by fatigue once there is any local reinforcement corrosion.

There is evidence to suggest that this approach is extremely pessimistic. Cairns (4) and others have found that the stress induced in deck slab reinforcement by wheel loads is substantially lower than predicted by elastic plate theory. As the fatigue life-stress range relationship is highly non-linear, this implies that the predictions for the fatigue life of reinforcement could be in error by several orders of magnitude. It is also possible that, even if some of the reinforcement had corroded right through or ruptured due to fatigue, a deck slab could continue to behave satisfactorily since it has been found (5, 6) that the local strength of bridge deck slabs is not dependent on reinforcement. These aspects are explored in this paper.

## COMPRESSIVE MEMBRANE ACTION

The reason elastic plate theory over-estimates wheel load stresses is that bridge deck slabs work by "compressive membrane action", otherwise known as "arching action", which is illustrated in Figure 1. This has been extensively researched in Ontario (5), Northern Ireland (6) and elsewhere. It has even been suggested that bridge deck slabs would have adequate strengths if completely unreinforced.

The author has recently conducted further research into compressive membrane action (7, 8). His tests differed from earlier work in two important respects. Firstly, only one of the two half-scale models was provided with the diaphragms previously considered necessary to provide the restraint needed to develop compressive membrane action. Secondly, the models were subjected to the whole of the, very severe, "HB" load (9) used in British bridge design practice. A major reason for these differences was that non-linear analysis suggested that the transverse global moments (the moments induced in the deck slab by the differential displacements of the beams) would significantly affect its local behaviour. The non-linear analysis also suggested that, under full global load, the diaphragms would be subjected to a small transverse compressive stress, implying that they could not be important to the restraint. The restraint came from the under-stressed steel and concrete surrounding the critical areas of the slab.

The analysis, supported by tests on the first deck, showed that (contrary to earlier suggestions) reinforcement was necessary to maintain the integrity of the deck slab; at least under the full design load. However, it was needed not to resist local moments but to provide restraint and to resist global transverse moments. This meant that the exact position of the reinforcement was relatively unimportant, which implied that the behaviour should be insensitive to local corrosion. It was therefore decided to incorporate a test to investigate this in the tests on the second model.

## CYCLIC LOAD TESTS

The large size of the model used, combined with financial and time constraints on the project, meant that it was not possible to apply enough loads cycles to realistically study metal fatigue in corroded reinforcement; this would have taken many months. It was therefore decided to investigate the effect of complete failure in several bars. The failure was simulated by cutting right through eight adjacent main bars. This was a very severe test; not only because it meant the slab was effectively completely unreinforced at its mid-span for a width equal to its span, but also because (even with intact reinforcement) the steel provided was equivalent to only approximately a third of that required by conventional design rules.

As the model was to be used to investigate the ultimate strength of a bridge deck with intact reinforcement, an area which would be only lightly stressed under the critical global load case was chosen to investigate the effect of local corrosion.

The bridge, which is shown under load in Figure 2, was subjected to two cycles of design ultimate HB load, 100 cycles of design service HB load and 5000 cycles of a lower load in several different positions. Most of these positions subjected the area with the cut bars to only modest stresses. However, one included a wheel applied immediately over the test area. Since a similar load case had previously been tested opposite hand, over intact reinforcement, this enabled the effect of the corrosion on the service load behaviour to be assessed.

A load of 1.1 times design ultimate load (1.3 times design service load) failed to produce any visible cracks in the relevant area of the deck. This was slightly surprising since the application of a 10% lower load opposite hand had cracked the equivalent area with intact reinforcement. It was also unfortunate as it had been intended to assess the behaviour of the cracked, and effectively unreinforced, concrete under cyclic loads. The strain in the bottom of the slab, 600 microstrain measured over a 100mm gauge length, indicated that a small increase of load would have cracked the concrete. However, the global load on the deck was now so high that it was feared an increase would have damaged other parts of the deck and upset the stress history analysis, which was important to other parts of the project. It was therefore decided to disconnect three of the four jacks so that the load on the area being investigated could be increased without damaging other parts of the deck. At approximately the same wheel load which had been applied previously, a crack became visible. The load was increased a further 10% before it was removed and the other two jacks were re-connected. The cyclic loads were then applied as for the other load positions.

After this treatment the deck was in good serviceable condition and, despite the initial over-load application, the cracks in the area with the cut bars were no wider than on the opposite side of the deck where the reinforcement was intact. The number of load cycles applied was small in terms of metal fatigue, but the reinforcement strains, both measured and estimated from the surface strains, indicated that the fatigue life of the remaining reinforcement, even if partly corroded, would be effectively infinite. In terms of the behaviour of cracked concrete, which is more sensitive to small numbers of large load cycles, the loading to which the bridge had been subjected was equivalent to a very long life. Exactly how long is difficult to estimate, but, since the loading applied included 100 cycles of design service load and the design service load is supposed to have only a 5% chance of occurring once in 120 years (10), the implication appears to be that it was equivalent to several thousand years.

## LOCAL FAILURE TEST

After the cyclic load tests had been completed the bridge was loaded to failure under the critical global load case. It failed by wheels punching through the slab, leaving the beams and the area of the slab with the cut reinforcement relatively undamaged. It was therefore possible to perform a local failure test over the cut reinforcement. Since the equivalent area on the opposite side of the deck was also intact, it was possible to obtain a direct comparison with an area with intact reinforcement.

The two areas were tested in turn using the rig shown in Figure 3. The load displacement responses are shown in Figure 4. Initially, the area with intact steel appeared to be less stiff than the area with cut steel. However, the former area was more extensively cracked by the previous test to failure so this may have no significance. The area with the cut bars did fail at a slightly lower load but the difference, 5%, is no greater than the normal variability of tests on concrete structures. The failure loads for both areas were very high; 4.8 times the required ultimate strength. This was some ten times the strength of the slab implied by elastic plate theory, even with intact reinforcement.

## USEFUL LIFE

The tests indicated that the local behaviour of bridge decks is largely unaffected by even severe local corrosion. However, analysis, supported by the global tests, suggested that some reinforcement was needed to provide restraint and to resist global transverse moments. Thus, there must be some limit to the amount of local corrosion which can be tolerated before a bridge deck slab becomes unsafe. The tests provide only a lower bound to this.

As the total reinforcement provided in the model was approximately a third of that required by current conventional design methods, the implication is that a slab with a random selection of two thirds of the bars corroded right through on any one section along the deck, plus local areas up to the slab span in width where all the steel had corroded right through, would exhibit satisfactory behaviour. Indeed the tests suggest that such a slab would still have a large reserve of strength. However, it should be noted that the behaviour is sensitive to the global behaviour of the beams; the failure under global loads occurred when the differential displacements of the beams became excessive (7, 8). Although loss of reinforcement would not greatly affect the displacement required to initiate failure, it would reduce the global transverse moments, and hence the total load, associated with such displacements. Thus, when assessing a deck slab which suffers from local corrosion (or which is believed likely to suffer from it in the future), it is necessary to allow for the resulting reduction in the distribution properties.

Even allowing for this, it seems unlikely that the reduction in static or fatigue strength of the deck slab resulting from local corrosion would be the critical factor in determining its life. Two alternative mechanisms are much more likely. The first is that more oxygen could get to the steel which would cause the black rust to turn to red rust and could also cause new general corrosion. Although general corrosion has less effect on reinforcement strength than does local, it can be more serious in a deck slab because it is far more expansive and so can cause delamination; that is, it can cause the whole of the cover concrete to separate from the rest of the slab. This is certainly undesirable and would be considered a serviceability failure, although many deck slabs designed to current rules would still have adequate strength in this condition.

The other mechanism which is likely to bring the useful life of the bridge to an end, is that the corrosion is likely to spread to the tendons in the beam. This is much more serious, as is evidenced by a recent failure (11), and the effect on life has been considered by Price and Aguilar (12). It should be noted, however, that the global and local behaviour of bridges are not independent (7, 8). Thus, the deck slab of a bridge could fail as a result of the differential displacement caused by local corrosion in the tendons of one of the beams or, conversely, a beam could fail as the result of the loss of distribution caused by corrosion in the deck slab. Thus, having estimated the extent of corrosion that is likely in a structure at some time in the future, it is only strictly possible to check safety either by using a non-linear analysis or a model test of the whole damaged structure.

CONCLUSIONS

Even with extensive local reinforcement corrosion, bridge deck slabs are remarkably strong and appear to have an effectively infinite fatigue life. It is thus very unlikely that the reduction in static or fatigue strength caused by local corrosion in the slab reinforcement would be the critical factor in determining life.

General corrosion in the reinforcement, causing eventual delamination, or local corrosion in the tendons are much more likely to be critical factors. However, the local corrosion in the reinforcement could lead to a deterioration in the distribution properties of the deck and hence in the tolerance to corrosion in the tendons.

ACKNOWLEDGEMENT

The test work considered in this paper was conducted at the British Cement Association whilst the author was an employee of that organisation. He would like to thank the Association for the opportunity to conduct the tests and for permission to publish this paper. He would also like to thank the many staff of the Association who assisted with the test work.

REFERENCES

1.      Pritchard B.P. and Chubb M.S. 'Concrete bridge integrity assessment'. Structural assessment; The use of full and large scale testing, London, Butterworth (1987), editors Garas F.K. and Clarke J.L.

2.      Department of Transport. Draft Departmental advice note 'Assessment of corroded reinforcing bars', London, (1987).

3.      British Standards Institution BS 5400: Part 10: 1980, Code of practice for fatigue, London, (1980).

4.      Cairns. 'Measurement of service strains in the deck slab of a highway bridge', Research Seminar, Wexham Springs, C. & C.A., (1986).

5.      Batchelor B. de V. 'Membrane enhancement in top slabs of concrete bridges' Concrete Bridge Engineering Performance and Advances, London and New York. Elsevier, (1987), editor Cope R. J.

6.      Kirkpatrick J., Rankin G.I.B. and Long A.E. 'Strength evaluation at M-beam bridge deck slabs'. The Structural Engineer, Vol. 62B No 3, September 1984.

7.      Cope R.J. and Jackson P.A. 'Prediction of overload response and strength of concrete bridges'. Re-evaluation of Concrete Structures. Danish Concrete Institute. To be published 1988.

8.      Jackson P.A. 'Compressive membrane action is bridge deck slabs', PhD thesis, Plymouth Polytechnic, (1989).

9.    British Standards Institution BS 5400: Part 2: 1978, 'Specification for loads', London, (1978).

10.   Department of Transport 'Background and synopsis to the proposed new loading', London, (1984).

11.   Woodward R.J. and Williams F.W. 'Collapse of the Ynys-y-Gwas Bridge, West Glamorgan'. Proceedings Institution of Civil Engineers, Part 1, No 84, August 1987.

12.   Price W.I.J. and Aguilar L.A. 'Assessment of deteriorating prestressed concrete bridges'. Assessment of Reinforced and Prestressed Concrete Bridges. Institution of Structural Engineers, September 1988.

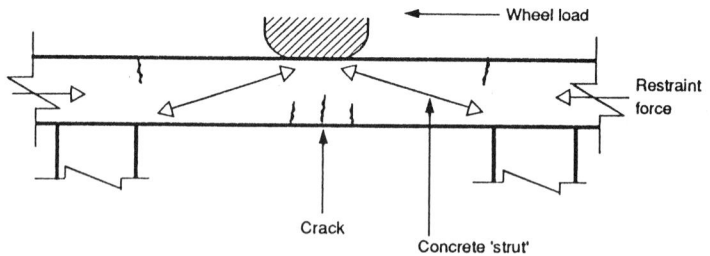

Figure 1    Compressive Membrane Action

Figure 2    Model Bridge Under Full H B Load

366

Figure 3    Single Wheel Test Rig

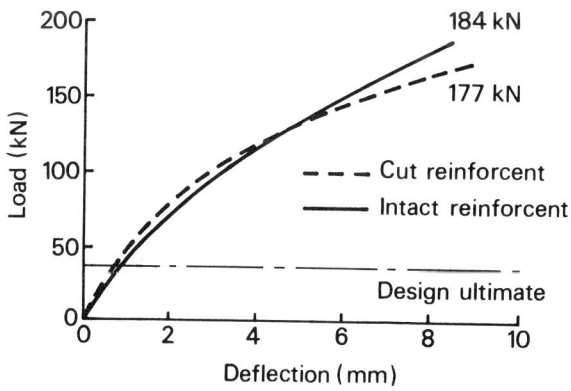

Figure 4    Load Displacement Response

# 42 Predicting the fatigue life of steel bridge decks

**J W Smith and M A Wastling,** Department of Civil Engineering, University of Bristol, UK

A method for calculating cycles of stress in steel
bridge decks under heavy traffic was developed. This
was achieved by using influence surfaces, derived by
finite element analysis, and simulating the passage
of heavy vehicles by a Monte-Carlo technique. The
predicted stress cycle counts compared favourably
with experimental observations on a real bridge using
an electronic continuous monitoring system. Finally,
an example is given to show how the stress cycle
counts may be combined with published fatigue data to
predict the fatigue life of welded joints.

INTRODUCTION

The structural efficiency of steel orthotropic decks ensures that they
are adopted for the longest suspension bridges and for many bascule
bridges where weight is an important consideration. The most common
form of construction in the UK is a 12mm steel plate stiffened by
trough shaped ribs running in the direction of traffic and welded to
the underside of the plate. The ribs are supported on transverse
girders or diaphragms and the plate is surfaced with approximately 40mm
of mastic asphalt.

However, high stresses occur in the deck in the close proximity of
wheel loads of heavy commercial vehicles and consequently the fatigue
strength of the welds is an important factor in their design.
Premature cracking of welds in actual structures has been reported.
Repair of welds in situ is expensive and disruptive to traffic.

Stress analysis of orthotropic decks has to take account of the
composite action between the steel deck plate and the surfacing
material(1) since it has been shown that there is a reduction in stress
as compared with stresses in unsurfaced decks(2). The properties of
bridge deck surfacing materials vary with temperature and vehicle speed
and data for various surfacing materials have been obtained by
Smith(3).

A simplified method for calculating the cycles of stress in steel decks
under heavy vehicles was developed by Smith and Wastling(4). This made
use of an average bridge temperature and treated wheel loads
separately. The results were conservative when compared with
observations on a full scale bridge. A more advanced method of
analysis is presented in this paper.

# THEORETICAL METHOD FOR STRESS CYCLE COUNTING

Influence surfaces for point loads:

The Muller-Breslau principle was used for calculating point load influence surfaces for bending and in-plane stresses at a critical location in a bridge deck. The principle is illustrated in Figure 1 where the influence line for bending stress in a simply supported beam can be determined by 'cutting' the beam at the point of interest (A) and introducing a rotation $\theta_a$. The Müller-Breslau principle states that the ensuing deflected shape is proportional to the influence line for bending at A. Therefore the bending moment at A due to a load $F_b$ at B may be calculated from

$$M_a = \frac{\delta_b}{\theta_a} F_b \qquad (1)$$

where $\delta_b$ is the vertical displacement of the deflected beam(5). Furthermore, the bending stress at A is directly proportional to $M_a$ and therefore the figure is a scaled influence line for stress.

This principle was extended to a three dimensional orthotropic deck structure by using the finite element method. The deck plate and stiffeners were modelled using thin shell quadrilateral elements possessing bending and in-plane displacement capability.

The method used to make the 'cut' and impose the appropriate displacement is shown in Figure 2. A single node(A) at the junction of 4 elements was replaced by 2 nodes. These nodes were coupled rigidly together in 5 out of the 6 possible degrees of freedom. The sixth unconnected degree of freedom was then available to impose a displacement. In this example relative rotation about the Y axis was imposed. This was done by applying moments in opposite directions on the nodes at A producing an arbitrary displacement $\theta_a$. The deflected shape of the plate in the z-direction then represents the influence surface for bending moment at A since

$$M_{ya} = \frac{\delta_{zb}}{\theta_a} \cdot 1 \qquad (2)$$

where $\delta_{zb}$ is the deflection at a point B and $M_{ya}$ is the bending moment about the Y axis at point A caused by a unit load at B. The bending stress is then directly proportional to $M_{ya}$.

A similar procedure can be adopted for in-plane stresses in the plate. The uncoupled degree of freedom is then selected to be displacement in the x-direction requiring opposing forces to impose a displacement $\delta_{xa}$. The influence surface is then defined by

$$F_{xa} = \frac{\delta_{zb}}{\delta_{xa}} \cdot 1 \qquad (3)$$

where $F_{xa}$ is the in-plane force at A due to a unit load at B. In-plane stress is directly proportional to $F_{xa}$.

Influence surfaces for stress in welds due to wheel loads:

The above procedure was used to obtain influence surfaces for bending and in-plane stress under the action of a unit load acting normal to

the deck plate. However, the stresses in a bridge deck are sensitive
to wheel contact area and therefore it is advantageous to obtain
influence surfaces for wheel loads of finite area. This was achieved
by performing a numerical integration of the vertical displacements of
the point load influence surface over a typical tyre contact area of
200 x 200mm. The finite element mesh for a typical orthotropic deck is
shown in Figure 3. The mesh was concentrated around the central
stiffener and influence surfaces were obtained for bending and in-plane
stress at a node close to the stiffener to deck plate fillet weld. The
combination of these two surfaces gave the influence surface for total
stress in the transverse direction. This is shown in Figure 4. The
surface stress is convenient because it can be compared with stress
gauge observations on a real bridge.

Monte-Carlo Simulation of traffic loading:

A procedure for simulating the passage of heavy vehicles over the steel
deck was developed as follows:

(1)    A wheel path was selected. It is known that moving vehicles
deviate from the mean wheel path according to a measurable
distribution(6). A deviation from the mean nearside wheel track was
selected from a suitable distribution at random.

(2)    A vehicle type was selected at random from the spectrum of heavy
vehicles published in BS5400:Part 10(6). The proportions of vehicles
in each class were obtained from a classification count at a real
bridge.

(3)    The axle spacings and wheel loads of each vehicle type are listed
in BS5400. The group of loads corresponding to the selected vehicle
was traversed over the influence surface along the selected wheel path
hence obtaining a stress history for the point chosen for analysis.

(4)    A "rainflow" cycle count was performed on the stress history(6).

(5)    The procedure was then repeated by generating new vehicles
randomly in succession.

The output from the analysis consisted of numbers of cycles of stress
range over a spectrum of magnitudes. A stress range is the algebraic
difference between the maximum and minimum stress in a cycle. Thus the
output was in the ideal form for fatigue assessment.


CONTINUOUS MONITORING OF A BRIDGE DECK

In order to confirm the theoretical analysis an electronic instrument
was installed within the box girder of an important motorway bridge. A
description of the instrument and its capabilities has been given in
earlier papers(4,7). The kernel of the instrument is an analogue peak
and trough detector which is connected to the output from a stress
gauge on the bridge. The peaks and troughs in the fluctuating stress
history under traffic are digitised and then reduced to a stress cycle
count by an on-board computer. The instrument was configured to
monitor six stress gauges simultaneously in real time.

The gauges were attached to the underside of the orthotropic deck as shown in Figure 5.

The experimental results obtained during an eleven hour classification count on the bridge were compared with the theoretical analysis. The stress in gauge 4 on the stiffener web was used as the basis of comparison here. Experimental and theoretical cycle counts are shown in Figure 6. The results are in reasonable agreement except for the higher stress ranges which were observed experimentally but not predicted theoretically. This is thought to be due to a small number of overloaded axles in the traffic stream. The damaging potential of the cycle counts were compared by evaluating an equivalent stress range per axle as defined by

$$\sigma_e = \left\{ \frac{\sum n_i \, \sigma_i^m}{n_e} \right\}^{1/m} \tag{4}$$

where  $n_i$ = the number of cycles of stress range  $\sigma_i$
$n_e$ = the number of axles
$m$ = the inverse slope of the S-N constant amplitude fatigue curve.

Equation (4) is derived from Miner's law of cumulative damage and an assumption that the S-N curve was of the form $N = K\sigma^{-m}$. The equivalent stress ranges are compared in Table 1 below.

Table 1      Equivalent stress range per axle - gauge 4.

| S-N slope | Stress Range $(N/mm^2)$ | |
|---|---|---|
| | Experiment | Theory |
| m=3 | 14.2 | 16.9 |
| m=5 | 19.6 | 20.6 |

It is evident that the theoretical procedure gives a good prediction of the damaging potential of the stress cycles.


EFFECT OF TEMPERATURE

It has already been mentioned that the asphalt surfacing acts compositely with the steel deck plate (1). Furthermore, since bituminous materials are generally viscoelastic the stiffnesses of asphalt surfacings vary with temperature and speed of loading. This may be taken into account in the analysis by modelling the surfaced deck plate by a simple plate with the same thickness as the steel but with an increased Young's modulus. The effective Young's modulus of the equivalent plate will vary with temperature and to a lesser degree on loading speed, and may be evaluated using the composite theory developed by Cullimore et al (1).

The greatest volume of traffic occurs during the day when bridge temperatures are generally above average. It would therefore be useful to obtain a relation between bridge temperature and the amount of accumulated fatigue damage. This was achieved by employing the

371

equivalent stress range per axle defined by equation (4). The
equivalent stress range per axle was evaluated from the experimental
observations over a full year of traffic using an S-N curve slope of m
= 5. The results for all six gauge points are shown in Figure 7. It
will be noted that the greatest stresses are in the deck plate at
points 5 and 6. The stiffener web stresses are smaller although the
fillet weld detail is relatively more susceptible to fatigue failure.

A useful point on each curve is the mean equivalent stress range per
axle. This may be evaluated from equation (4) but with the summation
taken over all stress cycles at all temperatures. The corresponding
temperature is the temperature that, if held constant, would result in
the same total fatigue damage over the full year of traffic loading.
This point is indicated for the deck plate stress gauge 5. The mean
equivalent temperature is approximately $22^\circ C$.

Finally, the theoretical method was used to derive a curve for
comparison with the experimental results of gauge 5. It is evident
that the theoretical method overestimates damage at high temperature
and underestimates at low temperature.


PREDICTION OF FATIGUE LIFE OF STIFFENER/PLATE WELD

The method of analysis described in this paper may be applied to the
calculation of the fatigue life of one of the welded joints in an
orthotropic deck. The stiffener web to deck plate fillet weld will be
taken as an example. The geometry of a typical weld is shown in Figure
8. It can be shown that the nominal stress, excluding the effects of
stress concentration, at the root of the weld is given by

$$\sigma_{root} = \sigma_a \left\{ \frac{T \cos(\theta/2)}{t} + \frac{3T}{t} \sin(90 - \theta/2) \right\} - \sigma_b \left( \frac{T}{t} \right)^2 \quad (5)$$

where $\quad \sigma_a = P/T = \quad$ axial stress in stiffener web

$\sigma_b = 6M/T^2 =$ bending stress in web

The influence surfaces for axial and bending stress at a temperature of
$30^\circ C$ were used to derive stress cycle counts caused by the actual
number of heavy vehicles crossing the bridge in the slow lane during
one year. Equation (5) was then used to calculate the number of cycles
$n_i$, of various stress ranges at the root of the weld. The results are
given in Table 2.

TABLE 2     Calculation of Fatigue Life of Welded Joint

| Nominal stress at weld root ($N/mm^2$) | $n_i$ $\times 10^5$ | $N_i$ | $\dfrac{n_i}{N_i}$ |
|---|---|---|---|
| 0-15 | 10.0 | $7.2 \times 10^{11}$ | $1.39 \times 10^{-6}$ |
| 15-30 | 5.5 | $2.3 \times 10^9$ | $2.39 \times 10^{-4}$ |
| 30-45 | 6.25 | $2.1 \times 10^8$ | $2.98 \times 10^{-3}$ |
| 45-60 | 4.0 | $4.4 \times 10^7$ | $9.09 \times 10^{-3}$ |
| 60-75 | 2.75 | $1.31 \times 10^7$ | $2.10 \times 10^{-2}$ |
| | | TOTAL | $3.33 \times 10^{-2}$ |

Fatigue endurance was estimated by assuming that the weld Class D design curve in BS5400 (6) was applicable. Hence the number of cycles to failure in a constant amplitude fatigue test, $N_i$, were recorded in the table and the total cumulative damage obtained from Miner's rule. The total in the table is the cumulative damage in one year. Hence the inverse of this implies a life of 30 years for a weld of this geometry. However, this estimate is conservative for two reasons. Firstly, weld metal is usually applied more generously than assumed. Secondly, the $30^{\circ}C$ influence lines are conservative relative to the mean equivalent temperature of $22^{\circ}C$ calculated for the actual bridge.

ACKNOWLEDGEMENTS

The authors are indebted to the Science and Engineering Research Council for supporting this investigation.

REFERENCES

1.	Cullimore, M.S.G., Flett I.D. and Smith J.W.  'Flexure of steel bridge deck plate with asphalt surfacing', Proceedings Int. Assoc. Bridge and Struct. Engng, P-57/83, 255-267 (1983).

2.	Smith J.W. and Cullimore M.S.G.  'Stress reduction due to surfacing on a steel bridge deck', Int. Conf. on Steel and Aluminium Structs, Cardiff, Elsevier (1987).

3.	Smith, J.W.  'Surfacing materials for othotropic bridge decks', Oleg Kerensky Memorial Conf. on Tension Structures, I.Struct.E., London (1988).

4.	Smith J.W. and Wastling M.A.  'Continuous monitoring of stresses in the deck of a steel box girder bridge under normal traffic', Seminar on Structural Assessment, Building Research Establishment (April 1987).

5.	Norris C.H. and Wilbur J.B.  'Elementary structural analysis', McGraw-Hill, New York (1960).

6.	BS5400:Part10: 'Code of Practice for Fatigue', Specification for steel, concrete and composite bridges, British Standards Institution, London (1980).

7.	Wastling M.A. and Smith J.W.  'An instrument for detecting arbitrary peaks and troughs of a fluctuating stress signal', Strain, British Soc. Strain Measurement, (August 1987).

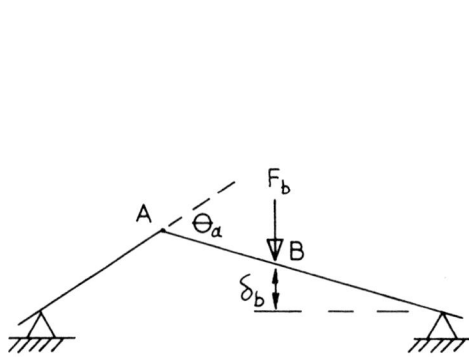

Figure 1  Müller–Breslau principle

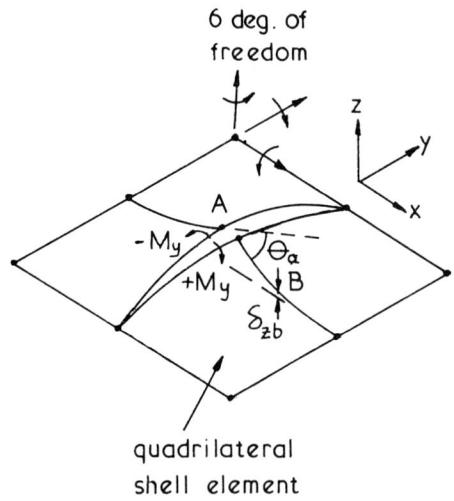

quadrilateral
shell element

Figure 2  Imposed rotation at A

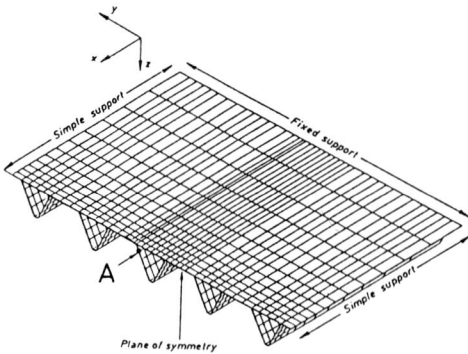

Figure 3  Finite element mesh

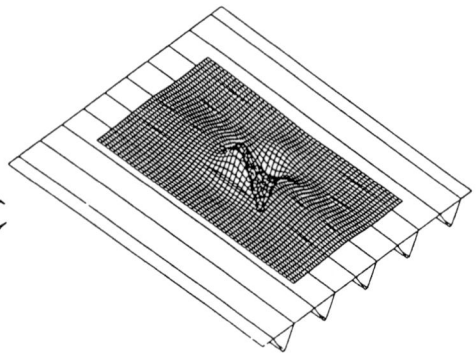

Surface stress at A

Figure 4  Influence surface

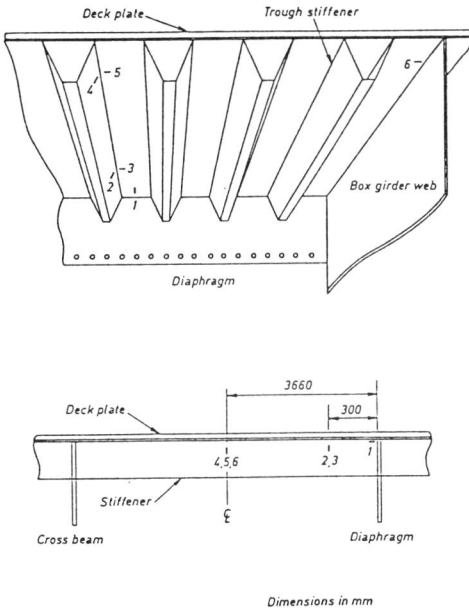

Figure 5  Location of gauges

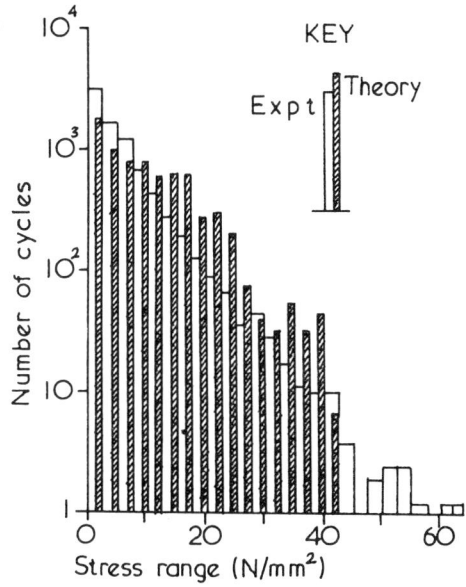

Figure 6  Stress cycle counts

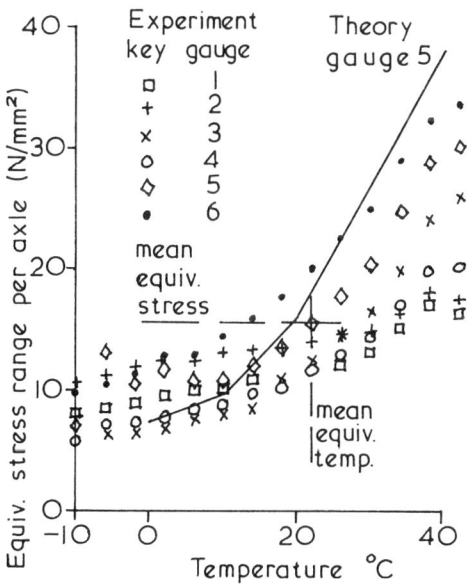

Figure 7  Stress range per axle

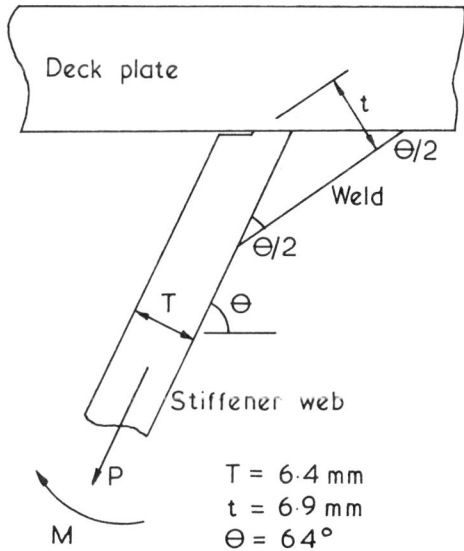

Figure 8  Weld geometry

DAMAGE DIAGNOSIS AND ASSESSMENT  OF REMAINING FATIGUE LIFE OF
OLD STEEL BRIDGES

K. Brandes

Federal Institute for Materials Research and Testing (BAM),
Berlin, West Germany

> The life of structures is limited to a certain
> extent because of technical reasons. The life of
> bridges may be restricted in terms of fatigue
> life or because of corrosion. Concerning old
> steel bridges, the fatigue life might be the
> governing quantity. However, if there are found
> cracks in an old steel bridge, the question
> arises for how much longer the bridge will re-
> sist the loading within certain limits of safety
> margins. This question had to be answered quite
> recently for a bridge in Germany crossing the
> Dortmund-Ems-Kanal (fig. 1).

INTRODUCTION

Bridges approach their theoretical design life after about 50
to 80 years if there are no drastic changes in their loading.
However, as known from experience, many bridges can survive
for more than 100 or even 150 years. We have a couple of difficulties in judging the remaining fatigue life of old steel
bridges which are obviously in good condition.

Figure 1: Highway bridge in north-west Germany. The span-width
of the riveted truss girder bridges amounts to 59 m

For the rating of old bridge structures, we recently extended the investigation by introducing crack growth tests to evaluate the sensitivity of the material to cyclic loading.
The procedure we apply for rating old steel bridge structures including all significant parameters shall be demonstrated by means of a highway-bridge of 59 m span crossing the Dortmund-Ems-Kanal near Osnabrück built in 1953 (fig. 1, fig. 2) [1].

## LOAD BEARING ELEMENTS OF THE BRIDGE

Two truss girders of 59 m span are the principal load bearing elements. Cross girders of 8.8 m span and longitudinal girders of 6 m and 5.5 m span are the secondary structural members, fig. 4.

Figure 2: View from underneath the bridge

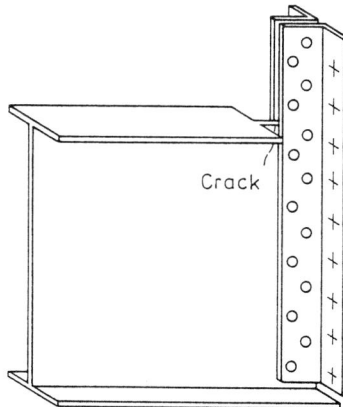

Figure 3:
Structural detail at the connection of the cross girders to the main truss girder and the location of cracks

377

## MOTIVATION FOR THE INVESTIGATION

During an inspection, we discovered long cracks at several cross girders near the connection to the main girders by angles (fig. 3). For this reason, an investigation of the bridge was initiated to diagnose the cause for the occurrence of the cracks and to evaluate the remaining fatigue life of the highly stressed members of the structure.


## INVESTIGATIONS

A rating of the structure with respect to its fatigue life includes different tasks:
- Evaluation of the fatigue load by modelling the traffic or by measurements
- Determination of stresses under static load and under traffic load
- Identification of the material of the structural members under consideration and their mechanical and fatigue behaviour.

During the investigation, we evaluated the stresses at several points at a cross girders and the central longitudinal girder as being the most critical elements, fig. 5, under defined static loading and under traffic flow, thus confirming the mechanical model of the structural performance and at the same time finding out a basis for assessing stress histories.

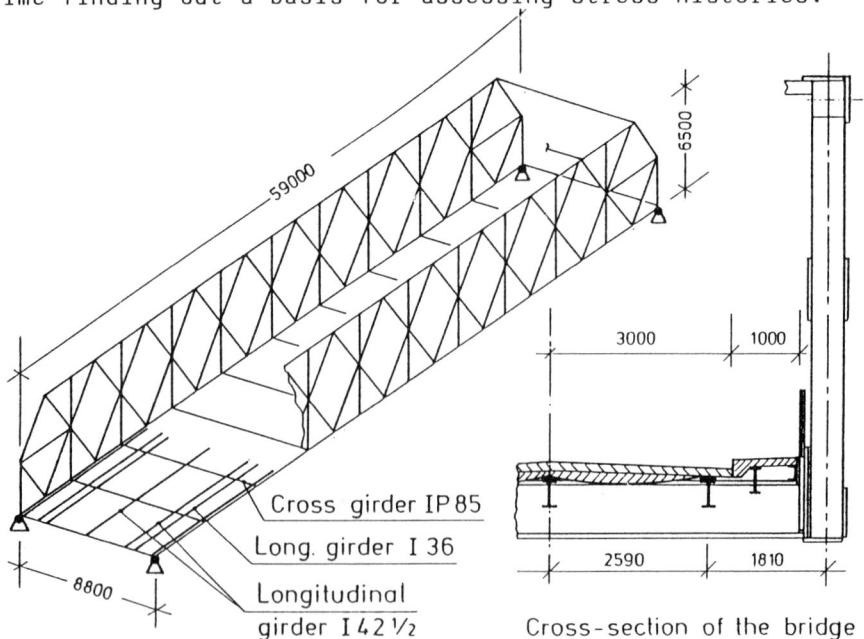

Figure 4: Load bearing elements of the bridge

The material was identified in standard tests. The material of the cross girders was Thomas-Steel, with its characteristic parameters of strength not quite coming up to the requirement for grade St 37. With respect to static loading, the stresses do not exceed the allowable stresses for this grade. However, concerning the fatigue strength, we decided to perform crack growth tests intending to confirm assumptions about the fatigue strength of the material.


RESULTS

The measurements under defined static loading by a two-axle truck confirmed that the measured stresses are smaller than the calculated ones, fig. 5. For the calculation, a grid-model was used, incorporating the interaction between the different elements and the elastic clamping at the main girders.

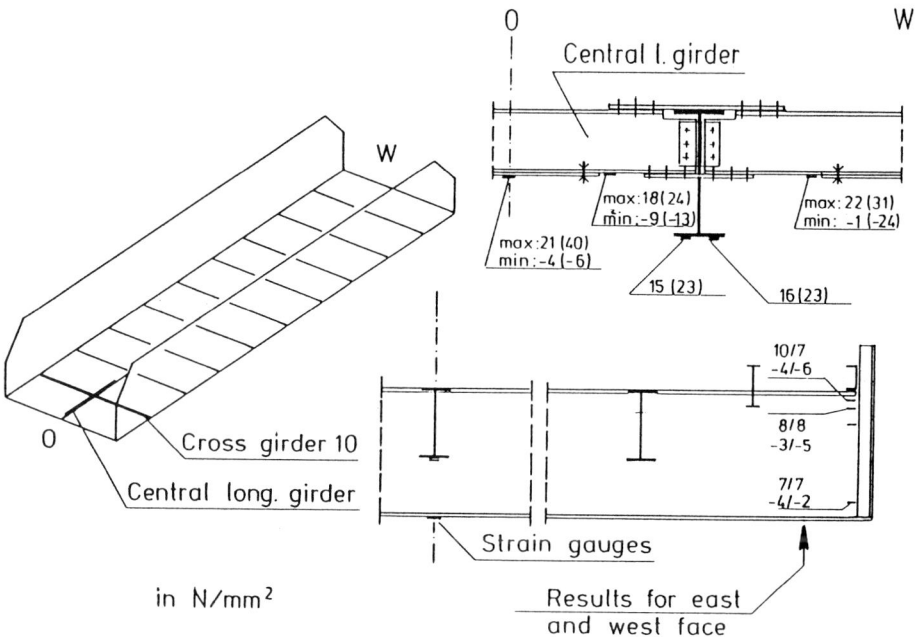

Figure 5: Results of the experimental and the numerical investigations. The experimentally evaluated stresses are compared with the calculated ones (in brackets)

In the crack growth tests on CTT-specimens (fig. 6), we observed a behaviour like the one well-known for grade St 37 steel, indicating that we could apply the standardized Wöhler-Diagrams. The result of one of the tests is plotted in fig. 7 with respect to the representation in terms of the Paris-equation (fig. 6) [3]:

$$da/dN = C \Delta K^n$$

$$K = \frac{\Delta F}{B\sqrt{W}} f(a/W), \quad \Delta F = F_u - F_l$$

a          crack length
N          number of cycles
ΔK         cyclic stress concentration factor
C;n        material parameters
B          thickness of the specimen (sheet)
f(a/W)     function depending on the geometry of the specimen

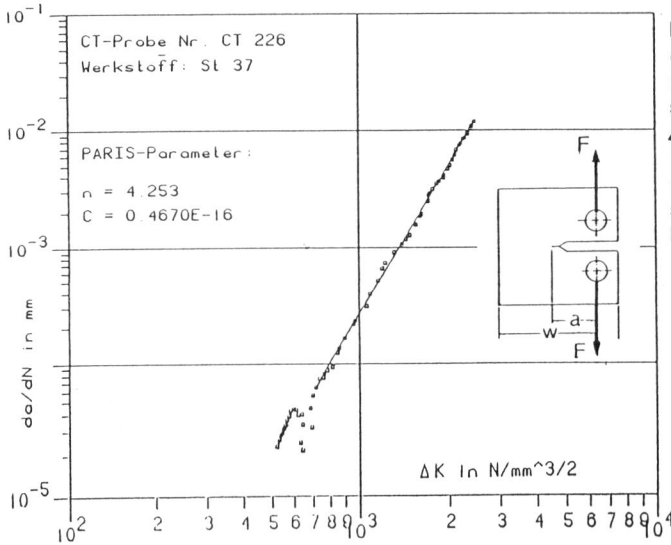

Figure 6: Diagram of crack growth rate versus cyclic stress intensity ΔK as elaborated for a compact tension (CTT) specimen (ASTM E 647-83)

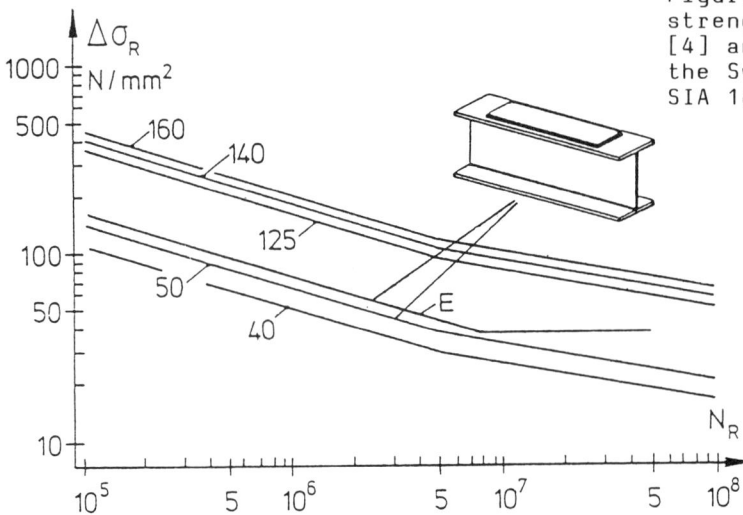

Figure 7: Fatigue strength curves [4] and curve E of the Swiss Standard SIA 161

As expressed by the Paris-equation, the crack growth rate is governed by the cyclic. stress intensity factor ΔK.

The remaining fatigue life was evaluated by applying the standardized "European Fatigue Strength Curves" as presented in EUROCODE No. 3 [4], fig. 7. A first estimate for the peak stresses at the measuring points was obtained by the measurements under traffic flow. During several hours of observation only a very few passing trucks produced stresses which exceeded the cut-off limits of the fatigue strength curves. An impression of measured time-histories during the passing of heavy vehicles is given in fig. 8. In all cases, only the mai peak exceeded the cut-off limit. Thus, there did not arise the question how to count cycles of different amplitudes during one passing. For only one point of the structure, the end of the coverplate at the longitudinal girder, the remaining fatigue life had to be evaluated underlying several assumptions about the load history and the future loading. For the past, the conservative estimation resulted in about 40 000 cycles of $\Delta\sigma = 45$ N/mm$^2$ (35 years). By doubling the present traffic, it would take about 50 years to reach the fatigue strength curve that means a probability of less than 5 % of occurence of a fatigue crack.

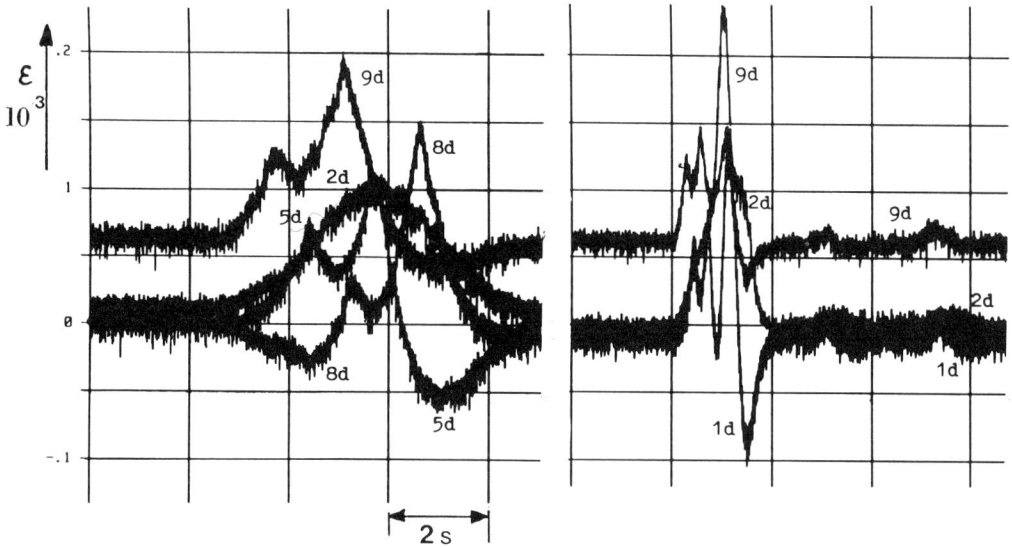

Figure 8: Strain histories of some measurement points for
truck passings.
Left: Two-axle truck of 20 t weight. Right: Large
five-axle truck.
Curves:
2d: Cross girder, center
9d: Longitudinal girder, first section, center
5d; 1d: Longitudinal girder, first section, at begin
of coverplate
8d: Longitudinal girder, second section, at begin of
coverplate

Whenever the crack growth behaviour of the material is known, there is the opportunity to evaluate the growing of a crack of a postulated length during an inspection time intervall. Assuming cracks like shown in fig. 9, we found that the lengthening of a crack of about 30 mm to 35 mm would take about 4 years or more. The inspection intervall is 3 years.

## DAMAGE DIAGNOSIS

The low level of measured stresses at an (uncracked) connection of the cross girder to the main girder, fig. 3, indicated that the cracks of the observed type were generated by another

381

mechanism. By measuring the deformations of the bridge, we observed that the entire roadway-plate acted as part of the lower truss of the framework of the principal truss, connected by a very weak element - the web of the cross girders (see fig. 4 and fig. 3). When passing the bridge, heavy trucks induced a marked 2.5 Hz-vibration of the whole structure stressing locally the web of the cross girder perpendicular to the web (see fig. 10). The cracks are generated by low-cyclic fatigue. After reaching a certain length of the crack, the mentioned connection weakens and the cracks do not grow any further. The construction has been improved by drilling bore-holes at the tip of the crack, thus stopping crack growth.

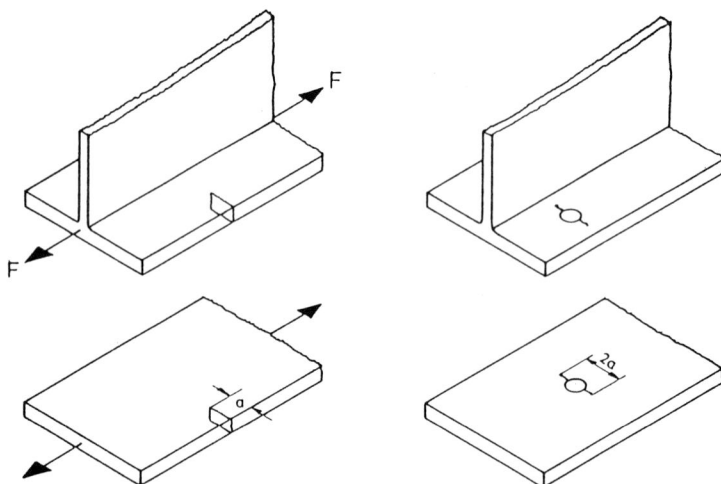

Figure 9: Crack pattern for evaluating stress intensity
factors and crack growing

Figure 10: Deformation pattern of
the principal truss girder system
(lengthening of the lower
truss of the framework)
with bending of the web
of the cross girders.
(The figure is
reduced to the
symmetric part
of the displace-
ments.)

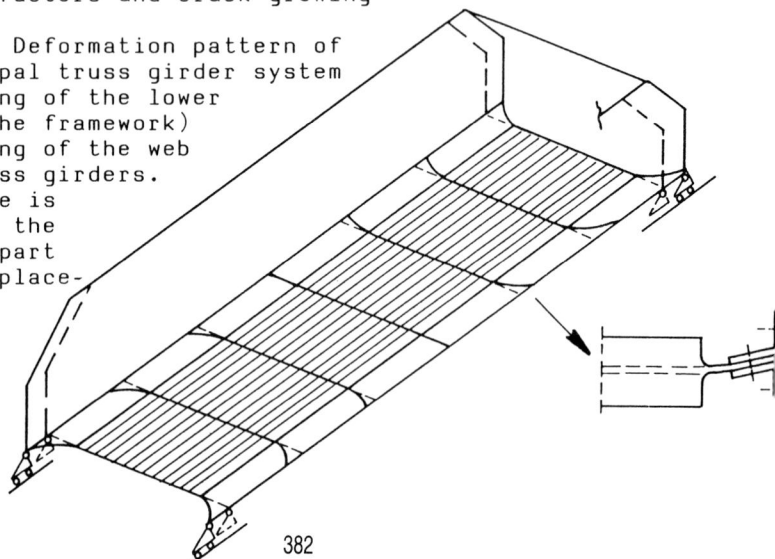

382

## CONCLUSION

Whenever a crack is found in a structure, this generally indicates that more cracks are present [2] and that the risk of
fatigue damage exists. However, by diagnosing the genesis of
cracks, one should be very carefully in relating it to fatigue
loading. The rating of a structure with respect to fatigue
life without measurement of strains (stresses) sometimes leeds
to insufficient results.
For old bridges the behaviour of the material is not accurately known and crack growth tests are recommended on the one
hand aimed to identify the material in question and on the
other hand to evaluate the growing of postulated or observed
cracks.

## REFERENCES

[1] Brandes, K.: Untersuchungen zum Rißfortschritt an einer
    Stahlbrücke. 13th Congress, IABSE, Helsinki, (June 6-10,
    1988), pp. 361-366

[2] Hirt, M.A.: Remaining Fatigue Life of Bridges. IABSE Sym-
    posium: Maintenance, Repair and Rehabilitation of Bridges.
    Washington, D.C., (1982)

[3] Heckel, K.: Einführung in die Technische Anwendung der
    Bruchmechanik. 2. Aufl., Hanser Verlag, München, (1983)

[4] EUROCODE No. 3: European code for Steel Construction,
    Comm. European Community, Brussels, (1981, draft)

# 44 A bridge can survive earthquakes to complete its intended economic life

**Ergin Atimtay,** PhD, PE, Middle East Technical University, Ankara, Turkey

The life of a bridge may be cut short mainly by
1) Abutment failure    2) Loss of span
3) Insufficient ductility.This paper describes how such
failures can occur and draws attention to precautionary
measures.Such measures deal with adequate seat length,
control of displacements,soil structure interaction,
liquefaction,piles and piers designed for ductility,and
preventing abutment failures.

DESIGN CRITERIA

The ultimate test for a structure is the nature itself.Therefore,it
is an absolute necessity to observe structures during and after natural
disasters and try to understand their performance to improve their design.

This fact has been very clearly realized after the 1971 San Fernando
earthquake (EQ) in California.Since then,observations of and research  on
bridges have resulted in a more realistic design approach today known as
AASHTO Guide Specifications for Seismic Design of Highway Bridges.

The life of a bridge may be cut short by
1) Loss of span  2) Insufficient ductility  3) Abutment failure
To prevent the occurance of the above modes of failure,careful
articulation of the bridge is essential as well a sound approach to
design.It is not an overstatement to say that an EQ resistant bridge is
one whose components have been meticulously detailed with EQ consideration
in mind.

For the above mentioned modes of failure,design measures and detailing
techniques accepted by the design profession will be presented.

Loss of Span
1. An evaluation of bridge failures observed in Japan,Alaska and
   California have shown that loss of span is a major cause.Loss of span
   may result from relative displacement effects,which arise from out-of-
   phase motion of different parts of a bridge,from lateral displacement

and/or rotation of the foundations and differential displacement of abutments.

2. Minimum support lengths at abutments,columns,and hinge seats must be provided (AASHTO 4.9.1).

3. Where the seismic risk is high and the superstructure segments are non-continuous,ties must be provide to keep the segments together. Particular attention must be paid for bridges with high columns or piers.

4. The design approach assumes that columns will yield when subjected to the design EQ.As yielding occurs,relative displacements will also increase.

5. If a response spectrum approach is being used,it must be realized that it does not by itself include the duration of the shaking,which may have a significant effect upon loss of strength once a structure yields.

6. The flexibility of the foundation may have a big contribution to the displacements.The flexibility of the foundation should be included in the elastic analysis by properly modelling the soil.If the foundation is not included in the analysis,it must be realized that displacements may be 50% larger for bridges founded on very soft soils.

7. Saturated uniformly sized granular foundation soils are subject to loss of strength during EQ which in turn may cause bridge failures. During the 1964 Alaska EQ,9 bridges suffered complete collapse and 26 suffered severe deformation or partial collapse mainly due to liquefaction of soils,leading to major displacements of abutment and piers.(3)

8. The best measure against liquefaction is to avoid deep,loose to medium-dense sand sites where liquefaction potential is high.Densification may be economically applied if dense or more compact soils are found at shallow depths.

Insufficient Ductility

1. Small-to-moderate EQ's should cause no significant damage.For large EQ's,plastic hinging in the columns is acceptable if adequate ductility is provided for.

2. The elastic design forces can be reduced by a factor (R-factor) for well-confined columns capable of deforming plastically.However, connections and foundations must be designed to accomodate the ground motion forces with little or no damage.

3. There is not much data available on the behavior of pile bents during actual EQ's.It is believed that there would be reduction in the ductility capacity of pile bents with batter piles.Batter piles are, of course,economical for resisting lateral loads,but they are also very rigid.

385

4. All piles should be adequately anchored to the pile footing or cap.

5. Unreinforced concrete piles must be avoided because they are brittle in nature and may create problems in free-field EQ motion.

6. It has been observed in past EQ's that concrete piles tend to hinge or shatter just below the pile cap.It is advisable to reduce the confinement spacing in this portion.

7. It must be ensured that piles should not fail below the ground level for safety purposes and ease of access for repair.

8. The design shear force on a column must be conservatively calculated, because a shear failure may result in a partial or total collapse of the bridge.

9. Vertical ties should be used in the footing to tie together the top layer of reinforcement to the bottom layer to prevent "delamination" in the footing concrete.(5)

## Abutment Failure

1. Abutment failures are usually observed as fill settlement or slumping displacements induced by high seismically induced lateral earth pressures,or the transfer of longitudinal or transverse inertia forces from the bridge structure itself.

2. Fill settlements were observed to be 10% to 15% of the fill height. Damage effects on bridge abutments in the Madang EQ in New Guinea reported by Ellison (4) showed abutment movements as much as 500 mm.

3. In free-standing abutments,inertia forces due to abutment itself should not be neglected.Such abutments rely on their mass for stability and the mass plays an important part in their behavior.

4. If batter piles or tie-backs restrain the abutment movements,attention must be paid to the fact that lateral pressures induced by inertia forces in the backfill will be greater than those given by a Mononobe-Okabe analysis.

5. Monolithic abutments where the end diaphragm is cast monolithically with the superstructure have performed well during EQ's.On the other hand, higher longitudinal and transverse superstructure inertia forces are transmitted directly into the backfill.Whereas damage may be heavier than that for free-standing abutments,with adequate abutment reinforcement the collapse potential is low.

## CONCLUSION

" The causes of earthquake are still not understood and experts do not fully agree as to how available knowledge should be interpreted to specify ground motion for use in design.To achieve workable bridge design provisions,it is necessary to simplify the enormously complex matter of earthquake occurance and ground motions.Any specification of a design ground shaking involves balancing the risk of that motion occuring against the cost to society of requiring that structures be designed to withstand that motion.Hence,judgement,engineering experience,and political wisdom are as necessary as scientific knowledge." (1)

REFERENCES

1. American Association of State Highway and Transportation Officials, "Guide Specifications for Seismic Design of Highway Bridges",1983

2. Applied Technology Council,"Tentative Provisions for the Development of Seismic Regulations for Buildings," ATC Report No.3-06,Berkeley, California,June 1978.

3. Seed H.B. and Idris,I.M.,"A Simplified Procedure for Evaluating Soil Liquefaction Potential",Journal of the Soil Mechanics and Foundations Division,ASCE,Vol.97,No.SM9,1971.

4. Ellison,B. "Earthquake Damage to Roads and Bridges-Madang,R.P.N.G.- Nov.1970", Bulletin,New Zealand Society of Earthquake Engineering, Volume 4,pp.243-257,1971.

5. State of California,Department of Transportation,"Bridge Memo to Designers Manual",California,May 1987.

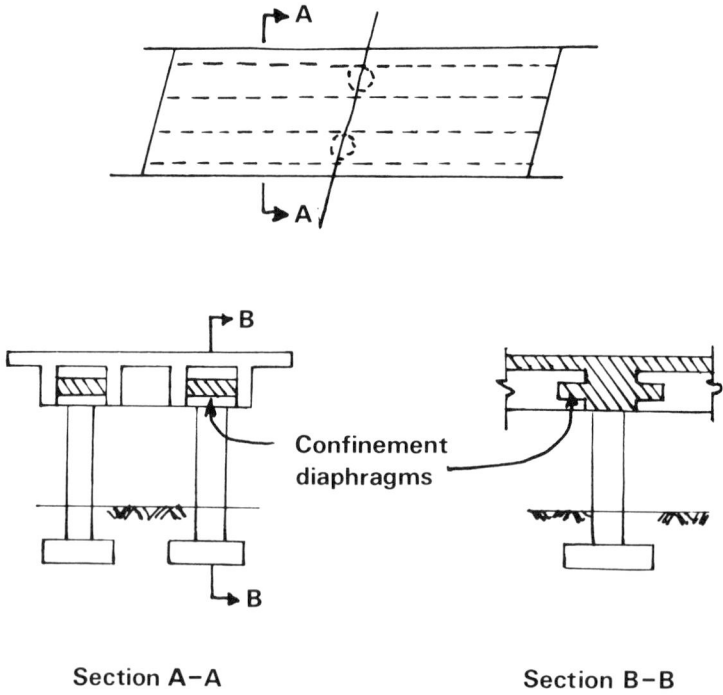

Section A–A                    Section B–B

Figure 1    Confinement diaphragms to prevent loss
            of span (5)

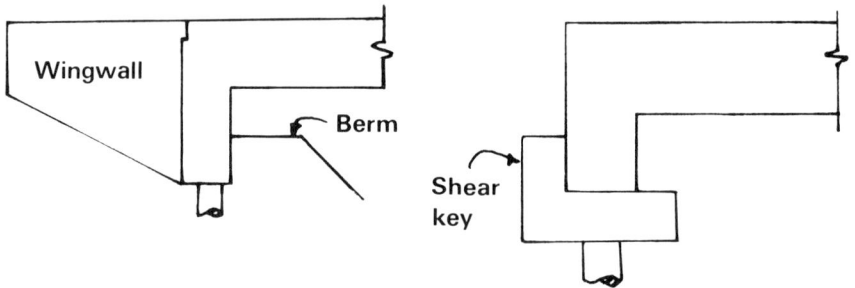

Figure 2    Monolithic abutments to control abutment
            damage (1)

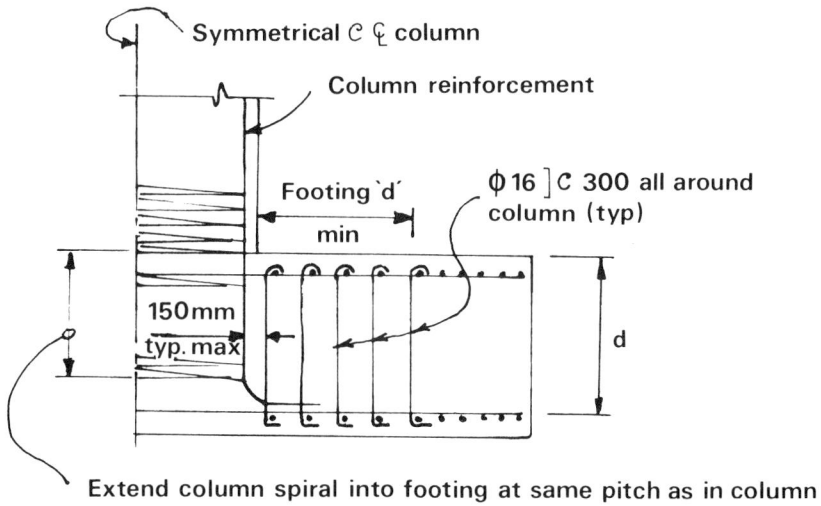

Figure 3   Footing reinforcement fixed column (5)

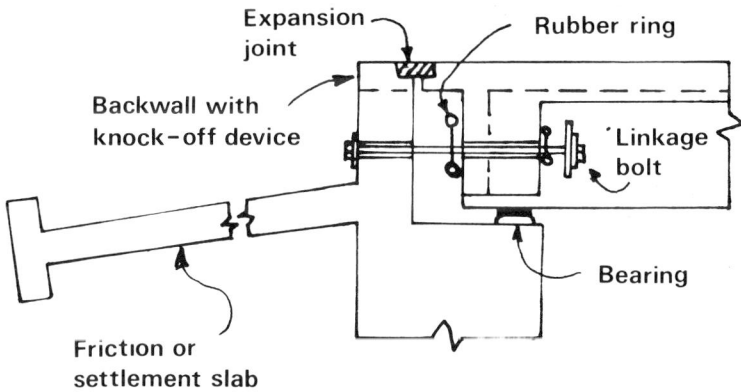

Figure 4   Abutment support detail (1)

# 45 The stress analysis of segmental prestressed concrete bridges during construction and long-term

**Tibor Jàvor**, Research Institute of Civil Engineering, Bratislava, CSSR

Great experimental observation was made during
construction of the over 1 km long prestressed
concrete box-girder segmental bridge using
embedded vibro-wire gauges.  The deflection
measurements were made by inductive transducers
and by geodetic methods.  The readings after
finishing the construction were recorded
directly and the results processed by an
Orion data logger with IBM PC/XT Controller,
by means of which the corresponding transform
-ation of measured strains into the stresses
was also made.  These values were plotted in
tables and graphically in diagrams with the
theoretical proposals and analysed.  Examples
are given in the paper.

## INTRODUCTION

The problem of the behaviour of structures and considering the influence
of the duration of load application is a problem that is drawing more and
more the interest of researchers.  Observation of states of stress is
made indirectly by evaluation of material constants and measuring of
relative deformations - strains in the structure.  Mutual correlation of
material constants and strains is in fact very difficult especially in
the case of concrete structures.  In essence for considerable
simplication of evaluation operation we consider the application of
Hookes Law.  It is desirable especially for long-termed observation to
take into consideration the influence of temperature and humidity
changes, of shrinkage and creep of concrete, as well as changes of the
modulus of elasticity of the bridge examined.  In the case of simple
statistically determinate structures we can assume that the stress at any
"n" time is given by

$$\upsilon_n^1 = \frac{{}^n\epsilon - \displaystyle\sum_{i \equiv 1}^{\substack{i = \\ j =}} \upsilon_j^1 \; \epsilon_{\frac{i+j}{2}}^{n - \frac{i+j}{2}}}{{}^1\epsilon_{\frac{m+n}{2}}^{n - \frac{m+n}{2}}}$$

where        $i = 0, a, b, \ldots, k, 1,$
              $j = a, b, c, \ldots, 1, n$

$${}^1\epsilon_{\frac{m+n}{2}}^{n - \frac{m+n}{2}}$$ = the measured deformation from unit load for concrete at the $\frac{m+n}{2}$ age and during loading period $n - \frac{m+n}{2}$,

$\upsilon_j^1$ = the unit stress at the $j$ time and

${}^n\epsilon$ = the calculated measured deformation in the direction of the strain gauge axis in the course of n period.

The resulting strain is obtained by superposition of partial stress components.


EXPERIMENTAL METHODS

The methods for measuring and checking of statical function of bridges are divided according to the kind of values measured as follows:

-      methods for measuring deflections and deviations,
-      methods for measuring strains,
-      methods for determination of various material constants.

The experimental analysis of prestressed box-girder bridges during construction as well as long-term, is made by the Czechoslovak embedded vibro-wire gauges. For long-term temperature measurements we used embedded vibro-wire thermo meters. Regarding the requirements of maximum automation of our measurements we applied the impulse wire gauges with only two-wire connection to the measurement equipment. In this case the gauge wire is set in damped oscillation by an 0.4 ms impulse duration. 100 oscillation periods are measured after a delay of T = 10 ms then we can determine the wire oscillation frequency from the relation f = 1/T. From this value, using the on calibrate gauge constant we determine the concrete strain depending on frequency, ie.

$$\epsilon = A.f^2 + B.f + C,$$

where A,B,C, are constants already determined separately for each vibro-wire gauge, during calibration.

For measuring purposes we use a measuring bus with a data logger, type HP 3050 with relevant digital voltmeter frequency reader, programmed switch unit, tuner and XY recorder. The system is oriented to the internationally codified IEC-BUS and controlled by controller HP 9826 S.

The measuring equipment enables measurements of gauge wire oscillation frequency by a velocity of 3 channels in one second. The measured values are stored in a magnetic cassette unit securing a long-term storage and measurement evaluation. The measurement results are recorded in real time, in table form or graphically, by line recorder. In our Institute we use very often also the modified Hewlett-Packard cybernetic system for model as well as structural analysis in situ. According to it the transducers of strains, deflections and temperatures are built in the tested structure during concreting of segmental elements. The sensor's outputs are connected to switch unit. By means of this switch unit the electric signals from a particular sensor are lead to the tuner and relevant multimeter, where the analog signal is transformed to a digital value. This value is lead by a switching box into the calculator memory for storing or further processing. We received the results in graphical or table form on the adjusted X/Y recorder or on the display unit. The digital clock and the time generator determined for programming the time course of measurements. Our system has at the present time three tuners by means of which is possible to measure 100 places. The system was enlarged for 400 measuring places for our large observation. For securing the mesurements by vibro-wire gauges the system was completed by the reader of frequencies.

DEFORMATION STATE DETERMINATION DURING PRESTRESSING OF BOX-GIRDER SEGMENTAL BRIDGES

At the Research Institute of Civil Engineering in Bratislava we have done experimental observation on prestressed concrete bridges during construction since 1954. We have examined various types of concrete bridges. In the case of observation of the state of stress for prestressed, cantilever concreted or bridges assembled of precast box elements - eg Figure 1, we determine the strains usually after prestressing or concreting, ie after particular working operations. In Figure 1 there is shown the course of strains of a cantilever assembled highway bridge determined in this way. The bridge was erected in 1965 and is observed long-term 23 years. The main span is 60m and each box-girder has for this span 2 x 9 precast segmental elements. The increase of strains of all box elements investigated had the same character. The largest deformations occurred in the upper fibre of the inner web of the cross section and reached as much as the double value in comparison with the strains of the similar fibre of external web.

The cantilever bridges with prestressed precast concrete elements had already larger deformations during prestressing in comparison with the monolithic concreted cantilever bridges. However one has to note that the desired prestress in the mentioned highway bridge with span 60.0 m was $-14.73$ $N/mm^2$/the measured prestress $-13.07$ $N/mm$ and in the monolithic cast in place bridge with span 70.0 m was designed $-8.30$ $N/mm^2$ and measured $-8.49$ $N/mm^2$. An other interesting observation was that in the cantilever precast bridge, the strains after 3 years were 100% higher than in the monolithic concreted bridge, but the deformations of the precast bridge stabilised in 4 years since the quality of the concrete is not bad.

An especially great experimental observation was made during construction of the prestressed concrete box-girder bridge with 17 spans from 30 m up to 70 m and piers of height 32.9 m.

The overall length of the superstructure is 1038 m. The superstructure is designed as a continuous box beam. It is divided into three separate units 245 m, 390 m and 403 m long respectively, mutually connected by hinges. It is mounted on fixed bearings of identical form. Longitudinal displacements from all loading factors are taken care of by the slenderness of pier walls or the use of rocking supports. The segmental superstructure is erected by an erection bridge 107 m long, weighing 250 t, which can place segments up to 80 t in weight in spans of up to 80 m, even horizontally curved. The segments are temporarily tensioned during their assembly by means of type Dywidag bars with a force of 588 kN. The permanent tendons consist of ropes with a prestressing force of 1600 kN each. The starting segments over the piers are stiffened by a precast cross beam. Our experimental research on the site should analyse the cooperation effectiveness of the precast cross beam and the starting segment during construction and determine the stress state in further segments in the course of construction. The construction during erection is shown in Figure 3 and the course of strain in the walls of the first segmental element during the erection is shown in Figure 4.

The strains at the corners of the cross beam over the pier are shown in Figure 5. During the concreting of the segments about 330 vibro-wire gauges and vibro-wire thermometers were embedded. The deflection measurements were made by inductive transducers and by geodetic methods, the inclinations by inclinometers Maihak and Huggenberger levels, some short measurements during the prestressing by Dywidag bars by resistance strain gauges and the local stresses by photostress method. The state of stress of the starting segment and the built-in cross beam was checked immediately after prestressing, then after the carriage transfer, during the further segment assembling and at one month intervals during the long-term observation.

THE LONG-TERM OBSERVATION OF SEGMENTAL PRESTRESSED CONCRETE BRIDGES AND THEIR SERVICE LIFE PREDICTION

The state of deformation of a structure is an important parameter of its service life and serviceability either from the point of view of traffic continuity or of the failure limit of the structure. We carry out the analysis of the deformation state experimentally:

- by current loading tests,
- by repeated static or dynamic loading tests, eg Figure 6 and
- by long-term observation of the bridge deformation.

While loading tests are performed according to valid standards and regulations, for long-term observation of concrete bridges there are no special standards with the exception of the "Recommendations" of RILEM drafted by the author of this paper. During long-term observation, the creep and the shrinkage are recommended to be checked on concrete samples situated in box-girder hollows and compared with samples stored in laboratory conditions where the changes of temperature and moisture are excluded. The long-term strain measurements of the mentioned bridge were continued after finishing the construction by automated equipment. The readings are recorded directly and the results are processed by an Orion data logger + IBM PC/XT Controller, by means of which the corresponding transformation of measured strains into the stresses is also made. These values are then plotted in tables and compared with the theoretical assumptions. The results obtained up to now show that the creep is stablized after five years and the influence of the summer and winter seasons upon the deformations of the structures are very expressive.

From the point of view of long-term measurement it is advantageous to replace the given measured curve by one or maximum two functions. We try to find such a substitute curve which satisfies both, deflection and strain for several similar bridges. The curves of strains and deflections show that the mathematical-statistic correlation leads to simple logarithmic functions, such as

$$Y = A \sqrt{t} + B.t \quad \text{or} \quad Y = A.\ln B.t,$$

which give satisfying curves for the periods after 5 to 10 years. The development of stresses in a concrete girder depends also on the season in which the structures has been concreted. It has been ascertained that the deformations of concrete placed in spring or summer, due to shrinkage, are partly compensated by elongation due to higher summer temperature. The best correlation function for this season is

$$Y = A + B \cdot \frac{T}{T + Q},$$

where A is the deformation in time O, B is the long-term deformation increment over the whole observation, Q represents the time in which half of the residual deformation will take place.

The prestressed concrete cantilever segmental bridges are an example of structure with concete of variable age, the creep analysis of which is highly exacting. History of deflection of one cast in place cantilever box-girder bridge of span 71.7 m compared with correlation function

$$Y = A \sqrt{t} + B.T$$

is shown in Figure 7 (A = -0.91, B = -2.53). The precast box-girder cantilever bridge of span 60 m differs from the cast in place bridges in assumption of lower shrinkage and creep during erection and traffic. The course of deflection of this bridge compared with the same correlation function is shown in Figure 8.

CONCLUSION

The described automated method of measurement by vibro-wire gauges and the appropriate method of data processing proved themselves very good. There is presented also an appropriate way of results interpretation as well as of their processing. The results from long-term observation of box-girder segmental bridges contributed to the analysis of their durability and can be used as basic material for designing similar bridges. The analysis of thermal effects, shrinkage, creep as well as losses of prestressing by appropriate transformation methods of strains into the stresses have been verified. The long-term measured deformations are compared with very simple correlation functins to predict the service life of the prestressed concrete bridges.

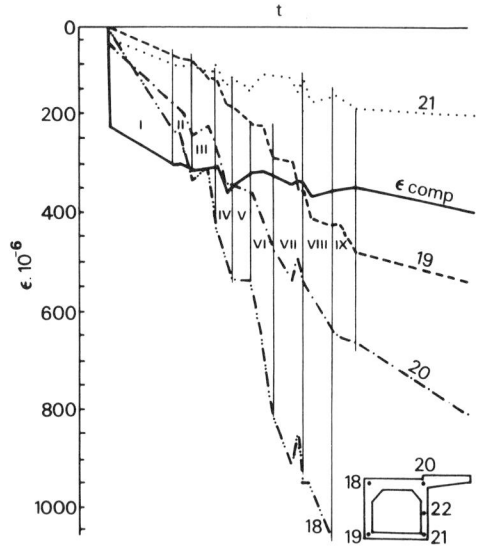

Figure 1   The cantilever precasted segmental bridge with course of
strain during erection

Figure 2   The segments are produced in stationary moulds, where the
vibro-wire gauges were embedded and when the concrete has
attained 70% of its strength, the segments were transported

Figure 3   The segmental bridge
during the erection

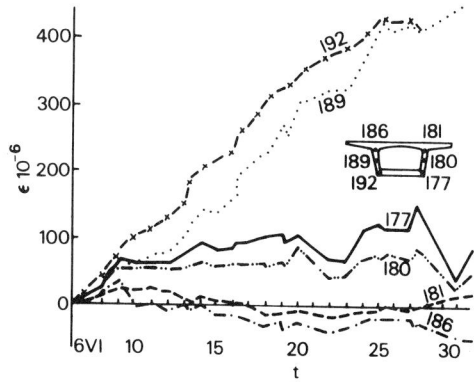

Figure 4   Course of strain in the
walls of the first segmental
box-girder elements during
the erection of the bridge

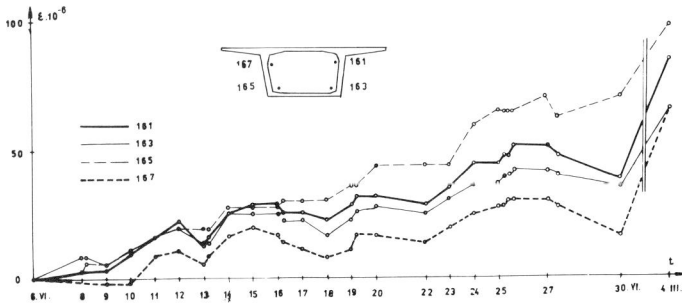

Figure 5   The course of strain at the corners of the
cross beam over the pier during erection

Figure 6 The repeated statical load-test of the prestressed concrete
segmental bridge observed by embedded over 330 vibro-wire
gauges and geodetically as well as by inclinometers

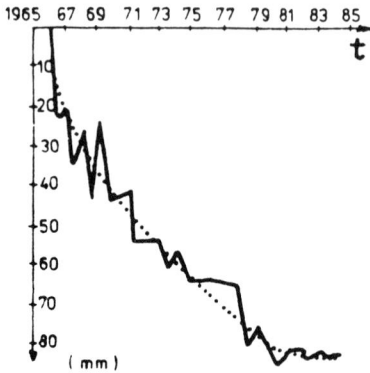

Figure 7 Course of deflection
during 20 years of the
cast in place cantilever
bridge and their
correlation function

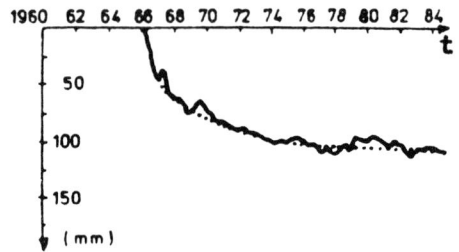

Figure 8 Course of deflection of the
precast box-girder cantilve
bridge compared with the
correlation function

# 46 Assessment of the long-term performance of segmental bridges

**P Waldron,** University of Bristol, Bristol, UK
**G S Ziadat,** G Maunsell and Partners, London, UK

A major prestressed concrete segmental bridge has been instrumented and monitored during construction and over the first two years of its service life. Results are presented from this study and from an accompanying laboratory investigation into both the short-term and time-dependent properties of the concrete used for bridge construction. A step-by-step analysis of the structure is described incorporating the various time-dependent material properties recommended in several design codes. By comparing the results of this analysis with those from the field and laboratory investigations, recommendations are drawn for the prediction of the long-term behaviour of segmental bridges.

## INTRODUCTION

Since 1984, three segmental bridges have been instrumented to monitor their time-dependent response and general performance during construction and early service life. The instrumented structures are the River Torridge Bridge, North Devon, and the Grangetown and Cogan Spur Viaducts, Cardiff (1). Accompanying laboratory tests have been carried out on specimens prepared from the concretes used for construction of the three structures. These tests were conducted in order to assess both the short-term physical properties and the long-term creep and shrinkage behaviour of the concrete. This paper will concentrate on the River Torridge Bridge since results are available over a longer period than from the two Cardiff bridges.

The principal aims of the investigation are (i) to evaluate both the short-term and long-term behaviour of segmental bridges, having regard for the time-dependent properties of the constituent materials and the prevailing environmental conditions in the UK; (ii) to compare the observed results with theoretical and empirical predictions using methods currently available for the design of segmental bridges subjected to time-dependent effects; (iii) to create a data base of field and laboratory results for future structural assessment and research; and (iv) to develop a step-by-step numerical technique suitable for the computer analysis of segmental prestressed concrete bridge structures.

## FIELD AND LABORATORY TESTS

The River Torridge Bridge, shown in Fig. 1, is a high level crossing carrying two lanes of traffic over the Torridge tidal estuary at Bideford. The 645 m long continuous deck consists of eight spans including three major spans of 90 m over the navigable channel. The superstructure takes the form of a single-cell, non-prismatic rectangular box girder varying in depth from 6.1 m at the supports to 3.1 m at midspan. Wide side cantilevers provide a total deck width of 13.3 m. The bridge was constructed from 251 precast concrete segments each approximately 2.5 m in length. Segments were match-cast using the short-line method and erected by means of a launching girder using the balanced cantilever technique. Segment casting commenced in June 1985 and bridge construction was completed in May 1987 (2).

Four segments were instrumented within Span 5, which is one of the central 90 m spans. Concrete strains were measured using a total of 122 vibrating wire gauges embedded within the concrete during construction. Both single gauge elements and delta rosette arrangements were employed placed on the median line of the web and flange walls in each of the four segments. A few additional gauges were also placed across the thickness of wall elements in two of the segments for the long-term assessment of shrinkage strains after erection. In addition, one of the four segments was instrumented fully for the measurement of temperature profiles across the concrete walls.

Field measurements of strain and temperature were recorded manually. Site security and the long distances between the four segments, especially before erection, made the use of automatic data logging equipment difficult. Readings from the bridge segments were taken at short intervals soon after casting, when shrinkage rates were high, and in the period during and immediately after erection. At other times readings were taken less frequently. In this way all major events including segment placing, temporary prestressing, final prestressing and changes in launching girder position, were picked up. An estimated 8000 readings were recorded during the 15 months period from casting up to completion of the bridge.

The physical properties of the concrete used in construction were measured using a large number of prism and cube specimens. Short-term measurements included compressive strength, modulus of elasticity, Poisson's ratio and coefficient of thermal expansion, at a number of different ages from 28-460 days.

Creep and shrinkage tests were conducted on prism specimens which were either fully sealed against moisture penetration, partially sealed (to take account of the surface area/volume ratio of the appropriate segment) or left unsealed. Shrinkage specimens were stored both indoors, within a controlled humidity and temperature environment, and outdoors, protected from direct rain and sunlight. Creep tests starting at different ages were conducted at a constant temperature of 23°C and 85% relative humidity, and under a constant compressive stress of 14.4 $N/mm^2$. This stress level represented approximately 25% of the 28 day characteristic cube strength. Early results from field and laboratory tests have been presented elsewhere (3, 4).


MATERIAL PROPERTIES

The concrete compressive strength was found to increase typically from 60 $N/mm^2$ at 28 days to 79 $N/mm^2$ after 365 days. Results for the concrete modulus of elasticity followed a similar trend by increasing from 30 to 39 $kN/mm^2$ over the same period. Poisson's ratio was found to be approximately 0.20 and the coefficient of thermal expansion was approximately 9.4 $\mu\epsilon/°C$ for all concrete ages.

Modulus of Elasticity

Table 1 shows the measured and predicted values of modulus of elasticity using available codes and recommendations commonly used for design. The American Concrete Institute recommendations ACI-78 (5) and the Concrete Society method CS-78 (6) are based on 28 day values and, as expected, yield results which agree closely with measured values at later ages. Methods based on characteristic compressive strength, such as the

European Codes CEB-70 (7) and CEB-78 (8), generally over-estimated the modulus especially at early ages.

Table 1    Measured and Predicted Values of Modulus of Elasticity $E_c$

| Age | Measured $E_c$ | Predicted $E_c$ (kN/mm$^2$) | | | |
|---|---|---|---|---|---|
| (days) | (kN/mm$^2$) | CEB-70 | CEB-78 | ACI-78 | CS-78 |
| 28 | 30 | 43 | 36 | 30 | 30 |
| 90 | 35 | 47 | 37 | 31 | 33 |
| 180 | 37 | 48 | 38 | 32 | 34 |
| 365 | 39 | 50 | 39 | 32 | 36 |

Shrinkage

Methods for the prediction of shrinkage available to designers can be divided into two categories. The first includes all methods which use strength, concrete mix composition and environmental conditions. These include the ACI-78, CEB-70, CEB-78 and CS-78 methods identified previously. The second type use short-term shrinkage data to predict long-term results using an exponential or hyperbolic type expression (9). These are rarely used in practice due to the obvious difficulties in performing meaningful laboratory trials in a realistic timescale during the design process. Table 2 summarises values of ultimate shrinkage predicted by the various methods. It can be seen that, in comparison to the measured long-term shrinkage strains in the segments, the CEB-70, CEB-78 and CS-78 predictions are low. The hyperbolic and exponential equations deduced using short-term results from indoor tests over-estimate ultimate shrinkage significantly. The ACI-78 method was found to provide the best estimate of ultimate shrinkage.

Table 2    Predicted Ultimate Shrinkage Strain (x $10^6$)

| ACI-78 | CEB-70 | CEB-78 | Hyperbolic | Exponential | CS-78 | Measured |
|---|---|---|---|---|---|---|
| 378 | 117 | 139 | 520 | 512 | 168 | 200-300 |

Creep of Concrete

Fig. 2 illustrates the various components of concrete deformation in typical prism specimens loaded at 180 days. It can be seen that shrinkage, basic creep (occurring in sealed specimens under hygral equilibrium) and drying creep (due to the presence of shrinkage in drying, partially sealed specimens) increase with time at a decreasing rate. The total strain at 700 days is $1100\mu\epsilon$. This comprises 38% elastic strain, 12% shrinkage and 50% total creep.

The effect of loading age on creep of concrete is illustrated in Fig. 3. Any delay in loading results in a significant reduction in total long-term creep.

As with shrinkage there are two approaches to the prediction of creep of concrete under constant load. The first requires knowledge of the mix, composition, strength of concrete and the operating conditions. The second adopts mathematical expressions containing coefficients which have to be determined experimentally by conducting short-term tests on the concrete being considered. Exponential and hyperbolic expressions, may be used but are not ideal for creep prediction since they have a limiting value.   Others which do not have a finite limit, such as the power and logarithmic expressions (9), may also be used.

Table 3 summarises the predicted values of ultimate creep strain for typical specimens loaded at 180 days from casting.   It can be seen that the CEB-70 code severely under-estimates ultimate creep.   The CEB-78 and ACI-78 methods are found to agree more closely with measured results, over-estimating them by approximately 25%.   The power expression gives unrealistically high predictions; the hyperbolic, exponential and logarithmic expressions are found to predict ultimate creep reasonably well.

Table 3   Predicted Values of Ultimate Creep Strain (x10$^6$) after 700 days

| ACI-78 | CEB-70 | CEB-78 | Exponential | Hyperbolic | Power | Log | Measured |
|--------|--------|--------|-------------|------------|-------|-----|----------|
| 623 | 232 | 609 | 488 | 471 | 1237 | 604 | 400-500 |

FIELD MEASUREMENTS

Typical longitudinal strains measured in the segment adjacent to Pier 4 are illustrated in Figs. 4 and 5.   From Fig. 4 it can be seen that shrinkage strains of 150-200$\mu\epsilon$ occur everywhere before the segment is erected.   The effects of prestressing more than counteract the bending stresses as successive segments are erected, as can be seen by the distribution of compressive strain around the cross-section in Fig. 5.   It is evident that the level of concrete strain continues to increase at all gauge positions after the completion of the instrumented cantilever.   This increase may be attributed principally to the action of creep since most shrinkage has taken place during the six months prior to erection.   Creep strains occurring from the completion of the cantilever until continuity is established with the adjacent cantilever, a period of approximately six weeks, amounts to approximately 25% of the total elastic strain exhibited due to the addition and stressing of further segments during cantilever construction.

STRUCTURAL ANALYSIS

A computer program has been developed to perform time-dependent analysis for segmental concrete bridges in the statically determinate balanced cantilever stage.   The step-by-step numerical procedure adopted predicts the strain history of the structure due to self-weight and construction loading by taking full account of the combined effects of creep and shrinkage of the concrete and relaxation of the prestressing steel.   It also considers the time-dependent nature of the construction procedure necessarily involved in the erection of segmental structures.   The method involves an accurate assessment of prestress loss at any section along the deck and then calculates the corresponding variations in strains and deformation at any section.   It accounts for the gradual stress reduction exhibited in the concrete due to prestress loss and due to the increase in its modulus of elasticity with aging.   Elastic and creep recoveries induced by the reduction in concrete compressive stress are also considered.

The bridge superstructure is first idealised as an assemblage of linear prismatic elements connected at nodes located on the centroidal axis of each cross-section.   A displacement method of analysis is used to calculate the level of prestress due to instantaneous deformations caused by changes in load.   These changes come about as a result of the addition of a new segment, tensioning operations or movement of the launching girder and other equipment used for erection.   The procedure is then extended to take account of the various time-dependent effects at different time steps.   Deflections and concrete strains around any cross-section along the member and induced deflections may then be predicted at any future age.

Three separate analyses were carried out for the instrumented cantilever of the River Torridge Bridge taking account of its detailed construction and loading histories. These were (i) an elastic analysis neglecting time-dependent effects; (ii) an analysis using CEB-70 recommendations to account for the creep and shrinkage of the concrete and relaxation of the steel, and (iii) a similar analysis but using ACI-78 recommendations to account for the time-dependent properties.

The CEB-70 and ACI-78 recommendations were chosen for this study since they are widely used in practice and because their loss functions are expressed simply in the form of multiplying factors. These are easy to evaluate and convenient to incorporate in a step-by-step procedure. Although the CEB-78 creep function showed good agreement with measured laboratory results, in practice it is not easy to evaluate and employ in a step-by-step analysis. Moreover, the approach creates a hypothetical division of strain into recoverable and irrecoverable components which does not conform with experimental observations (9).

Anchorage slip and friction losses along each tendon length were neglected in the analysis for the sake of simplicity. This may lead to an over-estimation of prestress forces and therefore concrete strains predicted are expected to be higher than measured strains. An over-estimation of creep strain is also expected in analyses (ii) and (iii) as a result of the method of superposition of virgin creep curves which was adopted for prediction of the creep strain under varying stress (9).

Results

Typical results for longitudinal strain in the top flange predicted by the three analyses are compared in Fig. 6 with those measured at various ages in the bridge. It can be seen that the elastic analysis under-estimates longitudinal strain by up to 25%, as might be expected. Predicted strains using ACI-78 recommendations appear to over-estimate measured results by up to 25%; the CEB-70 recommendations result in an under-estimation of strain. However, considering the over-estimation of strain due to the factors inherent in analytical modelling of the structure and in the assessment of creep strain development, results using ACI-78 are believed to show best agreement with the measured strains. Similar conclusions were deduced earlier from observed laboratory and field results.

By separating the various time-dependent effects and their influence on the development of concrete strain it is possible to determine their individual significance in the various analytical methods. For example, let us consider the top fibre strain for the segment adjacent to Pier 4 before continuity of the cantilever was established. Using the CEB-70 recommendations, elastic, shrinkage, creep and relaxation strains constituted approximately 50%, 26%, 23% and 1% respectively of total predicted strain. Using the ACI-78 approach the proportions were 74%, 4%, 20% and 2%. It can be seen that elastic strain is the largest component of total strain for both analyses but especially for the ACI-78 approach due to the relatively low prediction of elastic modulus. Creep and shrinkage strains are seen to share equal proportions of total strain using the CEB-70 results, as they are assumed to have similar rates of development with time. However, creep strain is seen to account more correctly for a much greater proportion of total strain than shrinkage for the ACI-78 results. Relaxation of the prestressing steel represents the lowest proportion of total strain for both analyses. This is because low-relaxation strand was employed in the structure which exhibits an ultimate relaxation of only 2% of initial load. Also the tendons were prestressed in stages before reaching the final load thus reducing intrinsic relaxation.

CONCLUSIONS

The ACI-78 and CS-78 recommendations were found to provide best predictions of elastic modulus. The CEB-70 and CEB-78 predictions greatly under-estimated measured shrinkage in the segments; ACI-78 predictions were found to over-estimate shrinkage to a smaller degree. Creep strain was over-estimated by the CEB-78 and ACI-78 methods, and greatly under-estimated by CEB-70 predictions. Shrinkage prediction methods using short-term results agreed reasonably well with measured results. However, those for creep prediction compared less favourably.

Neglecting time-dependent material behaviour can lead to an under-estimate of concrete strains by at least 25%. ACI-78 recommendations were found to yield the most accurate overall prediction of measured long-term behaviour. Short-term elastic behaviour contributed the major share of total predicted strains and deformations. Creep strain was found to have a significant effect; shrinkage strain contributed little to the concrete deformation in this structure due to the age of the segments at erection. Relaxation of prestressing steel had very little influence on long-term deformations because low-relaxation strand was used throughout.

REFERENCES

1.      Barr, B.I.G., Waldron, P. and Evans, H.R. 'Instrumentation of glued segmental bridges', Colloquium on Monitoring Large Structures and Assessment of their Safety, Intl. Assoc. of Bridge and Struct. Engg., Bergamo, (1987), 175-189.

2.      Ziadat, G.S. and Waldron, P. 'Measurement of time-dependent behaviour in the River Torridge Bridge', Vol. 1, Instrumentation and early results, Report No. UBCE/C/87/4, Dept. of Civil Engg., University of Bristol, (Nov. 1987), pp 124.

3.      Waldron, P., Ziadat, G.S. and Salfity, R.R. 'Early results from the instrumentation of a glued segmental bridge', Proc. I. Struct. E./BRE Seminar on Large and Full Scale Testing of Structures, London (April 1987).

4.      Ziadat, G.S. and Waldron, P. 'Monitoring and prediction of strain in a long-span segmental bridge', 13th IABSE Congress, Challenges to Structural Engineering, Helsinki, (June 1988).

5.      ACI Committee 209, Prediction of creep, shrinkage and temperature effects in concrete sctructures, 2nd draft, Am. Conc. Inst., Detroit, (Oct. 1978), pp 98.

6.      Parrott, L.J. 'Simplified methods for predicting the deformation of structural concrete', Report No. 3, Cem. and Conc. Assoc., London, (Oct. 1979), pp 11.

7.      CEB/FIP, International recommendations for the design and construction of concrete structures - principles and recommendations, FIP Sixth Congress, Prague, (June 1970).

8.      CEB/FIP, model code for concrete structures. Comite Euro-International du Beton - Federation Internationale de la Precontrainte, Paris, (1978), pp 348.

9.      Neville, A.M., Dilger, W.H. and Brookes, J.J. 'Creep of plain and structural concrete, Construction Press, (1983), pp 361.

Figure 1    The River Torridge Bridge

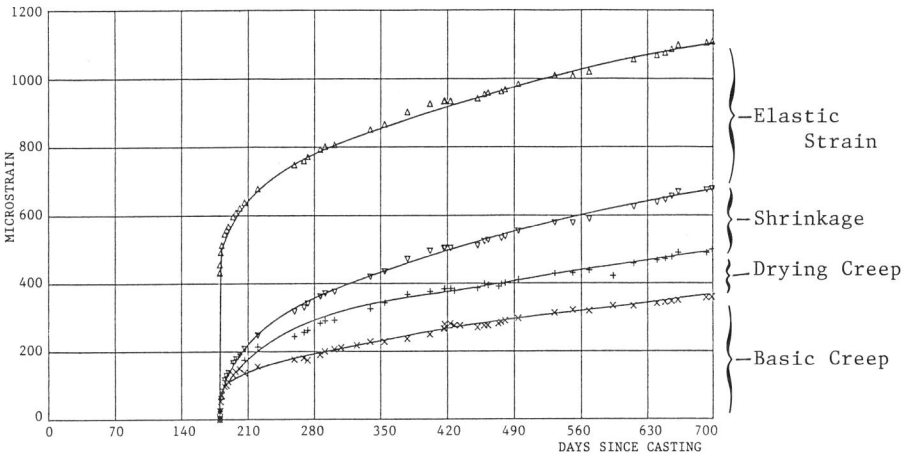

Figure 2    Components of Concrete Deformation

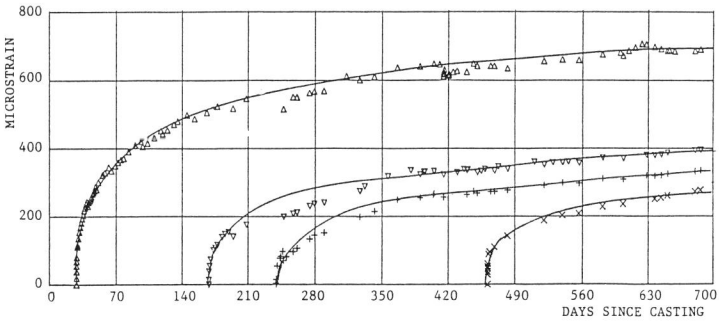

Figure 3    The Effect of Loading Age on Creep of Concrete

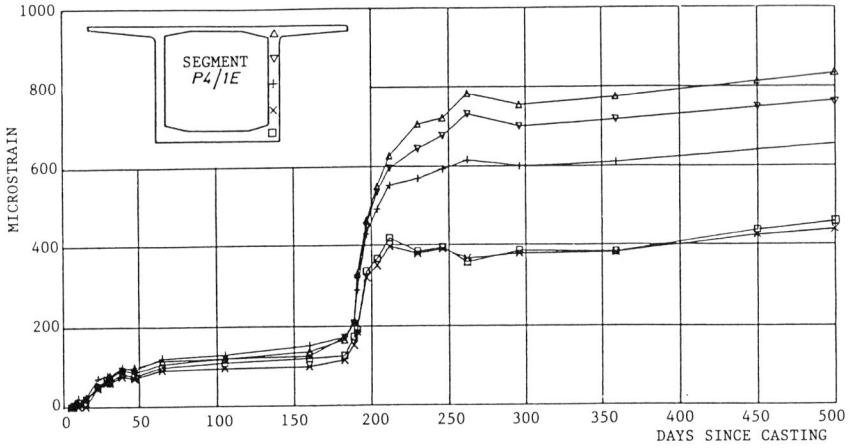

Figure 4    Distribution of Longitudinal Strain in one Web

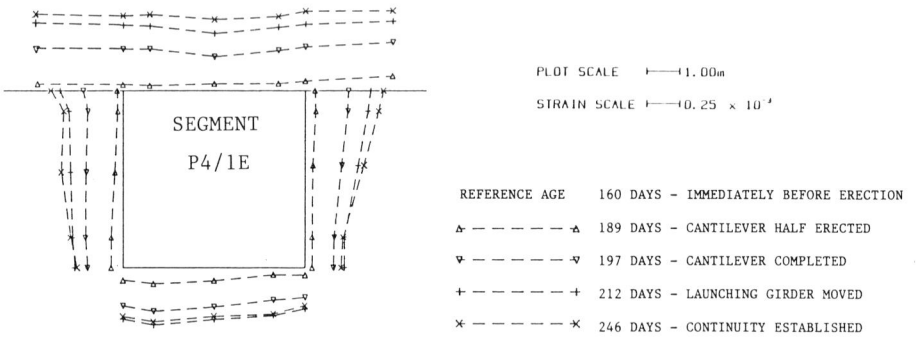

PLOT SCALE  ⊢——⊣1.00m

STRAIN SCALE  ⊢——⊣0.25 × 10⁻³

| REFERENCE AGE | 160 DAYS – IMMEDIATELY BEFORE ERECTION |
| Δ————Δ | 189 DAYS – CANTILEVER HALF ERECTED |
| ∇————∇ | 197 DAYS – CANTILEVER COMPLETED |
| +————+ | 212 DAYS – LAUNCHING GIRDER MOVED |
| ✻————✻ | 246 DAYS – CONTINUITY ESTABLISHED |

Figure 5    Distribution of Longitudinal Strain around Section

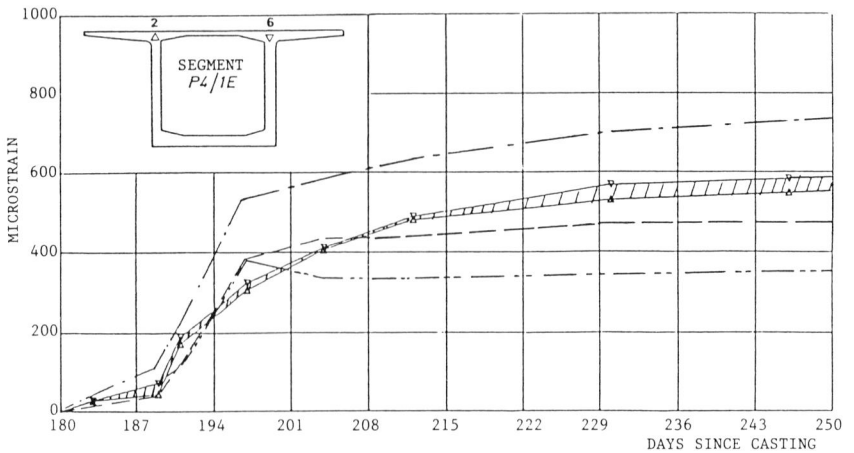

Figure 6    Comparison of Measured and Predicted Longitudinal Strains

406

R Lavender, Tower Hamlets Borough Engineers Service, UK

What consideration has been given to the replacement of the prestressing strands in the event of deterioration due to water and chloride attack.

The answer given by Mr Ziadat was that it was impossible to replace tendons on segmental constructed bridges, however, I consider that it is essential this condition is catered for as recent investigations on some early prestressed bridges has revealed some disturbing finds of severely corroded tendon strands.

Author's reply

A major cause of deterioration of segmentally constructed prestressed concrete bridges is the ingress of water and chlorides through the construction joints causing corrosion of the prestressing tendons. This problem has been highlighted recently in two bridges. The first structure is the Taf Fawr bridge in Wales which has been demolished and replaced recently after less than twenty years of service (1). A more dramatic case is that of the Ynys-y-Gwas bridge which collapsed under its own self-weight (2).

Both these bridges employed precast post tensioned elements with mortar/cast-in-situ concrete joints. Most precast concrete segmental bridges constructed recently contain epoxy glue rather than mortar at the joints to provide a better seal against ingress of water and contaminants (3).

However, if corrosion of the post tensioning steel does occur in bonded, grouted tendons, it is very difficult to detect since the tendons are embedded within the concrete section and therefore cannot be visually inspected. Moreover it is impossible to replace the tendons without partial or complete demolition of the structure.

In order to overcome this problem it is advocated that the use of corrosion protected unbonded tendons should be considered more seriously as an alternative for future design of prestressed concrete bridges (4). In this system the tendons can be extracted, inspected periodically and replaced if necessary without damaging the existing structure. Unbonded tendons have been successfully used in the construction of major segmental bridges including the Florida Keys bridge (USA) and Bubiyan bridge (Kuwait). However, care must be taken in the design of bridges using unbonded tendons to ensure adequate reliability since the British Standard for design of concrete bridges +(BS5400) does not cover unbonded tendon design.

## References

(1)  Corlett, M S., Demolition of a major prestressed bridge, Quarterly Journal of FIP, 1987/1, pp 11-14.

(2) Woodward, R J and Williams, F W., Collapse of the Ynys-y-Gwas bridge, West Glamorgan.  Proc. of I C E Part 1, No.84, Aug.,1987.

(3) The Concrete Society, Counter-cast segmental bridge construction, Technical Note No.19, 1980, 13pp.

(4) Sriskandan, K., Prestressed concrete road bridges in Great Britain: a historical survey, Proc. of I C E, Part 1, No.86, April 1989.

# 47 Condition survey of 21 bridges, A74(T), Dumfries and Galloway region

**Brian J Brown,** STATS Scotland Ltd, East Kilbride, Scotland

A condition survey of 21 bridges revealed low depths of cover, adequate cement contents but low strengths compared with current durability recommendations. Carbonation, however, was not generally a problem only occurring at localised areas of low cover. Chloride ingress was generally not a problem on deck soffits except at areas of leakage; supports were, however, effected by road spray and by leakage from above. Remedial measures were formulated based on the surveys.

## INTRODUCTION

21 bridges on the A74 trunk road between Beattock and the Scotland-England border in Dumfries and Galloway Region were investigated in 1985 (Figure 1). The work was carried out for Dumfries and Galloway Regional Council on behalf of the Scottish Development Department. All the bridges were constructed between 1959 and 1971 in reinforced concrete. The bridge types represented by the survey population included road, river and railway overbridges, road underbridges and accommodation bridges. Construction types included portal frames, piers with columns and capping beams and various types of decks (pre-stressed beams with infill concrete, post-tensioned and steel universal beams supporting r.c. slabs, continuous beam and slab decks and simply supported continuous r.c. decks).

## CONSTRUCTION

7 different sources were used for aggregate supplies dependent on site location and year of construction, all were local fluvial sand and gravels. The concrete produced was either site batched or supplied ready-mixed.
Based on year of construction, site location, aggregate source and concrete type, five bridge groupings can be recognsied, as summarised below.

1) Lockerbie Area, pre 1960 and 1962 construction, site mixed, aggregate from local borrow pits (excepting one bridge which was ready-mixed ex Barnhill).
2) Beattock Area, 1960 construction, ready-mixed supply ex Barnhill.
3) Ecclefechan Area, 1964 construction, site mixed concrete, gravel ex Hardgrove.
4) Southern section of A74, 1968-69 construction, ready-mixed supply ex Peth (occasionally Edenhall).
5) Gretna Area, 1971 constuction, ready mixed supply ex Lynefoot (occasionally Peth).

CONDITION SURVEYS

All of the bridges were assessed in a similar manner. Selected areas of deck and support structure representing visually good and poor areas were investigated. Work on-site included visual, depth of cover, half-cell potential, resistivity and carbonation surveys together with the taking of core and drill samples for laboratory analysis. Drill samples were taken at the surface (0-25mm) or as gradients at 25mm intervals and tested for chloride content. Core samples were visually appraised and tested for strength, density, microcracking and cement content and underwent petrographic examinations.

A report detailing the condition of the concrete, the corrosion condition of the reinforcement and the cause and extent of any defects was prepared for each bridge together with suggested remedial measures based on the condiiton survey results. A final summary report was then prepared collating the information from all 21 surveys and expanding upon that information so allowing an outline of an overall repair scheme to be formulated.

Details of some of the investigated parameters are given in the following paragraphs.

Depth of Cover

The cover to steel was measured as the lowest depth of cover within a set grid (generally 1m). The mean depth of cover (as measured) for all the bridges was 31.0mm with a standard deviation of 6.3mm; the range of average values for individual bridge elements was 21.2mm to 43.5mm. The lowest values recorded on individual bridge elements lie in the 0-30mm range with the highest values in the 25-60mm range. Analysis of the mean results for individual bridge elements is given in Table 1.

Table 1: Depth of Cover

|  | Deck Soffits | Piers | Abutments |
|---|---|---|---|
| Mean (mm) | 28.2 | 32.5 | 34.3 |
| Standard Deviation (mm) | 6.1 | 3.9 | 6.5 |

The variation in cover depth is thus large with standard deviations varying from 3.9mm for piers to 6.5mm for abutments. For a normal distribution 95% of results would be expected to be within two standard deviations of the mean. The mean should thus be approximately 8 to 13mm greater than the minimum cover (compare this with current specifications (1) that allow only 5mm between minimum and nominal cover). As the value recorded on-site is the lowest cover within a grid square and the statistics have been calculated on mean values for individual elements of the 21 bridges, the variation between mean and minimum may actually be greater.

The precise specification for cover requirements for the bridges was not established; however CP114:1957 (2) was current over the construction period. This standard recommended 1 1/2 inches (38mm) for all external concrete. It is apparent that the majority of the reinforcement did not comply with that recommendation. Current requirements as given in BS5400:Part 4:1984 (1) are for 35mm for deck soffits and 50mm for supports adjacent to the carriageway (assuming grade 40 concrete - see below) with a 5mm allowance between minimum and specificed value. Again the majority of the cover would not comply with these recommendations.

Compaction

Compaction was generally found to be good with very little in the way of honeycombing recorded and if found only in small isolated areas. The majority of the concrete was classified as having very good or good compaction with excess voidage (visually assessed) less than 1% and between 1 and 2% respectively. A number of cores (11 out of 77 examined) had only fair compaction with excess voids greater than 2%. Piers tended to be better compacted than abutments and deck soffits. Saturated densities were more consistent than other measured parameters with an overall mean of 2365kg/m$^3$ and a standard deviation of 37kg/m$^3$ (see Table 2).

Strength

Compressive strengths (measured as the Estimated In-Situ Cube strength) were variable with an overall mean of 44.8N/mm$^2$ and standard deviation of 11.3N/mm$^2$ for 21 bridges. Data for each construction phase is given in Table 2.

While it is recognsied that the strength of in-situ concrete is some 80% of the standard cube strength when both are tested at 28 days of age (3), the rate of strength gain between 28 days and 1 year is approximately 25% with minimal gain thereafter so that two factors broadly cancel one another out. On concrete several years old the in-situ strength is similar to the standard 28 days cube strength.

Comparisons between concrete quality and specification can be undertaken on the basis of mean values provided that the margin between mean values and minimum values is known. The minimum value is the characteristic value and the margin between this and the mean value is generally in the region of 10N/mm$^2$. The overall mean strength of 44.8N/mm$^2$ would indicate grade 35 concrete. On close examination the concrete from the Beattock and Gretna phases of construction would appear to have Grade 37.5 or 40 concrete whereas the other construction areas more correspond to grade 30 supplies. The grade 30 mixes may be equivalent to the 4500 lb/ft$^2$ (31N/mm$^2$) standard mix of CP114:1957 (2). BS5400:Part 4:1984 (1) relates minimum strengths to exposure conditions with grade 40 non air-entrained concrete the minimum for soffits and grade 50 non-entrained concrete the minimum for supports adjacent to the carriageway. It is thus apparent that the majority of the concrete does not meet current requirements.

Table 2: Material Properties

| Area | Compressive Strength | | Cement Content | | Saturated Density | | Carbonation k Factor | |
|------|------|------|------|------|------|------|------|------|
| | Mean | s.d. | Mean | s.d. | Mean | s.d | Mean | s.d. |
| Lockerbie | 41.6 | 13.1 | 339 | 69 | 2367 | 24 | 2.00 | 0.56 |
| Beattock | 50.3 | 11.4 | 406 | 95 | 2411 | 35 | 2.04 | 0.48 |
| Ecclefechan | 43.5 | 9.1 | 389 | 69 | 2358 | 34 | 1.63 | 1.04 |
| Southern | 42.0 | 10.5 | 379 | 43 | 2337 | 25 | 1.97 | 0.72 |
| Gretna | 48.2 | 10.7 | 402 | 37 | 2358 | 13 | 1.98 | 1.00 |
| All | 44.8 | 11.3 | 383 | 67 | 2365 | 37 | 1.91 | 0.77 |

## Cement Content

The mean cement content was high for all the structures at 383kg/m$^3$ with a standard deviation of 67kg/m$^3$. Data for each construction phase is given in Table 2.
The Beattock and Gretna phases which recorded the highest strengths also appear to have the highest cement contents.
BS5400:Part 8:1978 (4) currently requires a minimum cement content of 290kg/m$^3$ in soffits and 360kg/m$^3$ for supports adjacent to the carriageway to ensure durability when using 20mm aggregate. It is probable that all the bridges meet both requirements bearing in mind the accuracy of determining cement content by chemical analysis. The one exception was the piers to one of the Lockerbie overbridges which recorded values much less than 290kg/m$^3$ (and so reduced the mean value for the Lockerbie area).

## Carbonation

Carbonation depth is generally thought to increase proportionally to the square root of time. The proportionality factor (k) is dependent on a number of factors including water:cement ratio and moisture content of the concrete. k is known to vary considerably not only between structures but also between elements in any structural unit. The maximum k value can be expected to be between 2 and 3 times the mean in any batch of similar components (5).
Carbonation was measured as a maximum and minimum at each test location. Only the maximum values will, however, be considered. The range of individual maximum values was found to be 0 to 40mm. Mean maximums for bridge elements ranged from 2.9 to 18.4mm.

Assuming a 120 year design life for a bridge, the necessary k value to ensure that the carbonation front does not reach the reinforcement for the bridges in question is 2.83 (assuming an average level of cover of 31.0mm). If using the average depths of cover given in Table 1, the equivalent k values for deck soffits, piers and abutments are 2.57, 2.97 and 3.13 respectively.

The range of actual k values for each bridge structural element is 0.65 to 4.02 with an overall mean k of 1.91, standard deviation 0.77. Values for each construction phase are given in Table 2.

It is obvious that in the vast majority of the concrete, carbonation will not be the problem within the design life of that particular structure. However, areas of one deck in the Lockerbie area, the piers of one bridge in the Ecclefechan area, one deck in the Southern Section and one deck in the Gretna deck have predicted time to the start of carbonation induced corrosion much less than the time remaining of the 120 design life. In general, though, only areas of very low cover or very high carbonation (eg honeycombed areas) were found to be at high risk.

## Chlorides

The presence of chloride in concrete leads to a probability of corrosion of any embedded reinforcement. No chloride based admixtures were found to be present in the concretes investigated; however, all of the bridges were found to have chloride ingress derived from de-icing salts to varying degrees. Deck soffits were generally little affected by chlorides with values generally less than 0.2% Cl⁻ by mass of cement. Values increased dramatically in localised areas of persistent dampness via leakage of road salt laden water from the road above. At areas surrounding weep holes and at leaking joints, cracks and element intersections chloride values ranged from 0.24 up to 1.82% Cl⁻ by mass of cement. Values were generally lower at deck edges affected by run-off with values in the 0.14 to 0.35% Cl⁻ by mass of cement range.

Work on the supports adjacent to the carriageways indicated ingress of chloride via road-spray affecting the lower 2 to 3m of columns and walls with run-off from leaking deck joints or soffit defects (cracks etc) affecting the upper areas of columns and walls. Capping beams were similarly affected. Summarised results are given in Table 3.

Table 3:   Chloride content of Supports

| Height from G.L.(m) | Range of Chloride Content (% Cl⁻ by mass of cement) | |
|---|---|---|
| | Columns | Walls/Abutments |
| 0-1 | 0.06-1.53 | 0.03-2.42 |
| 1-2 | 0.08-1.10 | 0.06-0.77 |
| 2-3 | 0.08-0.53 | 0.06-0.33 |
| > 3 | 0.16-1.12 | 0.05-0.36 |
| Capping Beams | 0.06-0.79 | |

In supports (generally walls and abutments) beneath carriageways chloride ingress is from leaking wall or abutment intersections with the deck allowing salt laden water to flow down the face. Values are very variable and are dependant on the leakage pattern. Chloride contents ranged from 0.04 to 1.57% Cl⁻ by mass of cement.

## Half-Cell Potential

The vast majority of potentials were more positive than -200mV with very little more negative than -350mV. Summarised results are given in Table 4.

Table 4:  Half-Cell Potential

|  | | % no. of readings | |
| --- | --- | --- | --- |
|  | -200mV | -200 to -350mV | -350mV |
| Average Decks | 89.2 | 10.0 | 0.8 |
| Average Piers | 82.1 | 15.4 | 2.5 |
| Average Walls/Abutments | 52.3 | 27.3 | 20.4 |

In general areas of high chloride content, dampness and visual distress (spalling, exposed rebars etc) recorded high potentials as would be expected. Potentials are known to be affected by a large number of parameters (7) and the Van Deveer interpretive scheme given in ASTM C876 (6) (> -350mV 90% corrosion probability, < -200mV 10% corrosion probability, -200 to -300mV corrosion uncertain) may not be applicable to support structures. Certainly the polarisation effects from the damp soil at the base of columns, walls and abutments must be taken into account when interpreting results from the supports. The relationship between chloride content and half-cell potential was examined and showed in broad general terms that high chloride equates with high half-cell potential and vice versa. However, the relationship was by no means exact and the variability found is a reflection of the other parameters affecting potential especially moisture condition. The relationship found was similar to that reported by Vassie (8) for a single bridge deck.

## Corrosion Condition of the Reinforcement

Corrosion of the reinforcement was not judged on any single factor but on a combination of the visual, depth of carbonation, depth of cover, half-cell potential and chloride content survey results with due recognition of all factors that may effect a result. Much weight was placed on the shape and gradient of equipotential contours rather than on the numerical value of the half-cell potential. This coupled with the knowledge of the chloride and carbonation profiles within the concrete and especially the chloride content at the depth of the steel reinforcement and the visual condition of the structure allowed a valued judgement of the corrosion condition of the steel to be made.

SUGGESTED REMEDIAL ACTION

Based on the condition surveys of all 21 bridges outline remedial measures were formulated both on an individual structure basis and as an overall scheme.

Most of the bridges required effective waterproofing of deck joints to prevent leakage to the substructure. Minor remedial works such as localised patching to spalled areas, extending weep-holes clear of the soffit and prevention of run-off down edge beams was suggested for a number of bridges. All of the bridges except one required a protective coating (or similar) to prevent or retard further chloride ingress. Columns, walls or abutments required replacement of cover concrete subject to more detailed investigation in 14 structures especially in the lower 1.5 to 2m adjacent to the carriageway. The entire abutment walls of some underbridges with leaking joints might require replacement of the cover concrete. The entire soffit of one bridge deck was thought to require replacement subject to detailed economic and structural considerations.

Although the surveys of the 21 bridges were used to formulate suggested repair schemes in many instances the amount of data available was insufficient for a detailed consideration of quantities and further investigation of targetted areas would be required prior to and during actual remedial works.

COMPARISION OF DIFFERENT CONSTRUCTION PHASES

There is very little difference between the 5 phases of construction despite the fact that they have different aggregate sources and concrete supplies and are of different ages. All have defects to various degrees. Of greater effect on the overall longevity of the structures than their concrete and age are their inherent defects (such as leaking joints, weep-holes flush with the soffit, no dripstops on soffit edges etc) which all lead to salt laden water gaining local access to the concrete. If such leakages are remedied either by design or through maintenance and barrier coatings are applied to the spray zone of supports adjacent to the carriageway, then the problems of chloride induced corrosion of rebars will largely be overcome assuming adequate control over depth of cover and concrete quality at the time of construction.

ACKNOWLEDGEMENTS

The permission of the Scottish Development Department and Dumfries and Galloway Regional Council to publish this paper is gratefully acknowledged.

415

REFERENCES

1    BS5400:Part 4:1984, 'Steel, concrete and composite bridges.
     Code of practice for design of concrete bridges'.
2    CP114:1957, 'The structural use of reinforced concrete in
     buildings'.
3    BS6089:1981, 'Guide to the assessment of concrete strength in
     existing structures'.
4    BS5400:Part 8:1978, 'Steel, concrete and composite bridges.
     Recommendations for materials and workmanship, concrete, reinforce-
     ment and prestressing tendons'.
5    Currie, R.J., 1986, 'Carbonation depths in structural-quality
     concrete'. BRE Report.
6    ASTM C876-80, 'Standard test method for half-cell potentials of
     reinforcing steel in concrete'.
7    Figg, J.W. and Marsden, A.F. 1985, 'Development of inspection
     techniques for reinforced concrete: a state of the art survey of
     electrical potential and resistivity measurements for use above
     water level'. Offshore Technology Report OTN 84 205.
8    Vassie, P., 1984, 'Reinforcement corrosion and the durability of
     concrete bridges'. Proc.Instn.Civ.Engrs. Part 1, 1984,76,Aug.,713-
     723

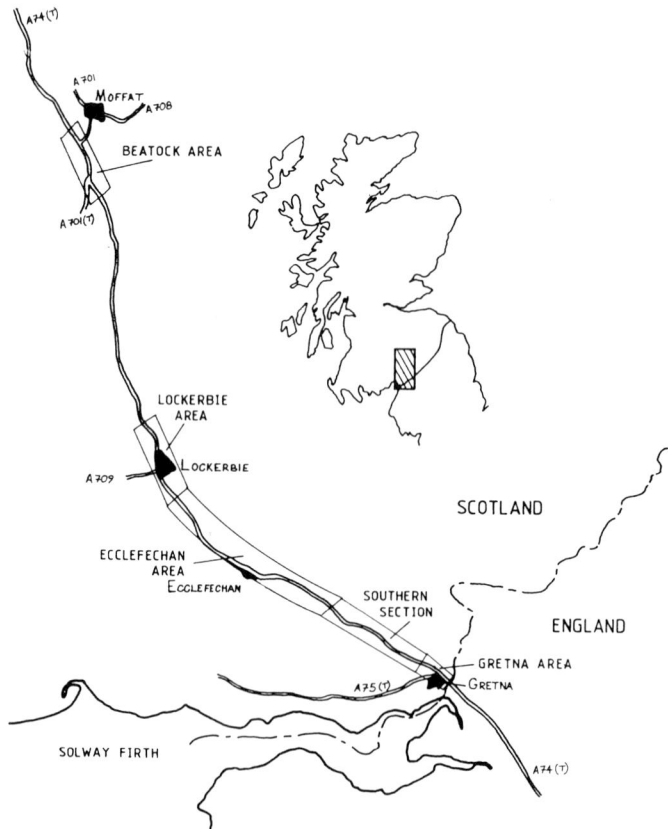

Figure 1  Location Plan

# 48 The performance of a population of large concrete silos

**A L Gilbertson,** WS Atkins, Epsom, Surrey, UK
**P Dawson,** Taywood Engineering, Southall, London, UK

This paper presents information about a population of large slipformed prestressed concrete silos which have been studied, surveyed and tested over a period of six years. The work is described and sufficient data is presented to give an understanding of the results. Finally, an explanation of how the results are being used is given and the basis for decision making is discussed.

The authors wish to thank British Sugar and their respective employees for permission to publish this paper and express their appreciation to their colleagues who have contributed to the work reported.

## 1.0    INTRODUCTION

Over the last 35 years, British Sugar plc have invested in the construction of 54 large prestressed concrete silos. Each silo can hold between 8,000 and 25,000 tonnes of granulated sugar and the silos have been built in groups to receive product directly from the sugar factories. This silo population is of particular interest because whilst the design of the silos has evolved to some extent, the silos are basically similar. They have however been built over a period of years by different Contractors in different locations. Details of the silos are given in Table 1.

Starting in 1983, as a result of operational difficulties and some concern about their major investment in silos, British Sugar decided to increase the level of maintenance activity and at the same time to enhance their understanding of how silos behave. Work has been undertaken by a number of parties (including the authors) and some interesting results have been obtained. In 1988, British Sugar published various papers about their silos (British Sugar plc Technical Conference, Eastbourne 1988) and both general information about the silos and detailed information about some of the studies undertaken can be found therein.

Table 1  Details of the Prestressed Silo Population

| Silo Site/No. | Date Built | Tonnage Capacity (each) | Founds | Roof | Prestress |
|---|---|---|---|---|---|
| A 1,2 | 1961 | 10,000 | Piled | Conical | Bonded |
| A 3 | 1982 | 17,000 | Piled | Conical | Unbonded |
| B 1,2 | 1956 | 8,000 | Piled | Flat | Wound |
| B 3 | 1966 | 10,000 | Piled | Conical | Bonded |
| B 4 | 1968 | 10,000 | Piled | Conical | Bonded |
| B 5 | 1983 | 14,000 | Piled | Conical | Unbonded |
| C 1,2 | 1961 | 8,000 | Piled | Flat | Unbonded |
| C 3 | 1983 | 20,000 | Piled | Conical | Unbonded |
| D 1,2,3,4 | 1972 | 12,000 | Rafts | Conical | Unbonded |
| D 5 | 1982 | 25,000 | Piled | Conical | Unbonded |
| E 1,2,3,4 | 1963 | 10,000 | Piled | Conical | Bonded |
| E 5,6 | 1980 | 10,000 | Piled | Conical | Unbonded |
| F 1 | 1966 | 12,000 | Piled | Conical | Bonded |
| F 2 | 1974 | 12,000 | Piled | Conical | Unbonded |
| G 1,2 | 1962 | 10,000 | Raft | Conical | Bonded |
| G 3 | 1969 | 11,000 | Vibro Raft | Conical | Bonded |
| G 4 | 1980 | 20,000 | Piled | Conical | Unbonded |
| H 1,2 | 1972 | 12,000 | Piled | Conical | Unbonded |
| J 1,2,3 | 1956 | 8,000 | Piled | Flat | Wound |
| J 4 | 1964 | 8,900 | Piled | Conical | Bonded |
| J 5 | 1983 | 14,000 | Piled | Conical | Unbonded |
| K 1,2,3 | 1976/78 | 12,000 | Piled | Conical | Unbonded |
| K 4 | 1981 | 12,000 | Piled | Conical | Unbonded |
| L 1,2 | 1965 | 12,000 | Piled | Conical | Bonded |
| L 3,4 | 1971 | 12,000 | Piled | Conical | Bonded |
| M1,2,3,4, 5,6 | 1968/73 | 12,000 | Piled | Conical | Bonded |
| M 7 | 1987 | 25,000 | Piled | Conical | Unbonded |
| N 1,2 | 1954 | 8,000 | Piled | Flat | Wound |
| N 3 | 1975 | 14,500 | Piled | Conical | Unbonded |

Notes  1.  Raft bearing on vibrostone columns in silty sand shown as "Vibro Raft" for G3.

2.  Rafts for D1,2,3,4 are on chalk.

3.  Unbonded prestress tendons are in plastic sleeves except C1,2 where they are in grease-filled metal tubes.

4.  Wound prestress tendons have a gunited protection.

5.  Bonded prestress tendons are grouted inside metal tubes.

This paper gathers together information specifically relating to the aims of the conference. The work we are reporting is as follows:-

(i)     General silo surveys and tests to investigate existing or potential problems.

(ii)    Particular concrete surveys and tests to provide specific data on the concrete strength of a particular group of silos.

(iii)   Measurement of the behaviour of a large silo to investigate the effects of sustained grossly eccentric emptying.


2.0     GENERAL SURVEYS AND TESTS

The majority of the silos have now been surveyed and tested. The level of surveying has varied depending on the past history and present condition of each silo. The work includes testing for carbonation and for chloride content, visual examination and crack mapping and investigation of cover and spalling due to corrosion of reinforcement.

(i)     Tests for Carbonation

Testing involved the drilling of three holes in a small triangle, breaking out the centre with a cold chisel and spraying on phenolphthalein to indicate the depth of carbonation. The results are presented in Table 2. Each silo was tested at a minimum of 12 locations. It can be seen that despite the age of many silos, carbonation has progressed only slowly. The greatest depth of penetration is at site N which is downwind of an industrial area and coal-burning power stations. For the majority of the silos the depth is only a few millimetres regardless of age and this is a reassuring fact with regard to the durability of exposed slip-formed structures.

Table 2   Tests for Carbonation

| Depth of Carbonation | Silos | Age Range |
|---|---|---|
| Min 3mm, Max 5mm | C1-2-3; E1-2,4; L1 to 3; M1 to 6; | 3 to 30 years |
| Min 3mm, Max 10mm | A1-2; L4; N1-2 (wall) E3; | 13 to 25 years |
| Min 3mm, Max 30mm | N1-2 (ring beam) N3; | 8 to 35 years |

419

(ii)   Tests for Chloride Content

Using powder samples taken directly from the drilling of holes in the silos, tests were made for chloride content. The results are presented in Table 3. In all cases, the chloride content was found to be low, indicating that chloride - accelerated corrosion would not be a problem.

Table 3   Tests for Chloride Content

| Chloride content as percentage chloride ion by mass of cement | Silos |
|---|---|
| 0-0.2% | C1 to 3; E1, 2; H1, 2; L1, 3, 4; M1 to 6; N1 to 3 |
| 0.2 - 0.4% | A1, 2; E3, 4; L2 |

(iii)   Visual Examination & Crack Mapping

These activities are not strictly "physical tests" but form a vital part of any testing programme. The silos varied widely in their performance and the main results are given in Table 4. The key factor which emerged is that problems are either related to initial construction difficulties, e.g. poor compaction and cold joints, or they occurred in a manner which it has not been possible to relate with any consistency to design, operation or age.

(iv)   Cover and Spalling Due to Corrosion of Reinforcement

Covermeter investigations of the external faces of the slipformed silo walls indicated that some cover was generally maintained, albeit varying considerably between 15mm and 70mm. In view of the generally low carbonation results, it is not surprising that corrosion/spalling problems are related to particular problem areas and Table 5 provides details.

Table 4  Visual Examination

| Condition | Silos | Comments |
|---|---|---|
| Very good condition | A3<br>M1 to 6 | But see also Table 5 |
| Severe cold joints (horizontal) | A1,2 | |
| Vertical cracks - external | A2; D1,2,3,4;H2; | Cracks in some cases noted soon after construction; others, noted many years after construction |
| Vertical cracks - internal | B1,2;D1,2,3,4; H2;L2;N1,2 | Each site shows distinctive crack patterns |
| Ring beam problems | A1,2;D4;E5,6 | Each site shows distinctive problems |
| Other comments | A1; H2<br><br>C1,2 | Large bulge on internal wall face<br>Failure of detail at wall seating |


Table 5  Particular Corrosion/Spalling Problem Areas

| Problem | Silos |
|---|---|
| Loose bars projecting from the rebar mesh | A1<br>H2 |
| Pilaster rebar low cover | H1,2 |
| Corrosion at prestressing anchorage zone | A1,2,3; B3,4; D1 to 4;. E1 to 4; H1,2; L1 to 4; M1 to 6 |
| Cracks greater than 0.3mm | A1,2; B1,2; D4; E1,2,3,4; L1,2,3,4; M2,6; N1,2. |

## 3.0    MEASUREMENTS OF SILO CONCRETE PROPERTIES

In the course of investigating the possibility of strengthening a pair of silos to improve their safe working capacity, a series of in-situ tests was carried out to determine the concrete strength and modulus of elasticity of the shells. The tests included cutting 100 mm diameter concrete cores from the full thickness of the shells and non-destructive testing (NDT) using ultrasonic pulse velocity (UPV) and Schmidt hammer (impact energy) techniques.

Sixteen concrete cores were cut from the full thickness of the walls of each silo, uniformly distributed vertically and horizontally. The NDT tests were made at 250 mm intervals on four vertical lines on each silo, with additional readings close by the core locations.

The equivalent cube crushing strengths, derived from the cylindrical cores varied widely, from 19 $N/mm^2$ to 59 $N/mm^2$ with an average strength of about 35 $N/mm^2$. An equivalent range of results was obtained from the UPV and Schmidt hammer measurements. Generally the Schmidt hammer results on and around the core samples reflected the core crushing strengths; the equivalent correlation was less apparent for the UPV results. It is considered possible that prior distortions to the shells had caused some microcracking in the concrete which influenced the UPV results.

Low concrete strengths on both silos were obtained from cylinder and NDT tests near the bottom and near the top. Both silos had been slip-formed and the results indicated a variability in the initial concrete mix, probably also associated with varying rates of slip progress which were achieved. This variability may be typical of this method of construction and should be taken into account in the design of structures where it is likely to be used.

The measurements of concrete modulus indicated an average value of about 30 $kN/mm^2$, after eliminating results from samples with obvious defects or embedments. This value is consistent with the average value of crushing strength.

A significantly higher crushing strength was obtained from limited tests on samples taken from the ring beams at the top of the silo shells. This may be because a higher strength concrete mix was used for these zones, or it may be because they were cast in-situ, rather than slip-formed.


## 4.0    SILO UNLOADING TESTS

A series of tests was carried out on one of a pair of 12000 ton capacity silos to investigate the effects of progressively increasing eccentric discharge. Measurements were made of the radial movement of the silo shell at five levels and of the vertical movements of the foundation slab supporting the silo.

The foundation slab rested on bored piles and it was initially supposed that these provided a substantially rigid support. The measurements demonstrated that this was not so. Not only did the silo foundation move vertically in response to the weight of ensiled material, by about 1 mm per 1000 tons, it also tilted towards the other silo when both were loaded with sugar and warped (ie curved about a horizontal axis) towards the other silo.

The movements and distortions of the silo foundation slab were matched by the movements of the silo shell. Its translation at various levels correlated with the tilt of the foundation slab and the ovalling of the shell corresponded to the warp of the foundation slab in both magnitude and direction.

The extent of ovalling was greatest at silo mid-height. This was expected as the shell is stiffened at the bottom by its connection to the foundation slab and at its top by a ring beam, which also supports the roof structure.

The ovalling also increased in proportion to the eccentricity of discharge. In all cases the minor axis of the oval broadly coincided with a line joining the silo axis and the discharge outlet. The differences between the diametrical lengths of major and minor axes was about 30 mm for discharge close to the silo wall. The tests clearly indicated that the shell ovalling was caused in part by the varying horizontal pressure distribution on the silo shell and in part by the shell being allowed to respond to a foundation driven ovalling arising from the warp of the base slab.

In a test where the discharge point was midway between the axis and wall of the silo a significant tendency for a tricorn shape to develop was detected with one of the 'flat' sides coinciding with the point of discharge. The horizontal bending moments deduced from the total distortion were greater during this test than for the test in which the discharge point was close to the silo wall.

Measurements of silo shell radial movements were complicated by the effects of atmospheric temperature. Not only did the shell appear to expand and contract in response to average ambient temperature, it also appeared to distort in response to irregular temperature distributions, for example when the silo was partially in the shadow of its neighbour. The consequent bending moments may not have been considered in the design of silos and it is recommended that future silo designs should make an appropriate allowance.


## 5.0 ENGINEERING APPLICATION

The information described above has enabled maintenance to be carried out in a controlled, appropriate manner such that the silos should continue to provide reliable product storage and to enable measures to be taken to reduce the risk of problems arising and to control them when they do arise.

(i)    General Silo Surveys and Tests

Although some problems are specific, a pattern has emerged showing areas of concern which relate to the silos as a population. For example, the silos walls are generally in good condition but the pilasters and particularly the prestressing tendon anchorages have frequently required attention. Because of the low depth of carbonation, coatings are not required to improve durability and there is no concern about chloride-accelerated corrosion. Where severe vertical cracking has occurred in the walls of particular silos, measures have been put in hand to monitor the situation.

(ii)   Particular Concrete Surveys and Tests

The results of these surveys and tests at one particular site clearly showed that the as-formed strength and condition of the concrete was not as good as had been expected from cube test records. Moreover, tests were carried out in areas of competent concrete; vertical cracks seen on the outside and associated internal fracturing in the walls were an additional source of weakness. Because of this, the provision of additional hoop prestress (which would increase the maximum compression in the concrete) was not pursued.

(iii)  Full-scale Silo Operation Tests

These tests confirmed laboratory test work which has demonstrated that under prolonged eccentric unloading conditions the silo walls are subjected to bending. Although the tests were extremely severe and the silo did not show any signs of distress which would be expected to affect its long-term performance, eccentric emptying is now prohibited until the silos are at least half-empty.

Application of the work outlined above has been based upon theoretical analyses and upon pragmatic Engineering Judgements - the large population of silos provides a particularly good basis for making such judgements. However, it must be stated that certain events have occurred on particular silos which have not been fully explained in terms of local conditions. For this reason, measures such as the imposition of central emptying are seen as prudent ways of improving the long-term reliability of the silos.

6.0    CONCLUSION

A variety of testing techniques have been employed both to assemble general information and to gain a detailed insight into particular aspects of the silo population. The tests have been carried out to meet specific objectives and the techniques employed have successfully supplied the required information. Interpretation of the information has assisted decision-making for both immediate maintenance activities and for consideration of measures to promote long-term durability and reliability.

# 49 Insight on structural safety and age of concrete buildings in Saudi Arabia

**Dr Habib M Z Al-Abidien,** Assistant Deputy Minister of Public Works, Riyadh, Saudi Arabia
Associate Professor, King Faisal University, Dammam, Saudi Arabia

This paper outlines the common causes of distress in concrete buildings in Saudi Arabia which are most likely applicable in other developing countries. In the Arabian Gulf Countries, during the construction boom of the last 15 years, many buildings did not receive the necessary quality control in construction which affected their life expectancy. Examples are presented to demonstrate some of the causes and major findings are summarized from investigations of more than 400 cases. The paper concludes with recommendations to alleviate dangerous cracking and improve the longevity of buildings.

INTODUCTION:

There is somewhat an informal acceptance that the life of a concrete structure is about 50-100 years with the exception of large civil structures such as dams that have a longer life expectancy. However, in some developing countries many concrete buildings started to age much sooner. This indicates that the age of many concrete structures in some developing countries will be less than their counterparts in other countries where construction methods are better and more up-to-date, quality is more controlled and where the maintenance is regular and systematic. This means that in these developing countries the major causes of distress in structures are poor construction quality and lack of supervision and regular maintenance that avoids problems before it is too late. This paper highlights the causes of distress in concrete structures as detected in Saudi Arabia. Cracks observed are structural and non-structural (1), (2), and lead to buildings shorter in life than hoped for by the investor.

NATIONAL STANDARDS:

In some developing countries, there is generally a lack of complete sets of national standards that meet the requirements of a particular country. Therefore, designers as well as constructors resort to foreign codes and standards, mainly European or North American, which may not be suitable for the conditions that exist in a certain country. For example, the skill of labor and methods of construction may not be as good as it is expected by the code or standard writer. Also, it is a rarity to find local building regulations at the provincial (state) or municipality levels; these consider in more scrutiny the particular local conditions such as environmental factors, materials availability, skill of labor among other factors that influence construction quality. There are considerable efforts by the Saudi Arabian Standards Organization (SASO) as well as the Ministry of Public Works and Housing (3) to develop standards that meet the requirements of this country to form a group of national standards. Until these standards are made and approved by the proper authorities the construction industry in this country will continue using mixed foreign standards and codes.

CONSTRUCTION MISTAKES:

Some of the construction mistakes that are common in Saudi Arabia and most likely in similar countries are given here. Lack of proper supervision that is capable of avoiding the wrongs before they become worse is a primary concern. Quality control tests for concrete and its constituents and correct procedures in selecting components – especially aggregates Ref. (4)-, are given little attention. Proper mixing and casting, hot weather concreting and concrete technology in general, are very often considered as unnecessary academic luxury. Weak formwork and poor choice of pour breaks and construction joints can be added to the frequent construction mistakes. Storing construction materials and equipment without taking the necessary precautions or removing forms and/or shoring prematurely causes overloading of concrete at early ages. Precast concrete members get exposed to excessive stresses by wrong handling. Contractors sometimes excercise the freedom not to follow the instructions of the engineers. In many occasions they commit infringements to please the owner by increasing the number of floors, reducing the size of structural elements or changing their location without proper re-design. Legal aspects of construction supervision were the subject of another paper by the author (5).

Some of the mistakes cited above can lead to very weak concretes. In Saudi Arabia a specified cylinder strength of f'c= 21 MPa (cube strength of 25 MPa) is considered very good concrete; in some remote rural areas actual strength of concrete may be as low as 10 MPa. and even less, which definately affects the stregnth, durability of the structure and its life expectancy particularly where harmful environments exist.

BUILDING ABUSES:

Building abuses that are common in this country are classified into three categories, namely the lack of maintenance, abuses during construction, and after construction. The first category includes the lack of checking of cracks and spalling and their effect on possible corrosion. The leakage of sanitary pipes and defects in water insulating layers is generally a cause of concern (see Fig 5). Abuses during construction include many examples such as making up differences in elevations by increasing thicknesses through adding concrete or sand; this happens sometimes when changing elevations between washrooms and other rooms is not considered in the design stage. In existing structures where repair works are performed, contractors sometimes store materials and equipment on roofs and floors exposing structural members to loads that are not designed for. Vertical expansion without redesign and changing building function e.g. from residential to school or offices are two examples of abuses after construction.

Safety Factors: The simplest definition of safety factor in design is that it is the ratio of calculated failure load to service load. Fig. 1 depicts how the safety factor of 3.5 for a column almost diminshes when the service load doubles and concrete strength becomes half the specified strength .

ENVIRONMENTAL CONDITIONS:

One of the main problems for the construction industry in the Middle East and Gulf regions has been the degradation of concrete because of the aggressive environment found above and below grade. The superstructure suffers in coastal areas mainly from the attack of chlorides whereas the substructure (foundation) is subject mainly to sulphates attack in many regions of the Kingdom. It is notewortly that in Saudi Arabia, the Ministry of Public Works specifies sulphate-resistant cement in concrete below ground unless the soil tests prove undoubtedly otherwise or through a qualified written request from the supervising agency to use ordinary Portland cement. This is because the presence of sulphates in most regions is the rule rather than the exception. The problems become more serious and develop faster in poor quality concretes; humid and hot weather creates the suitable environment for carbonation and oxidation in the superstructure. The deterioration of foundations could be aggravated further by open sewage where septic tanks filter fluids into soils adjacent to the foundation elements (footing, grade beams, sub-columns ... etc.)

DESIGN SHORTCOMINGS:

It was mentioned before that design codes used in this part of the world are a mixture of foreign codes which may not be suitable for a particular country and prevailing construction conditions. Other examples of design shortcomings include the lack of adequate details, neglecting soil testing to determine the proper foundations and instead a commonly thought to be safe figure is used (e.g.200 $KN/m^2$ bearing capactiy). Improper choice of materials such as using high strength steel with very weak concrete does not give good results. In some buildings the appropriate designers are not selected; some structural designs are performed by an architect or a draftsman; in other cases an engineer that is not experienced in certain types of structures conducts the work.

EXAMPLES:

Example 1: Figs. 2,3 and 4 are photographs of a collapsed building that suffered from a lot of abuses. This mosque which was 6 years old, had many problems that multiplied and caused the failure. Weak concrete of columns, sub- columns and grade beams. The concrete was subject to chemical attacks underground. Leakage from the roof was repaired by adding new layers on top of the existing roof which increased the thickness of roofing from 120 mm to 350 mm and consequently added substantial dead load. During the repair process the contractor stored a lot of construction materials on top of the roof.

Example 2: Hopsital in a Coastal City:
The building was twenty years old and had many structural cracks and concrete spalling in beams, slabs and some columns. The reinforcement was exposed and badly corroded in several parts of the building Fig. 5 shows the bad piping connections. The investigation revealed that the building had very poor concrete, the sanitary piping system was leaking and the water proof insulation in the kitchen and washroom floors were badly placed. Periodical maintenance was neglected. This building was demolished at the young age of 20 years which is less than half the expected life of such investment.

Example 3: A housing development in Al-Khafji:
This newly established city located on the Arabian Gulf had residential quarters of approximately 500 villas of one or two-storeys. The buildings of this private development exhibited tremendous deterioration in very short time after construction, so that in five years many buildings were on the verge of collapse. The most serious defects included weak concrete strength and extensive steel corrosion of floor and roof slabs (Fig. 6). Concrete samples that were tested chemically indicated very high contents of chlorides, and sulphates which are prime catalysts in steel corrosion. Chemical tests conducted on soils showed high percentages of sulphur-trioxide and sodium chloride which are detrimental to foundation concrete especially low strength concrete and concrete without sulphur-resistant cement. It is noteworthy that this development which is considered a total financial loss was constructed by unqualified contractors. Soil tests were not considered prior to construction and no control on materials and construction was performed.

Example 4: Communications Institute in Riyadh; this complex of several buildings was constructed from 1977-1980; problems started in two years after completion. Ground water table rose to about 30mm above basement slab and remained for a year causing settlements in slabs on grade because of consolidation of the backfilling. Cracks in some grade beams were observed. The site is at lower grades than surrounding areas that donot have drainage network which made the water collect in the site. Defects observed were numerous; block walls had 45$^{0}$ cracks, floor settlements near landscaped areas, corrosion in sub-column steel from high levels of ground water (Fig. 7) as well as structural (shear) cracks in beams of the restaurant and gymnasium buildings.

The investigation showed that concrete strength was low, aggregates were suspected to have potential of low silica-alkali reaction and that the backfilling around foundations and underground water contained high percentages of sulphates, chlorides and organic materials. The design check of buildings whose cracks were structural proved that concrete strength was . lower than specified and shear resisting steel was inadequate. Studies proved that the project can be saved by repairs. Where shear cracks existed external stirrups were recommended in the girders of the restaurant and gymnasuim, subcolumns and footings were to be strengthened by reinforced concrete jackets and foundations insulated. It was also recommended to expedite connecting the area to the city sewage network and to stop watering landscaped grounds. Drainage of surface water to open channels was also proposed.

CAUSES OF CONCRETE CRACKING IN SAUDI ARABIA- SUMMARY:

The investigation of more than 400 cracked buildings in Saudi Arabia indicated that the most important causes of distress are the poor quality of concrete, poor workmanship and lack of proper supervision. Other problems cited are related to soils, underground water and open sewage; poor execution of roof and floor insulations, and bad connections in water and drainage pipes. Extreme environmental conditions such as very high temperatures, very dry, very humid weathers and detrimental salts come next in order of importance followed by design and detailing errors. These aspects have been dealt with in more details in Ref. (6).

Statistics: The two most important findings regarding the age and safety of investigated buildings are the following. Firstly most of the cracks occurred between the age of 5 and 15 years, which is an indication that the buildings deteriorated in much less time than the life expectancy for concrete structures. Secondly, 67% of the cases were repaired or rehabilitated whereas 18% were demolished; the remaining 15% were put under observation to determine condition with time.

RECOMMENDATIONS:

Based on the present study the following recommendations are drawn to alleviate possibilities of dangerous cracking and collapse of buildings in Saudi Arabia and similar developing countries; these measures aim to improve the longevity of buildings:

(a) Providing suprvising agencies and contractors personnel with required knowledge of good construction practices, and requiring owners and developers to provide their own engineer or consultant to assure the adherence of contractor to requirements. Making citizens aware of the importance of quality assurance to protect their investments.

(b) Specifying quality control tests in all contracts, and performing control tests in aggregate quarries and crushers, and ready mixed concrete plants.

(c) Limiting the changing of building functions especially those that cause load increase and certainly never storing construction materials on floors and roofs during construction or maintenance operations.

(d) Conducting soils investigation for each site.

(e) Building permits should not be granted by municipalities without having the structural design checked properly by the municipality or its assigned counsultant who should also conduct periodical inspection of construction.

(f) Preparing national standards for construction and national codes for design suitable for the country or until these are available, well established standards of other countries may be adopted with adaptation for local conditions.

(g) Strength evaluation of structures and assessment reports must be approved by specialized governmental agencies or consultants nominated by such agencies.

ACKNOWLEDGEMENT:

The example investigations described in this paper were conducted by the Ministry of Public Works and Housing, Riyadh, Saudi Arabia. The efforts of the Ministry staff are gratefully acknowledged. Thanks are also due to Dr. Magdi Khalifa for his help in preparing the text.

REFERENCES:

(1) American Concrete Institute, Committee 224 'Causes, Evaluation and Repair of Cracks in Concrete Structures' ACI Journal, (May / June 1984) PP. 221-230.

(2) Concrete Society, 'Non-Structural Cracks in Concrete', Technical Report No. 22 (1982) Concrete Society U.K.

(3) Ministry of Public Works and Housing 'General Specifications for Building Construction' Riyadh, Saudi Arabia (1982).

(4) Al-Abidien, H. M. Z., 'Aggregates in Saudi Arabia: a survey of their properties and Suitability for Concrete', Materials and Structures, RILEM, (1987), 20, pp 260-264.

(5) Al-Abidien, H.M.Z. 'Towards Legalization of Construction Supervision' Ministry of Public Works, Riyadh, Saudi Arabia (1981)

(6) Al-Abidien, H.M.Z. 'Safety Assessment of Concrete Structures', Obekan Company for Printing and Publishing, Riyadh, Saudi Arabia (1987) (in Arabic).

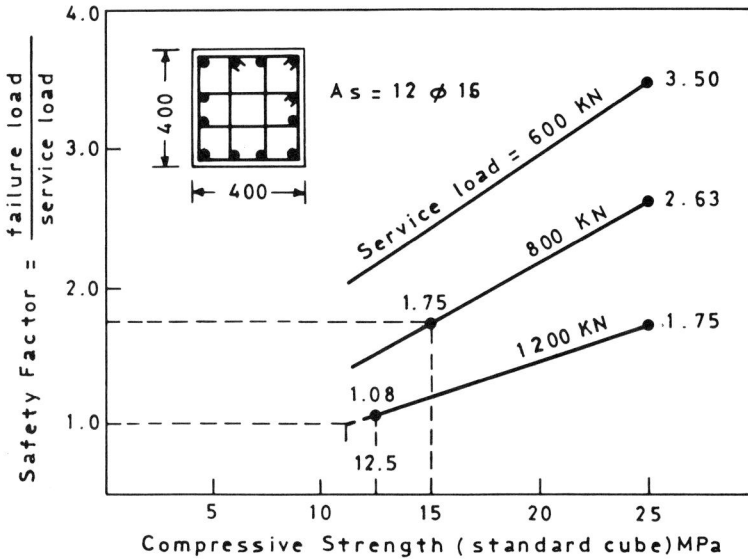

Figure 1    Safety Factor vs. Concrete strength at different
            Service Loads.

Figure 2    Mosque roof collapsed
            on ground

Figure 3    Mosque slab x-sec.
            showing added layers

431

Figure 4   sub-columns, reinft.
           corroded & buckled

Figure 5   Hospital with bad
           sanitary connections

Figure 6   Villa slab with extensive
           corrosion.

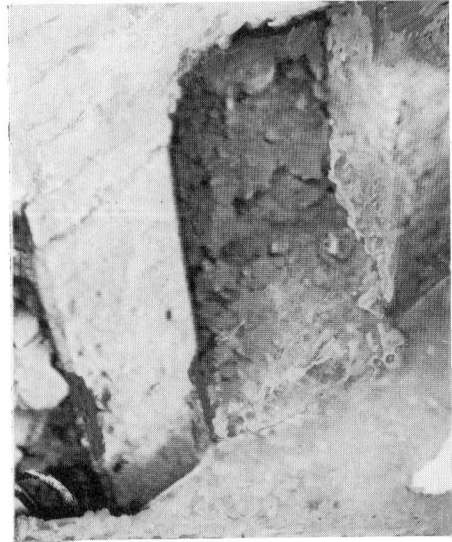

Figure 7   Ex. 4, foundation
           submerged.

**T M Chrisp, Mott MacDonald Group, UK**

The author has shown some very interesting slides and spoken with great openness of the problems of structural deterioration in Saudi-Arabia. He has indicated that a major element in these problems is inadequate construction supervision compounded by the lack of site quality assurance procedures.

In recent years the balance of responsibility for site supervision had changed, for example with the introduction of design and construct contracts. Does the author consider that this is a beneficial move for the industry? Could improved contract provisions and properly funded site supervision and testing play an important part in achieving better construction?

## Author's reply

I would like to draw the readers attention to the fact that this paper reports the unfortunate cases of construction where problems appeared. These represent a low percentage of the construction volume that was conducted in the same time period. The Kingdom of Saudi Arabia has many buildings that were professionally designed and properly constructed. Therefore it should not be construed that the paper represents the majority of concrete structures in the Kingdom.

The design-build concept in contracts is not applied in Saudi Arabia except in some complex large projects.

Construction regulations in the Kingdom require that the contractor must check the design adquacy before he builds which, if implemented properly, would benefit the contract. Improved contract provisions and properly funded site supervision and testing are very important, however, I would like to add that these would be more effective if conducted by indepedent organizations.

# 50 The future of Orlit houses in Northern Ireland

**D J Cleland,** The Queen's University of Belfast, UK
**D Lavery,** The Northern Ireland Housing Executive (now with University of Ulster), UK
**G I B Rankin,** The Queen's University of Belfast, UK

This paper describes the construction of Orlit precast concrete houses in Northern Ireland and the methods of assessment which have been employed to determine their condition. A test programme to examine the effect of loss of strength of HACC connections and of corrosion related damage on the performance of the concrete frames is reported. The paper also discusses how these factors have influenced an assessment of the remaining life of these structures.

## INTRODUCTION

### History of Problem

In the aftermath of the second World War there was considerable pressure for Local Authorities to provide new housing in large numbers over a short period of time. In view of this pressure and a shortage of skilled labour precast concrete housing systems were seen as a natural development. Several systems were developed by competing companies; easily the most common in Northern Ireland were the Orlit houses.

Orlit Northern Ireland Ltd. operated from 1945 to 1959 and for another five years through a receiver. The houses were of two types:
  (i)     Single storey 'Ulster Cottages'. These buildings were constructed with precast concrete storey-height panels
  (ii)    Framed Houses. Generally these were two storey but some apartments of up to four storeys and some single storey variations were also built.
This paper concentrates on those buildings in the second group.

In the early 1980's two problems with Orlit houses were uncovered. Firstly the insitu connections were found to have been frequently formed with High Alumina Cement Concrete (HACC) and secondly corrosion of reinforcement in the precast units was found to be present. Initially it was thought that the HACC problem was the most serious difficulty. This thinking was to change, however, as it began to emerge that the carbonation and chloride problems were sufficiently serious to bring into question the 30 year life span on which housing improvement calculations are normally based.

An initial report (1) completed in 1983 outlined the problems. The 1769
dwellings, owned by the Northern Ireland Housing Executive (NIHE), had
an estimated 'defect-free' value of £22m and replacement costs of the
order of twice that figure. The dilemma was exacerbated by the fact
that at least £8m had been spent on improving and upgrading Orlit
estates in the previous years and more schemes were in the pipeline.
The point at which the Orlit dwellings would cease to be serviceable
was, therefore, of interest in deciding a course of action.

Categorisation seemed feasible. It was decided that the category would
be indicative of the general condition of the buildings in the estate in
question and that they would be related to a defined period of time
during which the buildings would remain serviceable. This
categorisation would be greatly influenced by BRE's guidance (2,3).

## Form of Construction

In all the cases encountered in Northern Ireland the columns were
prefabricated in three sections. The upper sections of column were cast
integrally with a short length of first floor and roof main beam. The
three sections were bolted together as shown in Figure 1.

The columns are arranged in groups of three so that when connected to
the first floor and roof beams they form 2-bay, 2-storey portal frames
(Figure 1). The main beams spanning between the prefabricated column
heads are also prefabricated reinforced concrete units. When viewed in
section they are an inverted 'T' shape (or 'L' shape in the case of
gable beams). The beams are connected to the beam stubs on the columns
by means of an insitu concrete splice. The nibs of the main beams
support secondary beams placed at about 1000mm centres which span
longitudinally, i.e., parallel to the front and rear walls, between the
portal frames (Figure 2).

Timber joists and tongue and groove boarding (or 50mm thick concrete
slabs with a foamed slag screed in the case of apartments) span between
the secondary beams. Some of the houses have flat roofs constructed in
the same way as the floors and covered with felt. However many estates
have pitched roofs of timber purlin and rafter construction. The
external walls comprise two 50mm skins of precast units with a cavity in
excess of 150mm to accommodate the columns. In a few cases the internal
skins to the cavity walls are of ordinary blockwork.

## Ownership

There are twenty-three estates of two-storey Orlit houses and apartments
in Northern Ireland. Twenty-one of these came into the ownership of the
NIHE in 1973; 1913 dwellings in total of which about 144 were
subsequently sold to the occupiers. The estates vary in size
from 10 to 310 dewllings and are distributed throughout Northern Ireland.

## TESTING

When the Scottish Special Housing Association (4) made known their
concern about the quality of the HACC connections in October 1982
immediate action was taken to identify if such problems existed in

435

Northern Ireland. During that investigation it became apparent that Calcium Chloride and carbonation were also causes for concern and therefore a comprehensive investigation was initiated.

## Precast concrete elements

It was decided that if any attempt was to be made to define the extent and the nature of the problem further it would be necessary to remove ceilings in a sufficient number of houses to allow the problem to be properly appraised. A sample size of 5% of the Orlit dwellings was chosen but random sampling was not practical. Opportunities to inspect the frame only arose during vacancies caused by re-letting or during improvement work. In some estates where the turnover of dwellings was low the cooperation of selected tenants was necessary and therefore in some estates the sample size was not achieved but to compensate, in other estates more than 5% were inspected.

A standard survey report form which reflected the major areas of interest in the frame was used. It is divided into two parts; one part for recording detailed information on elements within the structure and the other a visual survey of the structure as a whole.

In all 98 houses were surveyed and in these 240 secondary beams, 147 main beams and 53 columns were inspected in detail (i.e., cover, depth of carbonation, etc. measured). Histograms of the carbonation and cover measurements for all elements are shown in Figures 3 and 4 respectively. These show that in 10% of cases carbonation has occurred at a depth of 30mm or more and in 25% of cases the carbonation depth exceeded 20mm. At the same time in about 28% of cases the cover to the reinforcement was less than 20mm. Some of the information is detailed in Table 1.

Table 1: Carbonation and Cover. Secondary and Main
beams. (As measured from the bottom of the beam)

|  |  | Av.Depth | Max. | Min. | Av.K.Value | Sample |
|---|---|---|---|---|---|---|
| Secondary Beams | Carbonation | 17.6 | 70 | 1 | 2.89 | 224 |
|  | Cover | 30.5 | 50 | 5 | − | 164 |
| Main Beams | Carbonation | 16.0 | 55 | 1 | 2.60 | 144 |
|  | Cover | 24.3 | 52 | 2 | − | 115 |

Currie (5) suggests an average depth of carbonation for Orlit housing of 13mm. The figure from the present study is about 17mm for all elements or 18mm for secondary beams. The K factor quoted in Table 1 is a permeability constant in mm/yr based on the idea that carbonation progresses at a rate proportional to the square root of time. This, too, is greater than the value quoted by Currie. There are two further factors which combine to indicate that the outlook for secondary beams in particular is not good:

(i)     If the bottom cover exceeds the mean value (approximately) the cover to the top of the bottom flange decreases to dangerous levels.

(ii)    Peaks in carbonation penetration occur at transverse cracks present in large numbers.

The distribution of Calcium Chloride concentration is shown in Figure 5 for all structural elements in the estates studied. It can be seen that the distribution is highly skewed with a mode value of 0.44% and a mean value of 0.49% chloride ion by weight of cement. An analysis of each of the element types, i.e. secondary beams, main beams and columns is summarised in Table 2. Again secondary beams are most critical - about 11% of the results were in excess of 0.6% chloride. This combined with low alkalinity due to carbonation would indicate a high risk of corrosion inside 25 years (2,3); a point borne out by the recorded incidences of corrosion.

Table 2: Chloride Levels. (All Elements, all Estates combined).

|  | Mean | Max. | Range | Mode | Sample |
|---|---|---|---|---|---|
| All Cases | 0.488 | 2.41 | 2.38 | 0.44 | 416 |
| S'ry Beams | 0.514 | 2.41 | 2.36 | 0.44 | 216 |
| Main Beams | 0.447 | 1.91 | 1.88 | 0.38 | 133 |
| Columns | 0.476 | 1.17 | 1.10 | 0.35 | 49 |

From all the examinations of reinforcement condition 8% were considered to be severely or very severely corroded and a further 18.5% were classified between moderate and severe.

Visual Survey

A summary of the visual survey of secondary beams is presented in Table 3. Clearly transverse cracking is easily the most prevalent defect with about 74% of secondary beams having at least one transverse crack. This type of cracking is most likely to have resulted from the casting and curing conditions. The side cracking, usually at the junction of the web and bottom flange, is associated with corrosion of the reinforcement. Three beams exhibiting some of the worst longitudinal cracking and spalling were removed for load testing in the laboratory.

Table 3: Number of secondary beams with visible defects

| Examined | cracked or damaged | with transverse | longitudinal soffit | side | rust staining | No. with spalling |
|---|---|---|---|---|---|---|
| 1085 | 828 | 808 | 50 | 83 | 49 | 25 |

Load Tests

The three secondary beams were tested in flexure and in shear and were found to satisfy the requirements of CP110:Part 1:1972 (6) in relation to recovery of deflections and had ultimate strengths at least 85% in excess of working load. Chloride contents measured at six locations in each beam were very variable but mean values for the three beams were 1.50%, 1.31% and 2.20%. The most severely cracked beam had the lowest

437

ultimate strength (marginally) and was found to be carbonated around the main reinforcement over the full length of the beam. Not surprisingly the reinforcing bar was surrounded by 1-2mm of corrosion scale.

## Insitu HACC Connections

Twelve estates were found to contain HACC connections. Since it was known that in practice the water/cement ratio of the insitu concrete varied considerably it was decided to use the pull-off test (7) to assess the insitu strength and the estimate of conversion was found by Differential Scanning Calorimetry.

In all 220 connections were examined in three surveys carried out at different stages (in one survey only conversion was measured). In most cases the pull-off test was used on the surface and after coring to eliminate any 'hard shell' effect from the results. Table 4 summarises the findings from which it can be seen that there is indication of an increase in conversion and decrease in strength between the time of the initial testing in 1982/83 and the more recent test programme in 1986/87.

Table 4   Mean conversion and strength of HACC connections

|  | 1982/83 | 1985/87 |
|---|---|---|
| Conversion in % | 71.6 | 80.5 |
| Equivalent compressive strength (surface) in $N/mm^2$ | 16.6 | 7.0 |
| Equivalent compressive strength (cored) in $N/mm^2$ | 7.7 | 5.5 |

## Load Tests

Load tests were carried out on main beams in three houses (in three different estates). Load was applied in the form of four point loads equally spaced along the span and the response measured in terms of mid-span deflection. The relative performance of each of the beams and the strength of the connections is shown in Figure 6.

The load/deflection relationships are complimentary to the connection strength with the lowest strength connections corresponding to the greatest deflection or poorest performance. There is some hint of a significant difference in performance when the compressive strength of the connections is less than about 10N/mm though even with very low strengths ($< 5N/mm^2$) the beams still satisfied the requirements of CP110:Part 1:1972 (6).

## CATEGORISATION OF ESTATES

Several studies were made to examine the viability of repair systems. The cost of essential repairs and rehabilitation, even using the systems specifically developed for precast concrete houses, was found to exceed the permitted yardstick; in some cases costing almost as much as a new house in Northern Ireland. Repair was therefore precluded as an option and so to obtain the best value for public investment a realistic assessment of the useful life remaining for each estate was desirable.

438

It was decided that a three tier system of categorisation would be appropriate. This would represent periods of remaining useable life:

Category A    15 years or more
Category B    10-15 years
Category C    0-5 years

In coming to a conclusion about the category of any estate a number of factors were taken into account. In order of significance these were:
  (i)    The condition of the precast concrete frame elements; particularly the secondary beams of which there are 28 in each house.
  (ii)   The age of the estate.
  (iii)  The existence and condition of HACC connections.
  (iv)   The condition of the external envelope.
  (v)    The presence of a flat roof.

Social conditions on the estate were not considered. Four estates were categorised as having 15 years or more useable life while another four were placed in Category C. The remainder were considered to be in the intermediate range, ie. Category B. The process was comparative and, to that extent can be substantiated. However the actual state of debilitation at any time in the future is less certain and the further predictions are in the future the less certain they will be. On the basis of experience gained over the past six years it is possible that they are unduly pessimistic.


CONCLUSIONS

On the basis of the fairly extensive survey of the Orlit housing stock the following conclusions can be drawn.
  (i)    Although the precast concrete elements are susceptible to reinforcement corrosion and many exhibit signs that this has already taken place there is no evidence that any of the estates present a hazard to the occupants at this time.
  (ii)   Where HACC has been used for the insitu connections in the frames the residual strength is low. This appears to affect the stiffness of the frame/beam but at this stage the overall frame capacity appears to be adequate, though in some cases only just.
  (iii)  The findings tie in with other work on the risk of corrosion in relation to chloride content and carbonation etc. However prediction of the rate of deterioration in the future is difficult since there is insufficient evidence on the performance of similar structures over 40 years old.
  (iv)   The present stock of Orlit houses will continue to be monitored and in due course more realistic categorisation of remaining usable life may be possible.


REFERENCES

1.   Lavery D., 'Internal report to NIHE Board identifying the existence of problems in Orlit Houses,' (Feb. 1983).

2.   Building Research Establishment, 'Durability of steel in concrete: Part 1 mechanism of protection and corrision,' Digest

No. 263. (1982).

3.    Building Research Establishment. ' Durability of steel in
      concrete: Part 2 diagnosis and assessment of corrosion - cracked
      concrete.' Digest No. 264. (1982).

4.    Scottish Special housing Association. 'Orlit House types:
      Guidence notes on concrete stitches.' (1982).

5.    Currie, R.J., 'Carbonation Depths in Structural quality concrete:
      an assessment of evidence from investigations of structures and
      from other sources.' IBRE (1986).

6.    British Standards Institution. 'CP110: Part 1: 1972 The
      structural use of concrete.' (1972).

7.    Murray, A.McC., and Long, A.E., 'A new method of assessing the
      strength of 'suspect' concrete.' Proc. Conf. on Structural
      Faults, Edinburgh. (1983).

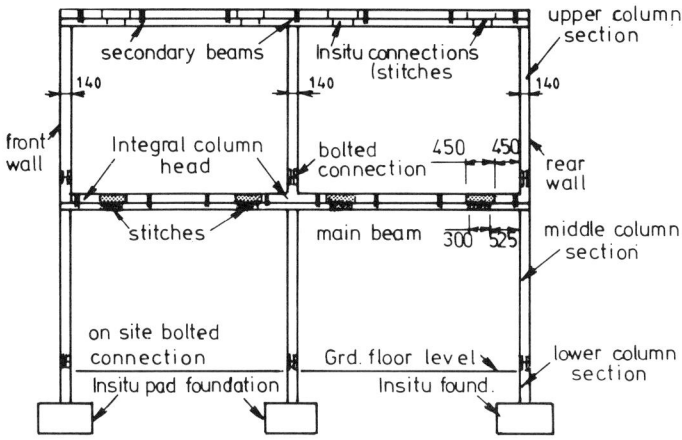

Figure 1 :   Typical Elevation of 2 Bay 2 Storey Portal

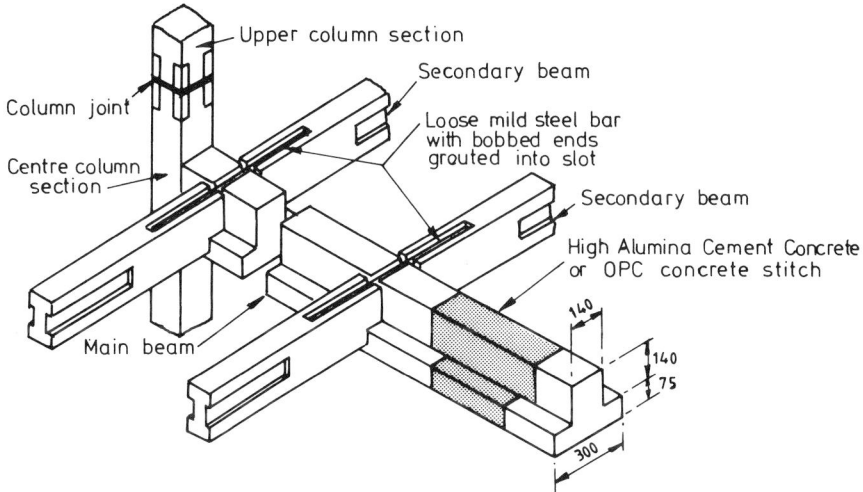

Figure 2 :   Detail of Secondary/Main Beam Connection

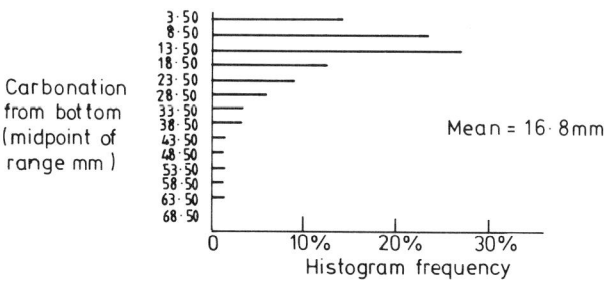

Figure 3   :   Carbonation in all elements (measured from bottom)

441

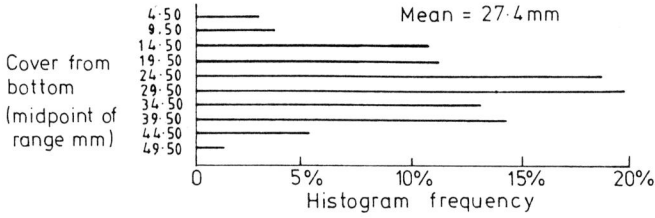

Figure 4 : Cover in all elements (Measured from bottom of section)

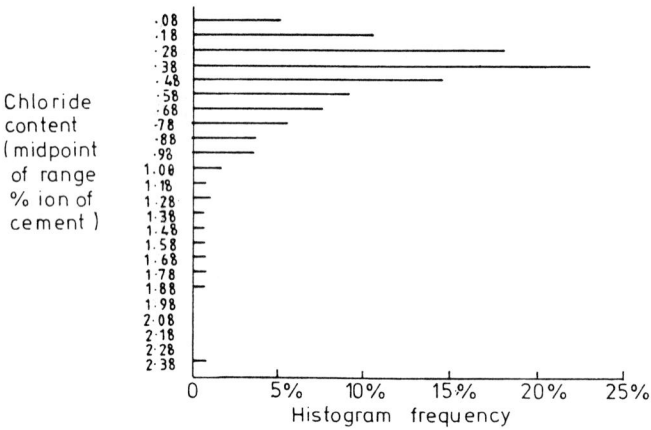

Figure 5 : Calcium Chloride Concentration (All Structural Elements)

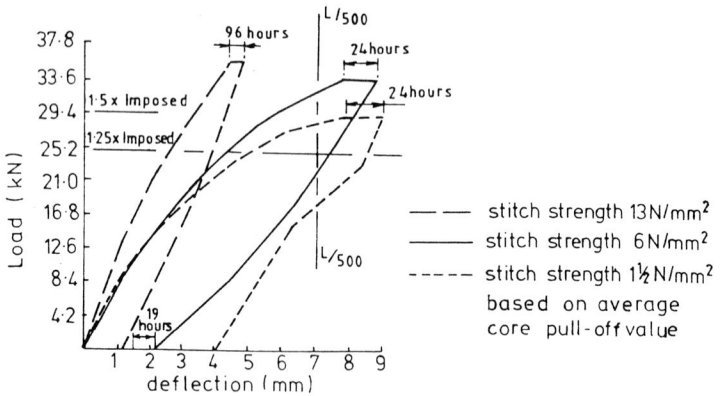

Figure 6 : Load & Central Deflection — Comparative Profiles

## J F A Moore, Building Research Establishment, U.K.

The author could perhaps cite the following relevant reference: 'The Structural Condition of Orlit Houses', BRE Report 1983. The data in this report, and that of Currie (author's reference 5), apply, essentially to large populations of dwellings and buildings generally, having wide geographical distribution and rather small sample size. It is to be expected that the relatively large sample from a closely defined population described in that paper might give results differing from the generality.

A final comment, not made at the meeting, is that no repair schemes are licensed by BRE-see section 'categorisation of estates', line 3. BRE is, of course, a member of the Assessment Committee of PRC Homes Ltd.

## Author's reply

Mr Moore is indeed correct. The purpose of the comparison was to determine to what extent, if any, the specific population of Orlit houses in Northern Ireland differed from the norm for all regions of the United Kingdom as reported in references (5) and (8).

8.  Building Research Establishment. 'The Structural Condition of Orlit Houses'. 1983.

# 51 PRC houses–the SSHA experience

**T Mitchell,** BSc, CEng, MICE, MIHT, MCIOB, The Scottish Special Housing Association, Edinburgh, Scotland

This paper describes the P.R.C. house types owned by the S.S.H.A. and currently listed as being defective under the Housing Defects Act, 1984. It gives an account of structural and material investigations carried out to determine high alumina cement in insitu concrete stitches, in 'Orlits',harmful chloride ion levels and carbonation depths in p.c. concrete components. The findings of a monitoring programme commenced in 1983 for the Orlit houses are discussed with a current assessment of the future service life against the prediction given five years ago.

WHAT IS S.S.H.A?

The Scottish Special Housing Association (hereinafter referred to as 'SSHA', or the Association) is a Government-sponsored body which designs builds and manages houses for the people of Scotland on a non-profit making basis. Set up in 1937 to provide houses in the Special (or "Distressed") areas, the Association's function was extended post war to provide houses for any local authority requiring assistance with their housing problems. In April 1989, the SSHA will merge with the Housing Corporation in Scotland to become 'Scottish Homes', a new government agency.

INTRODUCTION

The post war housing shortage resulted in government policy which promoted initiative in the precast concrete industry for the design and manufacture of houses based on production line principles. Site erection could be carried out by semi-skilled labour and completion times were greatly reduced.

However, research, prototype testing and contract production were carried out simultaneously and defects have become apparent after some thirty years of serviceable life. These defects have required building owners to become involved in costly structural repairs and in some cases, demolition has been carried out. The housing in this category is mainly in public ownership but in accordance with current legislation, many of these one-time rented houses are now owned by their former tenants.

444

The government has recognised the plight of these owners by introducing the Housing Defects Act 1984 which allows local authorities to buy back or grant the owners a sum not exceeding 90% of the cost of an approved repair.

## S.S.H.A.'s P.R.C. HOUSES

### Tarran - Clyde

This house type (known as the Newland in England) was built usually in the two storey semi-detached form with pitched roof. The prc structural elements are 400mm wide storey height, tray shaped panels, with prc corner columns. The structure is braced laterally at first floor with steel channel sections in external walls supported on tubular steel spine columns. The SSHA originally owned 247 on three sites.

### Unitroy

These were built in Central Scotland, in both two storey semi-detached and terraced forms with pitched roof. The prc structural panels are 900mm long,450mm high, laid with continuous vertical joints. The flanged ends of each panel are grooved to allow the casting insitu of a reinforced concrete column. The SSHA owned 168 on four sites.

### Whitson Fairhurst

These houses are constructed in both two storey semi-detached and terraced forms with pitched roof. The prc structure consists of 150mm square columns supporting p.c. beams on front and rear elevation with a spine frame running parallel supporting timber first floor and roof. Transverse p.c. beams connect these frames at first floor and roof level. The columns and beams are concealed in a cavity between either 75mm concrete blocks or external leaf of clay brick. The SSHA owned 626 of this type on six sites.

### Orlit

These houses are built in both semi-detached and terraced two storey cottages and flats, some authorities have three and four storey flats. The structural form is shown in figures No. 2, 3 and 4. The roofs are either of pitched timber construction or flat with p.c. concrete slabs as illustrated. Ground floors are either solid concrete or hung p.c. slabs as is the case in most SSHA stock.

## INVESTIGATION

### High Alumina Cement

'The SSHA Experience' commenced in 1980 after the reported failure of an insitu concrete stitch joint in the main frame of a local authority owned Orlit house in Edinburgh. After examination and laboratory analysis of a concrete sample from the stitch, it was confirmed that high alumina cement concrete (HACC) has been used and that the conversion factor was approximately 65%. The stitch was located at roof level. There was evidence of water penetration through the precast concrete slabbed flat roof at this location.

At that time there were an estimated 6000 Orlit houses located throughout Scotland. The potential seriousness of this failure and the possible reprecussions prompted the Scottish Development Department's Housing Research and Development Division to form an ad hoc committee to investigate the problem on a national scale. The SSHA, as a member of that committee, owned 1116 of these houses on 8 sites located throughout Scotland.

A programme of inspection and sampling concrete stitches was quickly carried out on 66 houses, 45 of which were unoccupied.

The samples were subject to Differential Thermal Analysis (D.T.A.) at Edinburgh University which confirmed that high alumina cement was present in 38 of the houses inspected and that the average degree of conversion was 54.6%, the highest being 60%.

All the stitch joints in one of the houses (i.e. 24 stitches) were subject to Ultrasonic Pulse Velocity Tests (UPV). This enabled the strength of the HACC to be determined relative to the strength of the Portland Cement Concrete (PCC) used in the precast concrete beams, adjoining the stitch. Because the stitch was congested with reinforcement, no core could be obtained to calibrate the UPV reading, however the relationship showed that the PCC was approximately three times stronger than HACC. Assuming the PCC to have a cube strength of $30N/mm^2$, then the strength of the HACC was approximately 10 $N/mm^2$.

Difficulty was experienced in checking the structural design of these pc concrete frames and it was confirmed in the Post War Building Studies Publication No. 25(1) that empirical methods were used to establish the structural sections which were proved by prototype testing at the Building Research Establishment.

Because of the uncertainty of their current serviceable condition, it was decided to subject the structure of three different houses to full scale load testing in accordance with clause 605 of CP 144:Part 2:1969. The roof beams of a portal frame were loaded in two separate houses, one with HACC stitches, the other with PCC stitches. The first floor beam in the third house on a different site had the HACC stitches.

The results of the test on the roof beams showed that both behaved in a similar manner, deflections were small, recovery was satisfactory and there were no visible signs of distress. The maximum deflection in the first floor beam was one third of the allowable. A stitch appeared to rotate slightly and lock which prevented both beams attaining full recovery on removal of the test load, however, since the deflections were so small,the test was deemed to have been satisfactory.

The report on the investigations into the HACC problem concluded that the stitches were structurally satisfactory but that every effort should be made to protect HACC stitches from water penetration which increases conversion and reduces strength.

In compliance with the recommendations and also for aesthetic reasons, the SSHA included in its modernisation programme, the superimposing of the flat roofs with timber pitched roofs, the additional loads being transmitted directly to the columns.

## Chloride and Carbonisation

The report of the ad hoc committee based on the Association's findings was passed to the Building Research Establishment (BRE) at Garston and after further investigation by BRE staff, a circular was sent by SDD in September 1982 to all public housing authorities in Scotland warning them of the potential HACC problems and also drawing attention to the possibility of high levels of chlorides in the P.C. concrete frame, found on a site in England by BRE staff.

The Association had already embarked on modernisation contracts for this house type and some were partially complete.

Further investigation of the Orlit stock commenced in conjunction with the SDD and BRE. Access to pc concrete main frames was more difficult as it necessitated stripping internal linings in void and occupied houses. Initially 141 samples from main frame components in 30 houses were analysed for chloride content, the results are shown in table 1.

Table 1 Chloride ion test results (Orlit main frames)

| % chloride ion # | Risk Category * | No. of Samples |
|---|---|---|
| less than 0.4 | Low | 59 |
| 0.4 to 1.0 | Medium | 69 |
| exceeding 1.0 | High | 13 |
| | Total No. of Samples | 141 |

# relative to weight of cement
* refer to BRE Digest No. 264 (2)

Carbonation depths detected using a phenolphthalein solution varied up to 30 mm in some cases. Visual inspection of the main frames showed beams with various degrees of cracks, spalling, corrosion and scaling of exposed reinforcement. BRE staff visited the sites and witnessed the problem which they pursued with other authorities throughout Britain.

Ceilings in void houses were stripped and it was evident that the smaller sectioned secondary beams were also affected by chlorides and carbonation. Some were badly cracked, others had spalled and exposed scaling reinforcement could be seen, see Fig. 1.

Feed back to BRE from SSHA and other authorities promoted the Government's Minister of Housing and make a statement in Parliament in February 1983 on precast framed and panelled houses of the post war era, warning of the potential problem.

It was concluded that many Orlit houses had a probable further service life of between 5 and 15 years but this was difficult to assess.

The Association decided that investigations would include the removal of all ceilings at first floor and roof level in flat roof constructions but this could only be carried out in unoccupied houses and those undergoing modernisation. Where secondary beams were not expected to give a further 10 to 15 year serviceable life they were removed in the modernisation contract and replaced with either timber or steel sections.

However, the occupied houses could not be inspected in similar way and it was decided to carry out the inspections of the main frame and secondary beams by borescope inspection. On several sites, columns could not be examined by this method because cavities had already been foam filled. However, so far, although containing chlorides and having varying depths depths of carbonation penetration, those seen had shown no signs of distress.

A grid system was set up with pro forma inspection sheets for each ceiling (pitch roof beams were inspected from roof space). Approximately twenty five 12mm diameter inspection holes were drilled in each ceiling to allow main frame and secondary beams to be viewed on elevation and all defects were recorded in accordance with a pre-arranged code. All secondary beams found to be potentially dangerous in the immediate future were removed and replaced.

Every occupied Orlit was inspected using the borescope method. At first difficulty was experienced in interpreting the view through the borescope but confidence was aided by stripping some ceilings after viewing to compare the results. The viewing holes were sealed with removeable plastic grommets which enabled easy access for further inspections.

Borescope inspections have been carried out annually. However in 1985 and 1986 only partial surveys took place on some sites to check those previous sightings which indicated more advanced signs of deterioration. Some further replacements of secondary beams have been necessary.

The other three PRC house types have undergone thorough investigation for defective concrete components. However, fortunately, none were found to contain harmful levels of chloride ions and carbonation depths were within acceptable limits, therefore deterioration from these conditions was not present.

Prestressed Concrete Floors:

It is now SSHA policy that prior to any modernisation contracts being prepared, a comprehensive examination of all concrete components is carried out.

Recently, during the structural examination of a four storey terraced block of traditionally built maisonettes, a hollow prestressed concrete floor at second floor level showed signs of abnormal deflection. These floors were constructed with 130mm thick prestressed concrete wide slab units. Samples were taken from every wide slab unit and high levels of chloride ion were found in the units, See Table 2.

448

Table 2   Chloride ion test results (Prestressed concrete units)

| % chloride ion [#] | Risk Category [*] | No.of Samples |
|---|---|---|
| less than 0.4 | Low | 29 |
| 0.4 to 1.0 | Medium | 72 |
| greater than 1.0 | High | 3 |
| | Total No. of Sample | 104 |

Greater than 0.8                                    17

# relative to weight of cement
* relative to B.R.E. Digest No. 264 (2)

Removal of the units was specified in the modernisation contract and every
unit with chloride ion exceeding 0.8% was broken to expose the single
strand type prestressing wires.   Pitting corrosion was evident at many
locations along the length of the wires although there was no evidence of
cracking on the exposed soffits.   Carbonation tests indicated penetration
of approximately 3 to 5 mm.   These maisonettes were constructed in 1966.
This type of defect is potentially more dangerous than in reinforced
concrete because of the possibility of sudden collapse.

Reinforced concrete lintels:

These components have also been found on many sites to contain high levels
of chloride ions and are affected by carbonation penetration.   Many show
visual signs of deterioration and have been patched during previous
maintenance inspections not knowing the reason for the defects occurring.
However concrete sampling and testing have confirmed the presence of
harmful chloride and these defective units are either replaced or patched
using approved remedial systems.

CONCLUSIONS

The Tarran-Clyde, Unitroy and Whitson Fairhurst house types were found to
be structurally satisfactory with an estimated further service life of at
least thirty years.   These houses are currently being modernised and many
have been sold.

The Orlit houses are structurally satisfactory, but require annual
monitoring because deterioration caused by harmful chlorides and
carbonation is continuing although slower than originally anticipated.
However, the degree of deterioration depends also on the location of the
component and on the age of the houses, with the older sites showing more
advanced levels of deterioration.

Where ceilings were stripped and defective beams replaced, a further 10
to 20 years serviceable life is predicted,but on sites where inspection has
been by borescope giving only lateral viewing,the original prediction of
15 years may be more accurate.

Two of the Association's Orlit sites are currently being demolished and new housing to suit the current needs is being constructed. These sites were due for modernisation but it was decided that this demolition policy achieved better value for money in the longer term.

The recently found problem with the prestressed floor units gives greater immediate cause for concern. A programme of inspection and sample testing from prestressed concrete floors is being carried out on all SSHA properties, prior to preparing modernisation contracts or sooner if abnormal deflections are observed.

Deterioration of lintels is being treated in a similar manner with replacement as necessary especially where window or door replacement or external insulation and rendering contracts are being prepared.

Table 3  General Information on SSHA Orlits

| Site Locations | No. of Houses | Site Completion Date | Current Status |
|---|---|---|---|
| IRVINE | 112 | 1953 | Modernised |
| AYR | 110 | 1952 | Demolished and Rebuild |
| HAMILTON | 130 | 1950 | Modernised |
| ABERDEEN | 176 | 1949 | Demolished and Rebuild |
| DUNDEE | 54 | 1948 | Part Modern-ised |
| KILMARNOCK | 50 | 1952 | Demolished |
| GLASGOW (North) | 292 ) | 1946–47 | Modernised |
| GLASGOW (South) | 202 ) | | Modernised |

REFERENCES

(1)  Post War Building Studies No. 25
     House Construction Third Report: London (1948)
     Published by H.M.S.O.

(2)  B.R.E. Digest No. 264   (August 1982)
     The durability of steel in concrete : Part 2
     Diagnosis and assessment of corrosion cracked concrete.

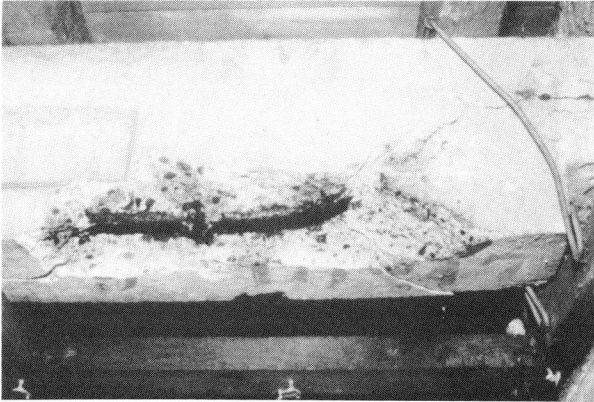

Figure 1 Defective Secondary Beam

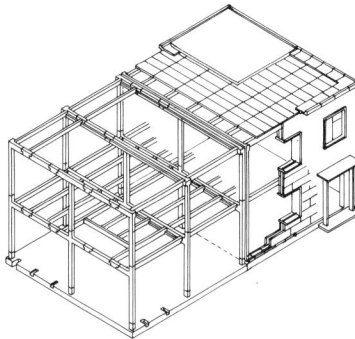

Figure 2 Orlit – General Construction

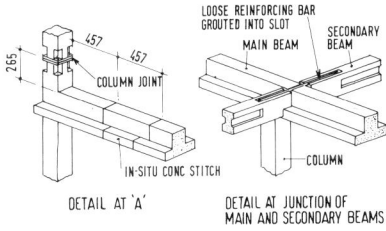

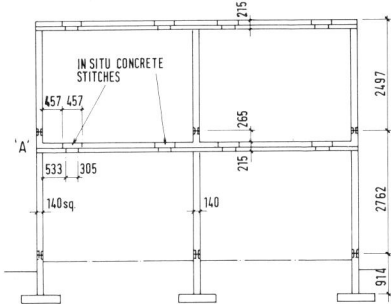

Figure 3 Orlit Details

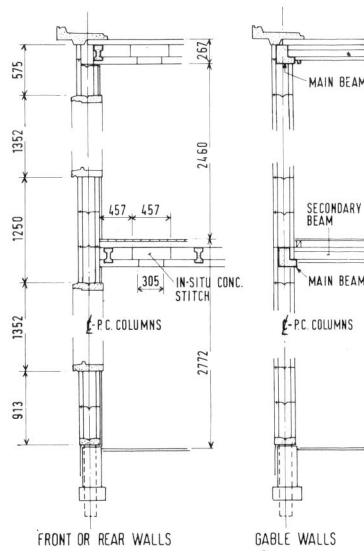

Figure 4 Orlit Details

S F Ray, Bingham Cotterell, UK

The author mentioned that secondary beams were the worst affected elements and that flat roofs were replaced by pitched roofs. Please would the author confirm if the roof beams were more seriously affected by reinforcement corrosion than the first floor beams. This might be the case because rainwater ingress was possible at roof level whereas condensation is perhaps the only source of moisture at the first floor level.

Please would the author also confirm if chloride attack or carbonation was the main cause of reinforcement corrosion at each level and if ways of preventing condensation reaching the members of the first floor level, where necessary, had been considered.

Author's reply

1.  Secondary beams were found at three levels in some orlit houses.

    (a) Those which supported p.c. concrete suspended ground floors were found to be the most seriously affected by corrosion of the reinforcement. The solum under the floors was not sealed thus the void contained below the floor area was very damp and unvented.

    (b) The secondary beams at first floor were least affected and also carried the least load.

    (c) The secondary beams at roof level of flat roofed houses were more affected than those at first floor level but like those at ground floor carried greater loading causing greater deflection and hence more flexural cracking. Condensation and rainwater penetration were also evident at this level.

2.  The beams were affected by both carbonation and chloride attack. Chloride levels varied and could also vary within the same beam (obviously added in flake form to the mix).

    Because of the flexural cracking, carbonation was more active at each of these locations and also where cover was not as originally specified.

    First floor level showed no signs of condensation as this is normally a warm zone. Access to any of the structural concrete components required a great deal of disturbance to the tenant.

P Barr, United Kingdom Atomic Energy Authority, U.K.

These papers have highlighted the importance of, firstly, monitoring structural condition; secondly, taking remedial action and thirdly, applying the lessons learned to future designs.

My own work with large industrial structures, whose performance is often of crucial importance to the safe operation of the installation as a whole, has highlighted the importance of ageing effects, and has impelled us to institute statistical assessment methods such as probabilistic safety assessment to attempt to predict overall structural behaviour. This, in turn, has necessitated a multi-disciplinary approach with structural engineers working alongside mathematicians, material scientists and physicists, particularly in studies of structural performance involving loadings beyond normal design basis resulting from certain types of severe accident of very low occurrence probability.

I foresee future work in this important area of study continuing and increasing, as well as the need to be able to forecast, with greater certainty, structural failure modes in regions of complex geometry.

With the increasing use of computer codes for the design and assessment of structures there is a tendency to assume that the uncertainties of structural performance have now been eliminated. I believe that even though such techniques are of undoubted utility and power, we still have much to learn, and not least in an understanding of the ways in which structures change with age.

# Index